W9-DHU-963

Ken-ya Hashimoto

Surface Acoustic Wave Devices in Telecommunications

Springer

Berlin
Heidelberg
New York
Barcelona
Hong Kong
London
Milan
Paris
Singapore
Tokyo

Engineering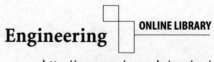

http://www.springer.de/engine/

Ken-ya Hashimoto

Surface Acoustic Wave Devices in Telecommunications

Modelling and Simulation

With 330 Figures

Springer

Professor Ken-ya Hashimoto
Chiba University
Faculty of Engineering
Department of Electronics & Mechanical Engineering
1-33 Yayoi-cho, Inage-ku
263-8522 Chiba
Japan
E-mail: ken@sawlab.te.chiba-u.ac.jp

Library of Congress Cataloging-in-Publication Data applied for.

Die Deutsche Bibliothek - CIP-Einheitsaufnahme

Hashimoto, Ken-ya: Surface acoustic wave devices in telecommunications: modelling and simulation/ Ken-ya Hashimoto. - Berlin; Heidelberg; New York; Barcelona; Hong Kong; London; Milan; Paris; Singapore; Tokyo: Springer, 2000 (Engineering online library) ISBN 3-540-67232-X

ISBN 3-540-67232-X Springer-Verlag Berlin Heidelberg New York

Springer-Verlag is a company in the BertelsmannSpringer publishing group.
© Springer-Verlag Berlin Heidelberg 2000
Printed in Germany

The use of general descriptive names, registered names, trademarks, etc. in this publication does not imply, even in the absence of a specific statement, that such names are exempt from the relevant protective laws and regulations and therefore free for general use.

Typesetting: Camera-ready copy by the author
Cover design: MEDIO, Berlin

Printed on acid-free paper   SPIN 10756695   62/3144/tr   5 4 3 2 1 0

Preface

Although the existence of the surface acoustic wave (SAW) was first discussed in 1885 by Lord Rayleigh [1], it did not receive engineering interest for a long time. In 1965, the situation changed dramatically. White suggested that SAWs can be excited and detected efficiently by using an interdigital transducer (IDT) placed on a piezoelectric substrate [2]. This is because very fine IDTs can be mass-produced by using photolithography, which has been well developed for semiconductor device fabrication, and proper design of the IDT enables the construction of transversal filters with outstanding performance.

Then, in Europe and America, a vast amount of effort was invested in the research and development of SAW devices for military and communication uses, such as delay lines and pulse compression filters for radar and highly stable resonators for clock generation. Research activities are reflected in the various technical papers represented by special issues [3–5] and proceedings [6]. The establishment of design and fabrication technologies and the rapid growth of digital technologies, represented by the microcomputer, meant that the importance of SAW devices for the military decreased year by year and most researchers in national institutions and universities left this field after reductions or cuts in their financial support. Then the end of the Cold War forced many SAW researchers in companies to do so, too.

On the other hand, Japanese researchers also paid much attention to SAW devices from the 1960s, but were mainly concerned with consumer and communication applications. Starting with intermediate-frequency (IF) filters for TVs [7], various filters and resonators were developed and mass-produced from the mid-1970s. After that, though, many companies left the competition because of price reductions and fixed market shares. The same reasons forced even successful companies to shrink their research sections. So Japanese research activities were diminished until the mid-1980s.

Rapid growth of the mobile communication market in the late 1980s changed the situation again. Many new-comers joined the surviving companies and very tight competition was restarted [8]. In Japan, more than 20 companies produce SAW devices now. The expanded market also stimulated research activities, and much innovative work has emerged in the last 10 years.

Through the development, the previous disadvantages of SAW devices, such as large insertion loss and low power durability, have been overcome. SAW devices have already been applied even to antenna duplexer of mobile phones, whose requirements are extremely severe.

In addition, until very recently, researchers believed that the practical frequency limit for the mass production of SAW devices is about 1 GHz. However, very rapid progress in microfabrication technologies imported from LSI production and in the refinement of the fabrication process enables mass-production of SAW devices in the 2.5 GHz range [9].

It is not unreasonable to expect that specifications for SAW devices will be more stringent in conjunction with rapid advances in digital communication technologies.

By the way, I entered the SAW field in 1977 as an undergraduate under the supervision of Professor Yamaguchi of Chiba University, and have continued my activities in the same laboratory for the last 20 years. My most recent main subject is the development of fast and precise simulators for SAW devices. There are many opportunities to partake in discussions with researchers who are interested in this work. During the discussion, they often asked me the questions: "Are there any textbooks about the subject?" or "Where is the discussion written up?" But, unfortunately, it is quite hard to answer them. This is because most books related to SAWs [10–18] were published before the rapid growth of the mobile communication market. They mainly concern transversal filters, and few descriptions are given of the resonator-related devices which are now widely used. Furthermore, devices employing SH-type SAWs and computer-based simulation technologies are scarcely detailed.

The aim here is to give an overview of the latest SAW technologies, such as the design and simulation of resonator-based devices employing SH-type leaky SAWs. Although the description is not highly mathematical, various theoretical foundations inherent in the development of precise simulation and design techniques are explained in detail.

Note that this book is not intended as a complete textbook. It is a supplement covering subjects not fully described in other books. From this reason, many important technologies, such as device fabrication, packaging, evaluation, optimal design, etc., are not discussed in this book.

This book is written such that readers can understand the content most easily. The explanations and descriptions are sometimes different from those in the original papers, and are sometimes even unusual. I do not doubt that there remain many errors and discrepancies. Any comments, questions, and error modification will be appreciated.

Of course, most of the content of this book is not my own achievement but is a gift of the tremendous effort of all SAW researchers around the world. Although I doubt if I am the perfect person to write such a book, much effort spent in the preparation has allowed me to do so.

The remaining original work was accomplished under collaboration with and/or supervision by Professor Masatsune Yamaguchi of Chiba University. Of course, I must express my special thanks to him for his support and guidance. Tireless effort and innovative ideas came from the staff members and students in our laboratory. Without their support, we could achieve nothing.

In addition, I must express my special thanks to Prof. Robert Weigel of Universität Linz who strongly suggested I publish this book and who kindly introduced me to Dr. Merkle of Springer-Verlag. Publication of this book would have been impossible without their understanding and encouragement.

I would like to thank all Japanese SAW researchers, represented by Professor Emeritus Kimio Shibayama of Tohoku University and Professor Yasutaka Shimizu of Tokyo Institute of Technology for their guidance and support. In addition, I appreciate many friends all over the world. Fruitful discussions and comments with them provided stimulation and encouragement.

Among them, I gratefully acknowledge Dr. Yoshio Sato of Fujitsu Laboratories, Inc., Mr. Hideki Omori and Mr. Yoshiro Fujiwara of Fujitsu Media Device Limited, Mr. Akihiro Bungo of Mitsubishi Materical Co. Ltd., and Dr. Clemens Ruppel of Siemens Corporate Technology for supplying much of the experimental data used in this book.

It is my great pleasure to express my special thanks to the following three persons. The late Professor Hiroshi Shimizu of Tohoku University showed me how university researchers should be. Dr. Fred Cho of Motorola Inc. encouraged me from an early stage of my career and trained me to survive in this world. Mr. Clinton Hartmann of RF SAW Components, Inc. is teaching me to be a true engineer. Discussions with him were always severe, but a lot of new innovative ideas came through them as a gift of his immense knowledge and experience. Thanks again.

Finally I thank my parents, my wife Kaoru and my daughter Hirono for their tireless support, encouragement and understanding.

Chiba University, January 2000   *Ken-ya Hashimoto*

References

1. L. Rayleigh: On Waves Propagating along the Plane Surface of an Elastic Solid, Proc. London Math. Soc., **7** (1885) pp. 4–11.
2. R.M. White and F.W. Voltmer: Direct Piezoelectric Coupling to Surface Elastic Waves, Appl. Phys. Lett., **17** (1965) pp. 314–316.
3. T.M. Reeder (ed.): Special Issue on Microwave Acoustic Signal Processing, IEEE Trans. Microwave Theory and Tech., **MTT-21**, 4 (1973).
4. L. Claiborne, G.S. Kino and E. Stern (eds): Special Issue on Surface Acoustic Wave Devices and Applications, Proc. IEEE, **64**, 5 (1976).
5. R.C. Williamson and T.W. Bristol (eds): Special Issue on Surface-Acoustic-Wave Applications, IEEE Trans. Sonics and Ultrason., **SU-28**, 2 (1981).
6. E.A. Ash and E.G.S. Paige (eds): Rayleigh-Wave Theory and Application, Springer-Verg, New York (1985).
7. S. Takahashi, H. Hirano, T. Kodama, F. Miyashiro, B. Suzuki, A. Onoe, T. Adachi and K. Fujinuma: SAW IF Filter on LiTaO$_3$ for Color TV Receivers, IEEE Trans. Consumer Electron., **CE-24**, 3 (1978) pp. 337–346.
8. K. Shibayama and K. Yamanouchi (eds): Proc. International Symp. on Surface Acoustic Wave Devices for Mobile Communication (1992).
9. T. Matsuda, H. Uchishiba, O. Ikata, T. Nishihara and Y. Satoh: L and S band Low-Loss Filters Using SAW Resonators, Proc. IEEE Ultrason. Symp. (1994) pp. 163–167.
10. B.A. Auld: Acoustic Waves and Fields in Solids, Vol. I & II, Wiley, New York (1973).
11. H. Matthews (ed): Surface Wave Filters, Wiley, New York (1977).
12. D.P. Morgan: Surface-Wave Devices for Signal Processing, Elsevier, Amsterdam (1985).
13. E.A. Gerber and A. Ballato (eds): Precision Frequency Control, Vol. I & II, Academic Press, Orlando (1985).
14. S. Datta: Surface Acoustic Wave Devices, Prentice Hall, Englewood Cliffs (1986).
15. G.S. Kino: Acoustic Waves: Devices, Imaging, & Analog Signal Processing, Prentice-Hall, Englewood Cliffs (1987).
16. C.K. Campbell: Surface Acoustic Wave Devices and Their Signal Processing Applications, Academic Press, Boston (1989).
17. C.K. Campbell: Surface Acoustic Wave Devices for Mobile and Wireless Communication, Academic Press, Boston (1998).
18. M. Feldmann and J. Henaff: Surface Acoustic Waves for Signal Processing, Altech House, Boston (1989).

Contents

1. **Bulk Acoustic and Surface Acoustic Waves** 1
   - 1.1 Bulk Acoustic Waves.................................. 1
     - 1.1.1 Elastic Waves in Solids........................... 1
     - 1.1.2 Wavevector and Group Velocity 4
     - 1.1.3 Behavior of BAWs at a Boundary 6
     - 1.1.4 Diffraction..................................... 8
     - 1.1.5 Piezoelectricity................................ 8
     - 1.1.6 Evanescent Fields 11
     - 1.1.7 Waveguides...................................... 12
     - 1.1.8 Behavior at the Boundary Between Waveguides 13
     - 1.1.9 Open Waveguides 15
   - 1.2 Waves in a Semi-infinite Substrate 17
     - 1.2.1 Excitation of L- and SV-type Waves 17
     - 1.2.2 Excitation of SH-type Waves...................... 18
     - 1.2.3 Leaky SAWs...................................... 20
     - 1.2.4 Leaky and Nonleaky SAWs 21
     - 1.2.5 SSBW 22
   - References 22

2. **Grating**... 25
   - 2.1 Basic Structure 25
     - 2.1.1 Fundamentals.................................... 25
     - 2.1.2 Reflection Center 26
   - 2.2 Behavior in Periodic Structures 27
     - 2.2.1 Bragg Reflection 27
     - 2.2.2 Energy Storing Effect 30
     - 2.2.3 Fabry–Perot Resonator........................... 31
   - 2.3 Equivalent Circuit Analysis 32
     - 2.3.1 Analysis.. 32
     - 2.3.2 Dependence of Reflection Characteristics on Parameters 35
   - 2.4 Metallic Grating 39
     - 2.4.1 Fundamental Characteristics 39
     - 2.4.2 SAW Dispersion Characteristics 40
     - 2.4.3 Approximated Dispersion Characteristics 42
   - References 46

3.   Interdigital Transducers 47
   3.1   Fundamentals 47
         3.1.1   Bidirectional IDTs 47
         3.1.2   Unidirectional IDTs 48
   3.2   Static Characteristics 52
         3.2.1   Charge Distribution 52
         3.2.2   Electromechanical Coupling Factor 53
         3.2.3   Element Factor 55
         3.2.4   Complex Electrode Geometries 56
         3.2.5   Effect of IDT Ends 59
   3.3   IDT Modeling 61
         3.3.1   Delta-Function Model 61
         3.3.2   Equivalent Circuit Model 65
         3.3.3   Other Models 66
   3.4   Influence of Peripheral Circuit 68
         3.4.1   Summary 68
         3.4.2   Smith Chart and Impedance Matching 70
         3.4.3   Achievable Bandwidth 73
   3.5 p Matrix 74
         3.5.1   Summary 74
         3.5.2   IDT Characterization by Using p Matrix 76
         3.5.3   Discussion on Unidirectional IDTs 77
   3.6   BAW Radiation 79
         3.6.1   Phase Matching Condition 79
         3.6.2   Radiation Characteristics 81
   References 84

4.   Transversal Filters 87
   4.1   Basics 87
         4.1.1   Weighting 87
         4.1.2   Basic Properties of Weighted IDTs 91
         4.1.3   Effects of Peripheral Circuits 92
   4.2   Design of Transversal Filters 97
         4.2.1   Fourier Transforms 97
         4.2.2   Remez Exchange Method 100
         4.2.3   Linear Programming 101
   4.3   Spurious Responses 103
         4.3.1   Diffraction 103
         4.3.2   Bulk Waves 107
         4.3.3   Other Parasitic Effects 110
   4.4   Low-Loss Transversal Filters 112
         4.4.1   Multi-IDT Structures 112
         4.4.2   Transversal Filters Employing SPUDTs 114
         4.4.3   Combination of SPUDTs and Reflectors 117
   References 120

5. **Resonators** 123
    5.1 One-Port SAW Resonators............................. 123
         5.1.1 Introduction 123
         5.1.2 Fabry–Perot Model 127
    5.2 Spurious Responses 130
         5.2.1 Beam Diffraction and Transverse Modes 130
         5.2.2 Transverse-Mode Analysis 131
         5.2.3 Effect of BAW Radiation 138
    5.3 Two-Port SAW Resonators 141
         5.3.1 Summary...................................... 141
         5.3.2 Fabry–Perot model 143
         5.3.3 Multi-Mode Resonator Filter 145
         5.3.4 Cascade Connection of Resonators............. 148
    5.4 Impedance Element Filters 149
         5.4.1 π-Type Filters 149
         5.4.2 Lattice-Type Filters 152
         5.4.3 Ladder-Type Filters 153
    References 160

6. **Selection of Substrate Material** 163
    6.1 Substrate Material and Device Characteristics......... 163
         6.1.1 Orientation 163
         6.1.2 Influence of Substrate and Electrode Materials 164
    6.2 Evaluation of Acoustic Properties by Effective Permittivity .. 167
         6.2.1 Effective Permittivity 167
         6.2.2 Approximate Expressions....................... 173
    6.3 Single Crystals 174
         6.3.1 Quartz....................................... 174
         6.3.2 $LiNbO_3$ 176
         6.3.3 $LiTaO_3$ 179
         6.3.4 $Li_2B_4O_7$ 182
         6.3.5 Langasite.................................... 183
    6.4 Thin Films 184
    References 188

7. **Coupling-of-Modes Theory** 191
    7.1 Fundamentals 191
         7.1.1 Collinear Coupling 191
         7.1.2 Periodic Structures 196
         7.1.3 Excitation 200
    7.2 COM Theory for SAW Devices 200
         7.2.1 Derivation................................... 200
         7.2.2 COM Equations in Other Forms 203
         7.2.3 Inclusion of Electrode Resistivity 204
         7.2.4 Examples 206

7.3   Determination of COM Parameters 214
      7.3.1   Perturbation Theory 214
      7.3.2   Wave Theory Based Analysis 216
      7.3.3   Analysis for Multi-Electrode IDTs 221
7.4   COM-Based Simulators 227
      7.4.1   SAW Device Simulation 227
      7.4.2   Inclusion of Peripheral Circuit 230
      7.4.3   Results of Simulation 231
References 235

8. Simulation of SH-type SAW Devices 237
    8.1   Physics of SH-Type SAWs 237
          8.1.1   Summary...................................... 237
          8.1.2   Propagation and Excitation on a Uniform Surface 237
          8.1.3   Behavior on a Grating 243
          8.1.4   Electrical Characteristics of IDTs.................. 246
          8.1.5   Effects of Back-Scattered BAWs.................. 248
          8.1.6   Influence of Grating Edge 249
    8.2   COM Theory for SH-Type SAWs 250
          8.2.1   COM Parameter Derivation 250
          8.2.2   Simulation..................................... 254
          8.2.3   COM Parameters for Rayleigh-Type SAWs 263
    8.3   Derivation of Approximate Dispersion Relations 266
          8.3.1   Derivation of Plessky's Dispersion Relation 266
          8.3.2   Derivation of Abbott's Dispersion Relation 267
    References 268

A. Physics of Acoustic Waves 271
    A.1   Elasticity of Solids...................................... 271
    A.2   Piezoelectricity.. 275
    A.3   Surface Acoustic Waves 278
    A.4   Effective Acoustic Admittance Matrix and Permittivity 282
    A.5   Acoustic Wave Properties in 6mm Materials 284
          A.5.1   Rayleigh-Type SAWs 284
          A.5.2   Effective Permittivity for BGS Waves 285
          A.5.3   Effective Acoustic Admittance Matrix.............. 287
    A.6   Wave Excitation....................................... 287
          A.6.1   Integration Path 287
          A.6.2   Electrostatic Coupling 288
          A.6.3   BGS Wave Excitation........................... 289
          A.6.4   SSBW Excitation 290
    References 291

B. Analysis of Wave Propagation on Grating Structures 293

   B.1 Summary 293

   B.2 Metallic Gratings 294

      B.2.1 Bløtekjær's Theory for Single-Electrode Gratings 294

      B.2.2 Wagner's Theory for Oblique Propagation 296

      B.2.3 Aoki's Theory for Double-Electrode Gratings 297

      B.2.4 Extension to Triple-Electrode Gratings 301

   B.3 Analysis of Metallic Gratings with Finite Thickness 304

      B.3.1 Combination with Finite Element Method 304

      B.3.2 Application to Extended Bløtekjær Theories 306

   B.4 Wave Excitation and Propagation in Grating Structures 310

      B.4.1 Effective Permittivity for Grating Structures 310

      B.4.2 Evaluation of Discrete Green Function 312

      B.4.3 Delta-Function Model 315

      B.4.4 Infinite IDTs 316

   References 318

Index 321

1. Bulk Acoustic and Surface Acoustic Waves

This chapter describes the excitation, detection and propagation of acoustic waves, including SAWs in various propagation media. Although I have tried to explain their physics as simple as possible, the content may seem difficult for readers who have dealt with SAW devices as an electronic element. But readers should not skip this chapter because it provides the technical background for the discussions given in the following sections.

1.1 Bulk Acoustic Waves

1.1.1 Elastic Waves in Solids

Bulk acoustic waves (BAWs) are elastic waves propagating in solids. They are categorized into a longitudinal wave shown in Fig. 1.1a and a transverse wave shown in Fig. 1.1b. The longitudinal wave is also called a pressure wave whereas the transverse wave is also called a shear wave.

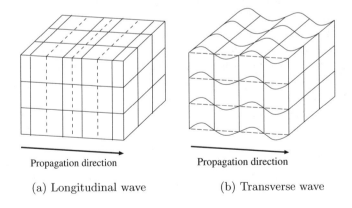

Propagation direction   Propagation direction

(a) Longitudinal wave   (b) Transverse wave

Fig. 1.1. BAW in solids

Even in the case where the propagation media are isotropic, their phase velocities, giving the transmission speed of the wavefront, are different from each other. That is, the phase velocities V_l and V_t for the longitudinal wave and shear wave are given by [1]

$$V_l = \sqrt{c_{11}/\rho}, \tag{1.1}$$
$$V_t = \sqrt{c_{44}/\rho}, \tag{1.2}$$

where c_{ij} is the stiffness constant and ρ is the mass density. Note that the longitudinal wave causes a volume change with propagation whereas the shear wave does not.

The longitudinal and shear waves are identified by their displacement direction which is called polarization: polarization of the longitudinal wave is parallel to the propagation direction whereas that of the shear wave is perpendicular to the propagation direction.

Elastic waves suffer attenuation of their amplitude with propagation. There are a few mechanisms which cause attenuation [1]:

(1) *Scattering loss due to inhomogeneity in the propagating media*: it is mostly temperature independent and is dependent upon the material preparation. This mechanism is dominant for most polycrystalline materials.

(2) *Scattering loss of coherent waves due to thermal lattice vibration*: it exists even in ideal perfect crystals and increases with temperature.

(3) *Energy transportation by thermal diffusion to neutralize temperature variation caused by the volume change*: since shear waves do not associate volume change with propagation, this mechanism occurs only in longitudinal waves.

These mechanisms become significant when the wavelength is comparable to the size of the scatterers. This means that the acoustic wave attenuation increases with frequency. For the same reason, materials with a higher velocity are generally preferable in order to reduce the propagation loss.

For the excitation and detection of elastic waves, piezoelectric materials are widely used. Since piezoelectric materials are inherently anisotropic, the properties of elastic waves are dependent upon the directions of propagation and/or polarization.

Let us consider common salt (NaCl), which has a cubic structure like stacked dice. It is clear that the elasticity when the dice are stressed perpendicular to the surface is different from that when they are stressed in the direction of the face diagonal. When the stress is applied obliquely, the tensile stress may cause shear strain, and the shear stress may cause compression. This means that, for most propagation directions in anisotropic materials, pure longitudinal and shear waves do not exist; these two waves are coupled. So they become quasi-longitudinal and quasi-shear waves, whose polarizations are close to being parallel and perpendicular, respectively, to the propagation direction. Generally, the quasi-longitudinal wave has a higher velocity than either shear wave.

Added to this, the elasticity of the shear strain is dependent upon the direction of the motion. Then, there are two quasi-shear waves called fast-shear and slow-shear waves. As is clear from their names, they have different phase velocities and polarizations. In general, for the specified propagation direction, there exists three independent acoustic waves. It is interesting to note that their polarizations are perpendicular to each other independent of the anisotropy of the propagation media [2].

The study of how acoustic wave velocities vary with different substrate orientations is vital as will be discussed in Chap. 7. However, when graphically displaying these results, the standard practice is to show the inverse of velocity or slowness ($S = V^{-1}$) because most useful analysis results depend on the wavenumber ($\beta = \omega S$ where ω is the angular frequency) which is linearly proportional to the slowness.

Figure 1.2 shows S of the (001) plane of Si as a function of the propagation direction. Since the fast-shear wave is always polarized toward <001>, the phase velocity is independent of the propagation direction, and the slowness surface composes a circle. On the other hand, the slow-shear and longitudinal waves change their velocities with the propagation direction, and the slow-shear wave becomes the slowest and the longitudinal becomes the fastest in the <110> direction. Note that the slowness shows four-fold symmetry associated with the symmetry of the crystal structure.

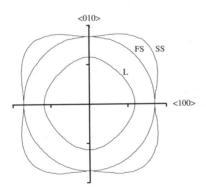

Fig. 1.2. Slowness surface of the (001) plane of Si. Unit scale is 1×10^{-4} sec/m

Figure 1.3 shows the slowness on the (001) plane of TeO$_2$. Due to very large anisotropy, the BAW velocities change considerably with the propagation angle. It is interesting to note that the longitudinal and shear wave velocities are mostly identical in the <100> direction. On the other hand, the shear wave velocity reduces to 650 m/sec in the <110> direction. Due to these very low velocities, TeO$_2$ has been applied for delay lines and acousto-optic devices [3].

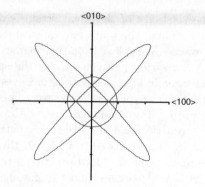

Fig. 1.3. Slowness surface on the (001) plane of $TeO_2(001)$ (5×10^{-5}s/m/div.)

Note that not only TeO_2 but also various single crystals belonging to the cubic 4mm crystal class, such as TiO_2 and $Li_2B_4O_7$ exhibit very large acoustic anisotropy.

For the discussion of surface acoustic waves, we sometimes use the polarization relative to the substrate surface. That is, shear waves having polarization perpendicular and parallel to the surface are called shear-vertical (SV) and shear-horizontal (SH) waves, respectively. Which is faster is dependent upon the anisotropy of the propagating medium.

1.1.2 Wavevector and Group Velocity

The wavevector β is useful to characterize wave properties. Its absolute value $|\beta| \equiv \beta$ is called the wavenumber, and gives the phase lag per unit length with propagation. It is related to the wavelength λ as $\beta = 2\pi/\lambda$. The direction of the vector gives the propagation direction of the wavefront as shown in Fig. 1.4. Since the phase lead per unit time is the angular frequency, the wavenumber is also called the spatial frequency.

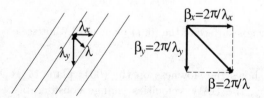

Fig. 1.4. Wavevector and wavelength

Let us consider oblique propagation of a wave in the coordinate system shown in Fig. 1.4. In this case, the phase shift due to propagation is given by the inner product of the wavevector $\boldsymbol{\beta}$ and the vector $\mathbf{x}$ corresponding to the propagation distance. That is,

$$\exp(-j\boldsymbol{\beta} \cdot \mathbf{x}) \rightarrow \exp(-j\boldsymbol{\beta} \cdot \mathbf{x}) = \exp(-j\beta_x x - j\beta_y y - j\beta_z z). \qquad (1.3)$$

In anisotropic materials, the direction and speed of the power flow often do not coincide with that of the wavevector. We refer to this phenomenon as beam steering, and the propagation speed of the power flow is called the group velocity. Plane wave propagation with beam steering is schematically shown in Fig. 1.5. In the figure, $\theta_p = \cos^{-1}(V_p/V_g)$ is the deviation in the directions of the phase velocity V_p and group velocity V_g, and is called the power flow angle.

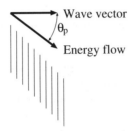

Fig. 1.5. Power flow and beam steering

By the use of the slowness surface, the relation between V_p and V_g is expressed as in Fig. 1.6 [4]. The power flow angle is mostly zero when the slowness surface is almost circular. But it sometimes approaches $90°$ in very anisotropic materials (see TeO_2 shown in Fig. 1.3).

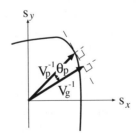

Fig. 1.6. Group velocity and phase velocity in the slowness surface

1.1.3 Behavior of BAWs at a Boundary

Let us consider oblique wave incidence to the boundary as shown in Fig. 1.7. Since the wavefront must be continuous at the boundary, the phase change

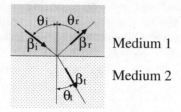

Fig. 1.7. Reflection and transmission for oblique incidence

per unit length, namely, the wavevector component parallel to the boundary, must be continuous at the boundary.

We set the wavenumbers for the incident, reflected and transmitted waves as β_i, β_r, β_t, respectively, and their angles as θ_i, θ_r, θ_t, respectively. Then the condition for the continuity of the wavevector is given by

$$\beta_i \sin\theta_i = \beta_r \sin\theta_r = \beta_t \sin\theta_t, \tag{1.4}$$

which corresponds to Snell's law in optics.

By using the relation $|\boldsymbol{\beta}| = \omega|S|$, (1.4) can be rewritten as

$$S_i \sin\theta_i = S_r \sin\theta_r = S_t \sin\theta_t, \tag{1.5}$$

where S_i, S_r and S_t are the slownesses of the incident, reflected and transmitted waves, respectively.

The relation (1.5) can be solved graphically by using the slowness surface shown in Fig. 1.8. That is, (1.5) is satisfied only when the projection of S_r and S_t on the horizontal axis is equal to that of S_i. It is clear from the figure that there is a situation where two kinds of waves are simultaneously reflected by one incident wave. Namely, an incident shear wave may generate both longitudinal and shear waves by reflection. In general, there may exist three kinds of transmitted waves and three kinds of reflected waves.

On the other hand, for an appropriate combination of materials, there is a situation where no transmission occurs as shown in Fig. 1.8b. This situation is called total reflection. The limiting incident angle for total reflection is called the critical angle.

Note that (1.5) determines only the propagation direction, and one may solve the other field problem to determine the transmission and reflection coefficients [5].

To characterize wave reflection, it is convenient to subdivide the shear wave into two components; one is the SH wave with polarization parallel to

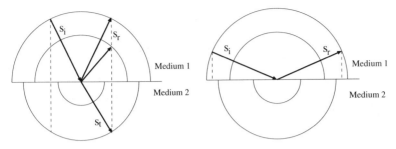

(a) When transmitted wave exists   (b) When total reflection occurs

Fig. 1.8. Reflected and transmitted waves in slowness surface

the surface and the other is the SV wave with polarization perpendicular to that of the SH wave.

Figure 1.9 shows the power reflection and coefficient for incidence of the acoustic waves to the free surface (vacuum) from polycrystalline Al. The result for SH wave incidence is not shown because it is always unity in this case. For longitudinal wave incidence shown in (a), the shear-vertical wave is generated by reflection for all incident angles. On the other hand, for shear-vertical wave incidence, the longitudinal wave is not generated when the incident angle is larger than $29°$. In this region, total reflection occurs.

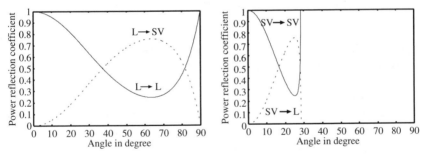

(a) Incidence of longitudinal (L) wave   (b) Incidence of shear-vertical (SV) wave

Fig. 1.9. Power reflection and transmission coefficients for incidence from Al to vacuum

It is interesting to note that the reflectivity for SV-type BAW incidence and L-type BAW reflection in Fig. 1.9a coincides with that for L-type BAW incidence and SV-type BAW in Fig. 1.9b when the horizontal axis is interpreted as the reflection angle. This relation is due to reciprocity [6].

1.1.4 Diffraction

Waves excited by a plane transducer with finite size will propagate as a plane wave for a certain distance. Then its wavefront spreads gradually with propagation, and finally it becomes circular (see Fig. 1.10). This phenomenon is called diffraction. The region where the wave propagates like a plane wave is called the Fresnel zone whereas the region where the wave propagates circularly is called the Fraunhofer zone.

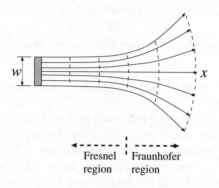

Fig. 1.10. Fresnel and Fraunhofer zones

The critical length x_c which distinguishes these regions is given by [7]

$$x_c = (1 + \gamma)w^2/\lambda, \tag{1.6}$$

where w is the width of the transducer, namely, the aperture, λ is the wavelength, and γ is a factor determined by the anisotropy of the media and is zero for isotropic materials.

To reduce the influence of diffraction, substrate materials with $\gamma \to \infty$ are desirable, where the acoustic wave propagates as an acoustic beam for considerably long distances. As will be shown in Sect. 4.3.1, $1+\gamma$ corresponds to the inverse of the curvature of the slowness surface.

1.1.5 Piezoelectricity

Let us consider the deformation of the trigonal structure shown in Fig. 1.11. When no external stress is applied, dipole moments in the structure are in equilibrium, and no net polarization appears (Fig. 1.11a). By the tensile stress, dipole moments displace and a net polarization appears (Fig. 1.11b). If a compressive stress is applied, the movement causes a net polarization with opposite sign (see Fig. 1.11c). Then the generated polarization is dependent not only upon the amplitude but also the direction of the applied stress.

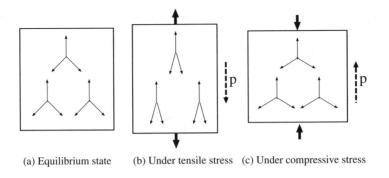

(a) Equilibrium state   (b) Under tensile stress   (c) Under compressive stress

Fig. 1.11. Trigonal structure under stress. Thin arrows indicate dipole moments

Namely, when the stress is small, the net polarization is linearly proportional to the stress.

Similarly, by applying an external electric field instead of a stress, it is possible to generate strain proportional to the electric field.

This characteristic, i.e., linear coupling between electricity and elasticity, is called piezoelectricity [8].

In contrast, let us consider the tetragonal structure shown in Fig. 1.12. Although movement of dipoles occurs because of the applied stresses, no net polarization appears in this case.

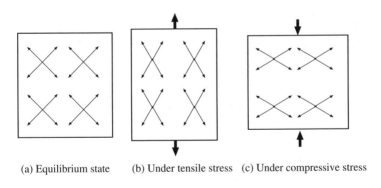

(a) Equilibrium state   (b) Under tensile stress   (c) Under compressive stress

Fig. 1.12. Tetragonal structure under stress. Thin arrows indicate dipole moments

This means that, for the existence of piezoelectricity, the structure must possess a certain crystallographic asymmetry [8]. In other words, all piezoelectric materials are anisotropic. And not only dielectric and elastic properties but also piezoelectric ones are critically dependent upon the propagation and polarization directions with respect to the crystal orientation.

Piezoelectricity is widely used for the excitation and detection of acoustic waves.

Let us consider the parallel plate capacitor shown in Fig. 1.13 where a piezoelectric plate is used as a sandwiched dielectric material.

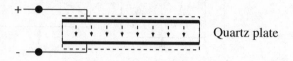

Quartz plate

Fig. 1.13. Strain caused by electric field in a piezoelectric plate

The applied electric field generates cyclic stress F in the plate through piezoelectricity, and the stress generates a displacement u as a reaction of inertia, elasticity and viscosity.

Then the equivalent circuit for the capacitor may be written as Fig. 1.14a. In the figure, $v = j\omega u$, C_0 is the static capacitance, M is the contribution of inertia, k is that of elasticity, η is that of viscosity, and the ideal transformer is to convert electrical voltage V to mechanical voltage, i.e., the stress F.

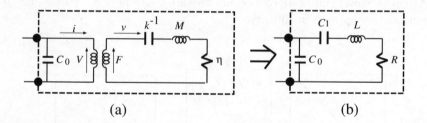

(a) (b)

Fig. 1.14. Electrical equivalent circuit for parallel plate capacitor with sandwiched piezoelectric material

When the driving frequency ω is chosen appropriately, a certain amount of current flows into the mechanical branch composed of k, M and η instead of the electrostatic branch, and a strong mechanical resonance can be excited and detected electrically. This feature is used for acoustic wave transducers and resonators.

Note that the equivalent circuit shown in Fig. 1.14a can be modified as Fig. 1.14b. This corresponds to the equivalent circuit for BAW resonators, and will also be used for SAW resonators which will be discussed in Chap. 5.

1.1.6 Evanescent Fields

Let us consider the total reflection shown in Fig. 1.8b. In this situation, although no transmission occurs, there exists a finite amplitude of vibration in medium 2. Namely, as shown in Fig. 1.15, the wave is reflected or regenerated after the incident wave energy is trapped near the boundary.

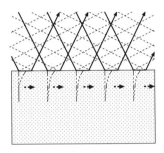

Fig. 1.15. Field distribution at total reflection

Equation (1.4) suggests that the wavenumber into the depth is given by

$$\beta_t \cos \theta_t = \beta_t \cos\{\sin^{-1}(\beta_i \sin \theta_i / \beta_t)\} = -j\beta_t \sqrt{(\beta_i \sin \theta_i / \beta_t)^2 - 1}. \quad (1.7)$$

Since it is purely imaginary, the transmitted wave component diminishes exponentially into the depth. This type of field dependence is called evanescent. It should be emphasized that this attenuation corresponds to the distribution of the stored energy, and not to the dissipation of its energy.

Due to the penetration of the wave energy into the medium, the wave can transmit through the medium as shown in Fig. 1.16 if its thickness is finite. And the transmission amplitude increases with reduction of the thickness. This phenomenon is called tunneling.

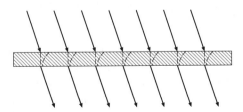

Fig. 1.16. Tunneling of evanescent field

1.1.7 Waveguides

When an impact is applied to one end of the plate with finite thickness shown in Fig.1.17, waves whose energy is concentrated within the plate can propagate, and are called waveguide modes, and this kind of structures is called a waveguide. Since the wave field is closed within the waveguide structure, this kind of waveguide is called a closed waveguide.

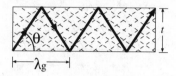

Fig. 1.17. Waveguiding by total reflection within a closed waveguide

Waveguide modes propagating in this structure can be expressed as the sum of plane waves generated by total reflection at the upper and bottom surfaces. Since a plane wave reflected twice propagates parallel to the original one, they interfere with each other. In order that the wave can propagate without cancellation, namely so that the waveguide mode can exist, the original and reflected waves must be in phase.

Let us consider a plane wave propagating obliquely with an angle θ in the medium. When the component of its wavevector β_B toward the thickness is $\beta_t = |\beta_B| \sin\theta$, and the phase shift due to reflection is ψ, the in-phase condition is given by

$$2\beta_t t + 2\psi = 2|\beta_B| t \sin\theta + 2\psi = 2n\pi, \tag{1.8}$$

where n is an integer. That is, the transverse phase shift during the round trip must be an integer times 2π. From this reason, this condition is called the transverse resonance condition.

What we can measure from outside the waveguide is the transmission response between input and output ports, where the phase lag is directly related to the wavevector component $\beta_p = |\beta_B| \cos\theta$ toward the propagation direction. Then, as a measurable quantity, the phase velocity $V_p = \omega/\beta_p$ of the waveguide mode is defined. Since $\beta_p < |\beta_B|$, $V_p > V_B$. Note that since the plane waves composing the waveguide mode propagate obliquely in the plate, its group velocity V_g, corresponding to the speed of power transfer, is lower than V_B.

When an excited ultrasonic pulse is detected after propagation for a distance L, the phase lag ϕ between the input and output ports is given by $\beta_p L = \omega L/V_p$. On the other hand, the time delay τ is given by L/V_g (see Fig. 1.18).

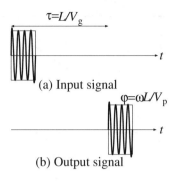

Fig. 1.18. Pulse response

As the simplest case, we will discuss the propagation of SH waves in an isotropic plate. In this case, the incident SH wave is totally reflected at the surfaces with $\psi = \pi$. Since $|\beta| = \omega/V_B$, (1.8) can be simplified to

$$\beta_p = \sqrt{(\omega/V_B)^2 - (N\pi/t)^2}, \tag{1.9}$$

where $N = n - 1$ is an integer.

Equation (1.9) suggests that multiple solutions exist when ω and t are sufficiently large. These solutions correspond to higher modes. For given ω, θ increases with N. Each mode only exists at frequencies higher than $NV_B\pi/t$ where $\beta_p = 0$. Namely, at the cut-off frequency, the mode corresponds to the thickness resonance in the plate. With an increase in frequency, the mode has finite β_p and propagates with small θ (see Fig. 1.19).

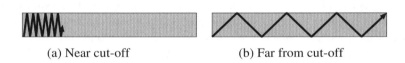

(a) Near cut-off   (b) Far from cut-off

Fig. 1.19. Mode propagation near and far from the cut-off frequency

Figure 1.20 shows the relationship between ω and β_p calculated from (1.9), which is called a dispersion relation. If the dispersion relation is given, the phase velocity V_p and group velocity V_g are determined by the gradient of the line from the origin and that of the tangent, respectively, as shown in Fig. 1.21 [9]. Thus $V_p \to \infty$ and $V_g = 0$ at the cut-off frequency, corresponding to the thickness resonance, whereas $V_p, V_g \to V_B$ at $\omega \to \infty$.

1.1.8 Behavior at the Boundary Between Waveguides

Below the cut-off frequency $NV_B\pi/t$, the wavenumber β_p given by (1.9) can be expressed as

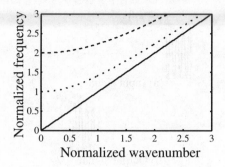

Fig. 1.20. Typical dispersion relation for closed waveguide

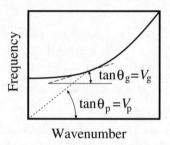

Fig. 1.21. Phase and group velocities

$$\beta_p = -j\sqrt{(N\pi/t)^2 - (\omega/V_B)^2}. \tag{1.10}$$

Thus β_p becomes imaginary, suggesting that the mode becomes evanescent.

Let us consider the case shown in Fig. 1.22a where the waveguide mode is incident upon the cut-off waveguide from the other one supporting only the lowest-order mode. Although the incident mode is totally reflected at the boundary, a phase shift occurs at reflection because the wave energy penetrates into the cut-off waveguide as an evanescent mode. This phenomenon is called the energy storing effect [10], which is expressed by the capacitance in the equivalent circuit shown in Fig. 1.22b.

In this case, the field of the mode penetrates into the waveguide in the form of $\exp(-|\Im[\beta_p]|x_1)$. The distance $1/|\Im[\beta_p]|$ where the amplitude decays by $1/e$ is called the penetration depth of the evanescent mode.

If the length of the waveguide is sufficiently short, the evanescent mode can pass through the waveguide as shown in Fig. 1.22c. Since the wavenumber is imaginary, no phase shift occurs with propagation. However, the phase shift due to the energy storing effect occurs, and it can be included in the equivalent circuit as shown in Fig. 1.22d.

Note that an energy storing effect due to higher modes also occurs, and its effect might also be taken into account as shown in Fig. 1.23.

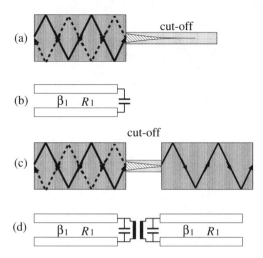

Fig. 1.22. Propagation of evanescent mode at cut-off and its equivalent circuit

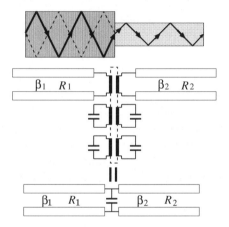

Fig. 1.23. Equivalent circuit for waveguides including higher evanescent modes

1.1.9 Open Waveguides

The waveguide can also be constructed by using an unbounded structure as shown in Fig. 1.24, where the inner and outer materials are chosen appropriately so that the total reflection occurs at the boundaries. This type of waveguide is called an open waveguide.

In this case, ψ in (1.8) changes with θ. Thus, even in simple structures, the dispersion relation cannot be expressed in a simple form. Figure 1.25 shows a typical dispersion relation for an open waveguide. In open waveguides, the cut-off frequencies are not determined by the thickness resonance condition

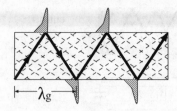

Fig. 1.24. Nonleaky mode in open waveguide

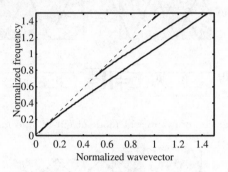

Fig. 1.25. Dispersion relation for open waveguide

where $\beta_\mathrm{p} = 0$, but is given by the condition where the propagation angle θ coincides with the critical angle θ_c for total reflection.

Note that the transverse resonance condition can be satisfied even when $\theta > \theta_\mathrm{c}$. Then the wave can propagate for a while with leakage of its energy to the outside as shown in Fig. 1.26. This type of mode is called a leaky mode. Its amplitude decays exponentially with propagation, and its wavenumber is complex.

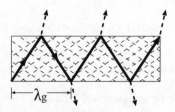

Fig. 1.26. Leaky mode

1.2 Waves in a Semi-infinite Substrate

1.2.1 Excitation of L- and SV-type Waves

Let us consider the case when a surface of semi-infinite substrate is hit by a hatchet. This impact excites BAWs into the substrate, which propagate isotropically as shown in Fig. 1.27. Since the BAW power spreads uniformly in all directions with equal speed, the law of energy conservation suggests the BAW power density decreases inversely proportional to the propagation distance r. Since the power density is proportional to square of the amplitude, it decrease as $r^{-0.5}$.

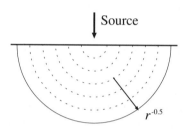

Fig. 1.27. Excitation and propagation of BAWs

Figure 1.28 shows typical radiation patterns of the BAWs launched from a line source; the SV- and L-type BAWs are not radiated exactly parallel to the surface because they cannot satisfy the surface boundary condition.

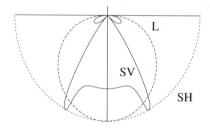

Fig. 1.28. Radiation pattern of BAWs launched from a line source on a plane surface

Both L and SV components excited close to the surface couple with each other through the boundary condition, and compose an eigenmode called a Rayleigh-type surface acoustic wave (SAW) [11].

Figure 1.29 shows the excitation and propagation of the coupled L and SV components composing the Rayleigh-type SAW [12]. In the figure, ϕ_c is the critical angle showing how the SAW fields grow up, and is given by

$$\phi_c = \cos^{-1}(V_S/V_B),\tag{1.11}$$

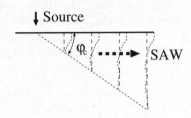

Fig. 1.29. Excitation and propagation of coupled L and SV components

where V_S and V_B are the SAW and BAW velocities, respectively. This means that the SAW fields do not penetrate into the bulk region satisfying $\phi > \phi_c$, where there only exist BAWs. This may be understood from the fact that acoustic fields do not spread with velocity faster than V_B.

Figure 1.30 shows the field distribution of a Rayleigh-type SAW. It is seen that the coupled field components are evanescent into the bulk and most of the wave energy concentrates within $1\ \lambda$ depth; they have an x_3 (into the depth) dependence in the form of $\exp(-\xi x_3)$, where

$$\xi = \beta_S \sqrt{1 - (V_S/V_B)^2} = \beta_S \sin \phi_c \tag{1.12}$$

is the decay constant into the bulk, and β_S is the wavenumber of the Rayleigh-type SAW. Note that $x_c = \xi^{-1}$ is referred to as the penetration depth.

Since the Rayleigh-type SAW possesses a phase velocity about 10% smaller than that of the SV-type BAW on the same substrate surface, (1.12) suggests that most of the energy concentrates close to the surface. Here, the difference between V_S and V_B is one of the measures which shows how strongly the SAW propagation is affected by the surface boundary condition or how much SAW energy is trapped close to the surface.

The Rayleigh-type SAW exists on all solids including anisotropic materials, and is mainly composed of L and SV components. Of course, it exists on piezoelectric materials. Since the wave energy is trapped at the surface mainly due to the mechanical boundary conditions, the penetration depth of the wave is scarcely affected by the electrical boundary condition.

1.2.2 Excitation of SH-type Waves

The SH component can satisfy the surface boundary condition, and propagates along the surface as well as into the bulk. This means that from a

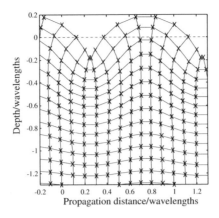

Fig. 1.30. Field distribution of a Rayleigh-type SAW

line source, the SH-type BAW is launched in every direction including the direction exactly parallel to the surface with the same amplitude (see Fig. 1.28). As with SV- and L-type BAWs, the amplitude of the SH-type BAW decays as $1/\sqrt{r}$. In other words, there is no mechanism which causes coupling between wave components. Thus, no SH-type eigenmode exists on the surface of semi-infinite isotropic materials.

On piezoelectric substrates, BAWs may couple with the electric potential Φ. If the SH field component and Φ couple with each other, they compose another type of surface acoustic wave called an SH-type SAW, and it satisfies both mechanical and electrical boundary conditions on the surface. For example, it is well known that, in some materials with a specific crystallographic symmetry, a pure SH-type SAW propagates accompanied by Φ, which is called the Bleustein–Gulyaev–Shimizu wave (BGS wave) [13-15].

The field distribution of the SH-type SAW is also determined by (1.12), and the SH-type SAW velocity is smaller than that of the SH-type BAW. The velocity reduction reflects how strong the piezoelectricity, i.e., the electromechanical coupling, is near the surface; with a decrease in the velocity, the SH-type SAW energy tends to concentrate close to the surface (see (1.12)). That is, the SH-type SAW is excited and propagates close to the surface by the guiding mechanism basically based on the electromechanical coupling. Namely, with an increase in the piezoelectricity of the SH-type SAW the effects of the surface electrical boundary condition becomes significant, and wave energy tends to concentrate at the surface.

It should be noted that a metallized surface concentrates the propagation of piezoelectric SH-type SAWs much closer to the surface than at a free surface.

Note that SH-type SAWs also appear by depositing materials with low shear wave velocity. This type of eigenmode is called a Love wave [16, 17].

As will be described in Sect. 8.2.2, SH-type SAWs also appear as a result of periodic surface roughness, i.e., a grating on the propagation surface. This type of mode is called a surface transverse wave (STW) [18, 19].

1.2.3 Leaky SAWs

Since piezoelectric materials are anisotropic, both the SH component and Φ generally couple with the L and SV components. So both Rayleigh-type and SH-type SAWs discussed above may possess three mechanical components and Φ; the Rayleigh-type SAW propagates as a "quasi" Rayleigh-type SAW and the SH-type SAW as "quasi" SH-type SAW.

Assume the situation where the velocity of a slow-shear SH-type BAW is smaller than that of the "quasi" Rayleigh-type SAW. As can be seen from (1.12), the SH component no longer concentrates its energy close to the surface. This means that the Rayleigh-type SAW energy leaks into the bulk with propagation and its amplitude on the surface decays exponentially; this type of surface acoustic wave is called a leaky SAW (of Rayleigh-type) or a pseudo SAW [20]. Similarly when the velocity of the "quasi" SH-type SAW is larger than that of the SV-type BAW, it becomes leaky.

If coupling with the slow-shear component is negligible, the propagation loss becomes very small, and the leaky SAW can be treated as a nonleaky SAW. Note that the propagation loss is critically dependent upon the surface boundary condition which generates the coupling among the components.

Figure 1.31 shows the excitation and propagation of a leaked BAW [12]. In the figure, ϕ_c is the critical angle given by

$$\phi_c = \cos^{-1}(V_B/V_S). \tag{1.13}$$

This angle specifies the limited region of the field growth and corresponds to the propagation direction of the leaked BAW. Hence, at a specified x_1, the amplitude of the leaked BAW increases with x_3 because the leaked BAW energy flows toward the direction determined by ϕ_c and the total energy is the accumulation of the energy leaked from the propagating leaky SAW.

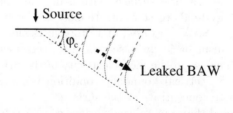

Fig. 1.31. Excitation and propagation of leaked BAW

It may be of current interest to note that there is a situation where a SAW is not of a leaky nature even when the SAW velocity is larger than the slow-shear BAW velocity. This is occasionally encountered when the slow-shear BAW component is completely decoupled from other components. For example, the BGS wave mentioned above is just this case, and the decoupling is due to the crystallographic symmetry for the substrate orientation.

1.2.4 Leaky and Nonleaky SAWs

Generalization of the above discussion suggests that any coupling of two components among four, that is, L, SV, SH, and the potential Φ, generates SAWs, and whether the SAW is leaky or nonleaky is determined by whether the other components are faster than the SAW or not (see Fig. 1.32).

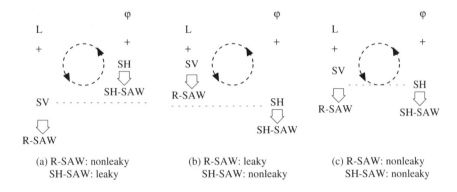

    (a) R-SAW: nonleaky   (b) R-SAW: leaky   (c) R-SAW: nonleaky
       SH-SAW: leaky   SH-SAW: nonleaky   SH-SAW: nonleaky

Fig. 1.32. Which wave becomes leaky, Rayleigh-type SAW or SH-type SAW?

Rayleigh-type SAWs are composed of SV and L components, and become leaky if V_S is faster than the SH-type BAW velocity. Since V_S is about 90% of the SV-type BAW velocity, Rayleigh-type SAWs are usually nonleaky unless the anisotropy of the propagating media is large (case (a)).

On the other hand, SH-type SAWs are composed of SH and Φ components, and become leaky if V_S is faster than the SV-type BAW velocity. Since V_S is smaller than the SH-type BAW velocity and the difference is determined by the piezoelectricity, SH-type SAWs are generally nonleaky if the SH-type BAW is slower than the SV-type BAW (case (b)).

It is interesting to note, as will be discussed in Sect. 6.3.3, that there is a situation where both the Rayleigh-type and SH-type SAWs are nonleaky (case (c)). This occurs when the velocities of the SH-type and SV-type BAWs are in close proximity and the piezoelectricity is sufficiently strong [21].

It is also noticed for 4mm materials such as $Li_2B_4O_7$ where Poisson's ratio becomes very small at a particular rotation angle. That is, in this situation, the coupling between the SV and L components becomes very small, and there exist SAWs mainly composed of L and Φ [22]. This type of SAW is called a longitudinal-type leaky SAW, and is paid much attention because of its extremely high velocity, up to 6500 m/sec.

1.2.5 SSBW

BAWs excited by the input IDT are sometimes detected by the output IDT as a spurious response. Usually they arrive at the output IDT after reflection at the bottom surface of the substrate, and they can be suppressed by sandblasting the bottom surface. There is another kind of BAW which arrives at the output IDT without reflection as shown in Fig. 1.33. This type of BAW is called a surface skimming bulk wave (SSBW) [23] or a shallow bulk acoustic wave (SBAW) [24]. It is clear that this cannot be suppressed by sand-blasting.

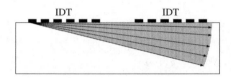

Fig. 1.33. SSBW

At particular crystal orientations for quartz, only the SH-type SSBW of 5000 m/sec is predominantly excited by the IDT and the Rayleigh-type SAW decouples with the piezoelectricity. In the late 1970s to the early 1980s, the SSBW on the substrates was investigated aggressively for the application to high frequency and temperature stable devices [25].

Lately, Thompson and Auld suggested that the SSBW can be trapped by a grating and composes the eigenmode [18]. This wave is called a surface transverse wave as described in Sect. 1.3.2, and is now paid much attention instead of the SSBW.

References

1. B.A. Auld: Acoustic Waves and Fields in Solids, Vol. I, Chap. 3, Wiley, New York (1973) pp. 57–98.
2. B.A. Auld: Acoustic Waves and Fields in Solids, Vol. I, Chap. 7, Wiley, New York (1973) pp. 191–264.
3. J.F. Rosenbaum: Bulk Acoustic Wave Theory and Devices, Altech House, Boston (1988).
4. B.A. Auld: Acoustic Waves and Fields in Solids, Vol. I, Chap. 5, Wiley, New York (1973) pp. 135–162.
5. B.A. Auld: Acoustic Waves and Fields in Solids, Vol. II, Chap. 9, Wiley, New York (1973) pp. 1–61.

6. D.P. Morgan: Surface-Wave Devices for Signal Processing, Appendix B, Elsevier, Amsterdam (1985).
7. G.S. Kino: Acoustic Waves: Devices, Imaging, & Analog Signal Processing, Prentice-Hall, Englewood Cliffs (1987).
8. B.A. Auld: Acoustic Waves and Fields in Solids, Vol. I, Chap. 4 Wiley, New York (1973) pp. 101–134.
9. B.A. Auld: Acoustic Waves and Fields in Solids, Vol. II, Chap. 10, Wiley, New York (1973) pp. 63–219.
10. R.E. Collin: Field Theory of Guided Waves, Chap. 8, McGraw-Hill, New York (1960) pp. 314–367.
11. L. Rayleigh: On Waves Propagating Along the Plane Surface of an Elastic Solid, Proc. London Math. Soc., 7 (1885) pp. 4–11.
12. T. Tamir and A.A. Oliner: Guided Complex Waves, Proc. IEE, 110, 2 (1963) pp. 310-334.
13. J.L. Bleustein: A New Surface Wave in Piezoelectric Materials, Appl. Phys. Lett., 13 (1968) p. 412.
14. Y.V. Gulyaev: Electroacoustic Surface Waves in Solids, Soviet Phys. JETP Lett., 9 (1969) p. 63.
15. Y. Ohta, K. Nakamura and H. Shimizu: Surface Concentration of Shear Wave on Piezoelectric Materials with Conductor, Technical Report of IEICE, Japan US69-3 (1969) in Japanese.
16. B.A. Auld: Acoustic Waves and Fields in Solids, Vol. II, Chap. 5, Wiley, New York (1973) pp. 135–161.
17. A.E.H. Love: Some Problems of Geodynamics, Dover (1967).
18. B.A. Auld, J.J. Gagnepain and M. Tan: Horizontal Shear Surface Waves on Corrugated Surface, Electron. Lett., 12 (1976) pp. 650–651.
19. Y.V. Gulyaev and V.P. Plessky: Slow Surface Acoustic Waves in Solids, Sov. Tech. Phys. Lett. 3 (1977) pp. 220–223.
20. H. Engan, K.A. Ingebrigtsen and A. Tonning: Elastic Surface Waves in α-Quartz: Observation of Leaky Surface Waves, Appl. Phys. Lett., 10 (1967) pp. 311–313.
21. K. Hashimoto and M. Yamaguchi: Non-Leaky, Piezoelectric, Quasi-Shear-Horizontal Type SAW on X-Cut LiTaO$_3$, Proc. IEEE Ultrason. Symp. (1988) pp. 97–101.
22. T. Sato and H. Abe: Longitudinal Leaky Surface Waves for High Frequency SAW Device Application, Proc. IEEE Ultrason. Symp. (1995) pp. 305–315.
23. M. Lewis: Surface Skimming Bulk Waves, SSBW, Proc. IEEE Ultrason. Symp. (1977) pp. 744–752.
24. K.F. Lau, K.H. Yen, R.S. Kagiwada and K.L. Gong: Further Investigation of Shallow Bulk Acoustic Waves Generated by Using Interdigital Transducers, Proc. IEEE Ultrason. Symp. (1977) pp. 996-1001.
25. K.H. Yen, K.F. Lau and R.S. Kagiwada: Recent Advances in Shallow Bulk Acoustic Wave Devices, Proc. IEEE Ultrason. Symp. (1979) pp. 776–785.

2. Grating

This chapter discusses the grating structure as the most fundamental element of SAW devices. In addition to the fundamentals, modeling based on the equivalent circuit and basic properties are included. The behavior of metallic gratings is also discussed by using a simple and effective theory developed by Bløtekjær et al. [1, 2].

2.1 Basic Structure

2.1.1 Fundamentals

SAW devices employ various periodic grating structures shown in Fig. 2.1 such as reflectors and interdigital transducers (IDTs), etc.

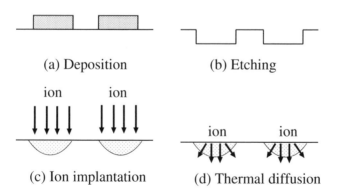

(a) Deposition   (b) Etching

ion   ion

(c) Ion implantation   (d) Thermal diffusion

Fig. 2.1. Grating fabrication

For the fabrication of the grating structure, various techniques including ion implantation (c) [3] and thermal diffusion (d) [4] have been applied. The layered structure (a) fabricated by the deposition of appropriate materials is most commonly used. The groove structure (b) fabricated by the etching of substrates [5] is also used when reduction of the SAW propagation loss is crucial.

For the analysis of SAW propagation in grating structures, the equivalent circuit model shown in Fig. 2.2 is frequently used. In the figure, β_i and R_i are the wavenumber and acoustic impedance, respectively, for the SAW in the region i. In this model, SAW reflection occurs at the impedance discontinuity R_1/R_2 that is modeled to exist at each electrode edge.

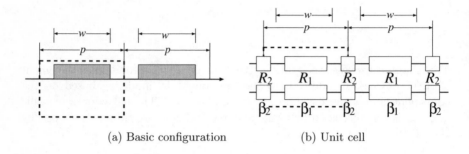

(a) Basic configuration   (b) Unit cell

Fig. 2.2. Equivalent circuit model

Note there is no physically clear definition of the acoustic impedance for the SAW. Since SAW devices are usually characterized by electrical quantities measurable at electrical ports, there is no need to know the actual field strength. Hence only the ratio R_1/R_2 is required for the analysis.

For BAWs, the acoustic impedance is given by $R = \rho V_{\mathrm{B}}$, where ρ is the mass density and V_{B} is the BAW velocity. This relation may be used for the SAW as an analogy.

Added to the reflection caused by the mechanical discontinuity at the surface, a reflection due to the electrical discontinuity at the surface also occurs when a metallic grating is deposited on piezoelectric substrates. This type of reflection will be discussed in Sect. 2.4.

2.1.2 Reflection Center

Let us consider reflection from the strip shown in Fig. 2.3. A portion of incident SAW with amplitude A_{in} is reflected due to the impedance mismatch at the strip ends.

From the equivalent circuit also shown in Fig. 2.3, the SAW amplitudes A_- and A_+ reflected at the left and right ends, respectively, are given by

$$A_- = r_- A_{\mathrm{in}} \exp\{-2j\beta_2(L - w/2)\}$$

and

$$A_+ \cong r_+ A_{\mathrm{in}} \exp\{-j(2\beta_2 L + \beta_1 w)\},$$

where r_- and r_+ are the reflection coefficients at the left and right ends given by $r_- = -r_+ = (R_1 - R_2)/(R_1 + R_2)$. For the derivation, $|r_\pm| \ll 1$ is assumed. Then the reflection coefficient Γ_{s} per strip is given by

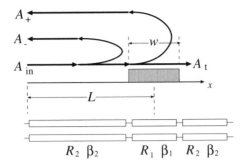

Fig. 2.3. SAW incidence and reflection at a strip

$$\Gamma_\mathrm{s} = \frac{A_+ + A_-}{A_\mathrm{in}} \cong 2jr_- \sin(\beta_1 w) \exp(-2j\beta_2 L). \tag{2.1}$$

This suggests that the reflection by the strip behaves like that occurring at its middle. The equivalent reflection point is called the reflection center.

Since $|\Gamma_\mathrm{s}|{=}2|r_- \sin(\beta w)|$, $|\Gamma_\mathrm{s}|$ takes a maximum value when $w/\lambda \cong 1/4$. Note $\angle\Gamma_\mathrm{s} = \pm\pi/2$ and the sign is determined by that of r_-.

2.2 Behavior in Periodic Structures

2.2.1 Bragg Reflection

Consider wave propagation in a periodic structure with periodicity p shown in Fig. 2.4.

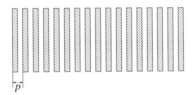

Fig. 2.4. Periodic structure

If the structure is infinitely long, every period is equivalent to every other. So if eigenmodes called grating modes exist, their field distribution $u(x_1)$ for each period must be similar to each other. Namely, $u(x_1)$ must satisfy the following condition, called the Floquet theorem [6] :

$$u(x_1 + p) = u(x_1) \exp(-j\beta p), \tag{2.2}$$

where β is the wavenumber of the grating modes.

If we define a function $U(x_1) = u(x_1)\exp(+j\beta x_1)$, the Floquet theorem suggests $U(x_1)$ must be a periodic function with periodicity p. This means $U(x_1)$ can be expressed in the Fourier expansion form of

$$U(x_1) = \sum_{n=-\infty}^{+\infty} A_n \exp(-2\pi j n x_1/p). \tag{2.3}$$

Then we get

$$u(x_1) = \sum_{n=-\infty}^{+\infty} A_n \exp\{-j(\beta + 2n\pi/p)x_1\}. \tag{2.4}$$

This relation suggests that the field in the periodic structure with periodicity p is expressed as a sum of sinusoidal waves with discrete wavenumbers $\beta_n = 2\pi n/p + \beta$. In other words, wave components with spatial frequency $\beta + 2n\pi/p$ are generated by the spatial modulation of the incident wave with β. We refer to $2\pi/p$ as the grating vector.

Wave fields scattered at each period may interfere. Usually, over wide frequency bands, they cancel each other, and the total reflected field will become negligible. In this case, the wave can be accurately described in (2.4) by a single term with $n = 0$. However, in a certain frequency band, the scattered waves are in phase, they will sum up and cause very strong reflection. In this case, multiple terms in (2.4) must be used for an accurate description of the wave field. This situation is called Bragg reflection.

When λ_S denotes the SAW wavelength in the grating, the phase matching condition is given by

$$2p = n\lambda_S \qquad \text{or} \qquad 4\pi p/\lambda_S = 2n\pi. \tag{2.5}$$

This relation is called the Bragg condition (see Fig. 2.5). By using the

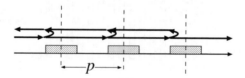

Fig. 2.5. Bragg reflection

wavenumber $\beta_S = 2\pi/\lambda_S$, (2.5) can be rewritten as

$$\beta_S = n\pi/p \qquad \text{or} \qquad |\beta_S + 2n\pi/p| = |\beta_S|. \tag{2.6}$$

This relation can be regarded as follows. Various components with the wavenumbers $\beta_S + 2n\pi/p$ are generated by the spatial modulation of the incident SAW with the wavenumber β_S. Then Bragg reflection occurs, and a

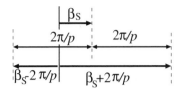

Fig. 2.6. Bragg condition

component with wavenumber $\beta_S - 2\pi/p$ will grow only when the component has the same phase velocity as the incident SAW (see Fig. 2.6).

By defining the equivalent SAW phase velocity V_S in the grating, the Bragg frequency is given by

$$\omega \cong n\pi V_S/p, \tag{2.7}$$

where the condition of (2.6) is satisfied.

Bragg reflection also occurs between different kinds of waves. Figure 2.7 shows generation of BAWs due to back-scattering of the incident SAW.

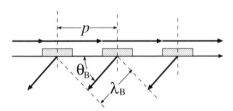

Fig. 2.7. BAW radiation by back-scattering of incident SAW

For this case, the phase matching condition is given by

$$2\pi p/\lambda_S + 2\pi p \cos\theta_B/\lambda_B = 2n\pi, \tag{2.8}$$

where λ_B is the BAW wavelength, and θ_B is the BAW radiation angle. By using the BAW wavenumber $\beta_B = 2\pi/\lambda_B$, it can be written as

$$\beta_S + \beta_B \cos\theta_B = 2n\pi/p. \tag{2.9}$$

As well as (2.6), this relation can be regarded as follows: among the various components with $\beta_S + 2n\pi/p$ generated by the incident SAW with β_S, only the component with $\beta_S - 2\pi/p$ will grow only when $|\beta_S - 2\pi/p|$ coincides with the BAW wavenumber $\beta_B \cos\theta_B$ parallel to the surface (see Fig. 2.8).

Next, let us consider oblique incidence of the SAW to the grating as shown in Fig. 2.9. By using the wavevector, the Bragg condition for this case is simply represented by Fig. 2.10. From this figure, it is clear that the condition is given by

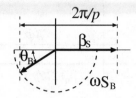

Fig. 2.8. Bragg condition for BAW diffraction caused by an incident SAW

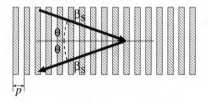

Fig. 2.9. Oblique incidence of SAW

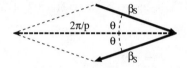

Fig. 2.10. Bragg condition for oblique SAW incidence

$$\beta_S \cos \theta = n\pi/p. \tag{2.10}$$

This indicates that the Bragg frequency increases with θ if β_S is independent of θ.

2.2.2 Energy Storing Effect

Equation (2.9) suggests that back-scattering to the BAW of the incident SAW occurs only when

$$\omega > \frac{2n\pi}{p}(V_S^{-1} + V_B^{-1}), \tag{2.11}$$

where V_S and V_B are the SAW and BAW velocities, respectively. Since $V_S < V_B$, back-scattering does not occur at frequencies lower than the Bragg frequency with $n = 1$ for the SAW.

Note that this is due to the mutual cancellation of BAWs scattered by the grating structure, and the BAW scattering itself is not suppressed (see Fig. 2.11a). Thus the energy of the scattered BAW is stored near the strip edges, and this causes the reduction in the SAW velocity. This phenomenon is called the energy storing effect [7]. In the equivalent circuit, this effect is expressed by adding capacitors C_e in parallel at the electrode edges (see Fig. 2.11b).

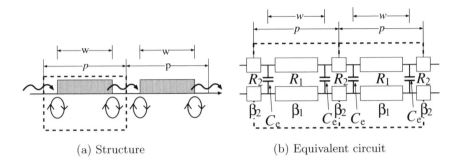

(a) Structure   (b) Equivalent circuit

Fig. 2.11. Energy storing effect

On the other hand, when $\omega > 2n\pi/p \times (V_S^{-1} + V_B^{-1})$, the energy storing effect is weak. This is because, at this frequency range, the scattered BAW energy is efficiently radiated into the bulk, and interference becomes small.

Note that since the interference occurring under the periodic grating structure is broken at discontinuous regions such as the edges of the grating structures, part of the scattered BAW energy is radiated into the bulk of substrate. Hence, the energy storing effect is made weak.

2.2.3 Fabry–Perot Resonator

Consider the optical resonator shown in Fig. 2.12, where two mirrors are placed in parallel at a distance L. When the reflection and transmission coefficients of each mirror are Γ and T, respectively, the law of energy conservation requires $|\Gamma|^2 + |T|^2 \leq 1$.

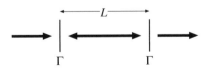

Fig. 2.12. Fabry–Perot resonator

For normal incidence, the transmission coefficient $\hat{T}$ and reflection coefficient $\hat{\Gamma}$ for the total structure are given by

$$\hat{T} = \sum_{n=0}^{+\infty} \Gamma^{2n} T^2 \exp\{-j\beta L(2n+1)\} = \frac{T^2 \exp(-j\beta L)}{\Gamma^2 \exp(-2j\beta L) - 1}, \qquad (2.12)$$

$$\hat{\Gamma} = \Gamma + T^2 \sum_{n=1}^{+\infty} \Gamma^{2n-1} \exp(-2jn\beta L) = \Gamma \frac{(\Gamma^2 - T^2)\exp(-2j\beta L) - 1}{\Gamma^2 \exp(-2j\beta L) - 1}.$$

(2.13)

If Γ is real, $|\hat{T}|$ takes maxima $T^2/(1-\Gamma^2)$ and minima $T^2/(1+\Gamma^2)$ periodically, and they appear when $\beta L = m\pi$ and $\beta L = \pi(m+1/2)$, respectively. Thus, by employing mirrors with $|\Gamma| \cong 1$, very narrow bandpass filters with large out-of-band rejection can be constructed. This type of resonator is called a Fabry–Perot resonator.

These maxima and minima are due to multiple reflection between the mirrors. Since the ratio between their amplitudes is given by $(1+\Gamma^2)/(1-\Gamma^2)$, this relation can be used for the experimental determination of Γ.

When the grating reflector is employed instead of the mirrors, the behavior of the resonator is mostly the same. The only one difference is that the frequency dependence of the reflector also affects the total frequency response.

2.3 Equivalent Circuit Analysis

2.3.1 Analysis

SAW propagation in the grating structure is analyzed by using the equivalent circuit shown in Fig. 2.13, where the energy storing effect is taken into account by C_e.

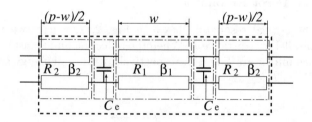

Fig. 2.13. Equivalent circuit

The two-port matrix $[F]$ for a period is given by

$$[F] = [F^{(2)}]^{1/2}[G][F^{(1)}][G][F^{(2)}]^{1/2},$$

(2.14)

where

$$[F^{(i)}] = \begin{pmatrix} \cos\theta_i & jR_i\sin\theta_i \\ jR_i^{-1}\sin\theta_i & \cos\theta_i \end{pmatrix},$$

(2.15)

$$[G] = \begin{pmatrix} 1 & 0 \\ j\omega C_\mathrm{e} & 1 \end{pmatrix},$$

(2.16)

$\theta_1 = \beta_1 w$ and $\theta_2 = \beta_2(p - w)$ are the phase shift in the electroded and unelectroded regions, respectively.

Since the structure is symmetric, $[F]$ can be expressed in the form

$$[F] = \begin{pmatrix} \cos\theta_e & jR_e\sin\theta_e \\ jR_e^{-1}\sin\theta_e & \cos\theta_e \end{pmatrix},$$

(2.17)

where $\beta_e p = \theta_e = \cos^{-1}(F_{11})$ is the effective wavenumber given by

$$\theta_e = \cos^{-1}(f_{11}f_{22} + f_{12}f_{21}),$$

(2.18)

and $R_e = \sqrt{F_{12}/F_{21}}$ is the effective acoustic impedance given by

$$R_e = \frac{f_{11}f_{12}}{j\sin\theta_e} = \sqrt{\frac{f_{11}f_{12}}{f_{22}f_{21}}},$$

(2.19)

where

$$f_{11} = \cos(\theta_1/2)\cos(\theta_2/2) - (R_2/R_1)\sin(\theta_1/2)\sin(\theta_2/2)$$
$$-\omega C_e R_2 \cos(\theta_1/2)\sin(\theta_2/2),$$
$$f_{22} = \cos(\theta_1/2)\cos(\theta_2/2) - (R_1/R_2)\sin(\theta_1/2)\sin(\theta_2/2)$$
$$-\omega C_e R_1 \sin(\theta_1/2)\cos(\theta_2/2),$$
$$f_{12} = jR_2[\cos(\theta_1/2)\sin(\theta_2/2) + (R_1/R_2)\sin(\theta_1/2)\cos(\theta_2/2)$$
$$-\omega C_e R_1 \sin(\theta_1/2)\sin(\theta_2/2)],$$
$$f_{21} = jR_2^{-1}[\cos(\theta_1/2)\sin(\theta_2/2) + (R_2/R_1)\sin(\theta_1/2)\cos(\theta_2/2)$$
$$+\omega C_e R_2 \cos(\theta_1/2)\cos(\theta_2/2)].$$

By substituting into (2.18), we obtain

$$\cos\theta_e = \cos(\theta_1 + \theta_2) - \frac{1}{2}\left\{\frac{(R_1 - R_2)^2}{R_1 R_2} - \omega^2 C_e^2 R_1 R_2\right\}\sin\theta_1\sin\theta_2$$
$$-\omega C_e R_2 \sin(\theta_1 + \theta_2) - \omega C_e R_2(R_1/R_2 - 1)\sin\theta_1\cos\theta_2.$$

Assuming $R_1 \cong R_2$ and $\omega C_e R_2 \ll 1$, it can be simplified as

$$\cos\theta_e \cong \cos(\theta_1 + \theta_2) - \psi\sin(\theta_1 + \theta_2) - \Delta\psi\sin\theta_1\cos\theta_2$$
$$-\frac{1}{2}(\Delta^2 - \psi^2)\sin\theta_1\sin\theta_2$$
$$\cong (1 + T^2/4)\cos(\theta_1 + \theta_2 + \psi) - (T^2/4)\cos(\theta_1 - \theta_2 - 2\eta), \quad (2.20)$$

where $\Delta = R_1/R_2 - 1$, $\psi = \omega C_e R_2$, $T = \text{sgn}(\Delta)\sqrt{\Delta^2 + \psi^2}$, and $\eta = \tan^{-1}(\psi/\Delta)$. Equation (2.20) suggests that $|\cos\theta_e|$ becomes larger than unity, namely, θ_e becomes complex when $\theta_1 + \theta_2 + \psi \cong m\pi$. Since this system is lossless, a complex wavenumber means the corresponding mode is evanescent. Thus if the grating structure is infinitely long, the incident wave is totally reflected (Bragg reflection).

By applying the same approximation, we also obtain

$$R_e \cong R_2 \sqrt{\frac{\sin(\theta_1 + \theta_2 + \psi) + T\sin(\theta_1 + \psi/2 - \eta)}{\sin(\theta_1 + \theta_2 + \psi) - T\sin(\theta_1 + \psi/2 - \eta)}}. \tag{2.21}$$

It should be noted that Bragg reflection occurs not only at a particular frequency but also in a certain range of frequencies where θ_e and R_2 become complex. We call this frequency range the stopband, and the other frequency range the passband. In addition, we sometimes use the term "Bragg frequency" for the frequency giving maximum reflection.

From (2.17), the matrix $[F]^N$ for the cascade-connection of N unit cells is given by

$$[F]^N = \begin{pmatrix} \cos N\theta_e & jR_e \sin N\theta_e \\ jR_e^{-1} \sin N\theta_e & \cos N\theta_e \end{pmatrix}. \tag{2.22}$$

Thus, the reflection coefficient for the finite grating structure shown in Fig. 2.14 is given by

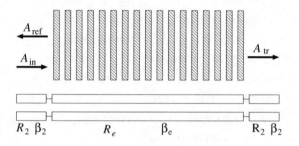

Fig. 2.14. Finite grating structure

$$\Gamma = \frac{j(R_e/R_2 - R_2/R_e)\tan N\theta_e}{2 + j(R_e/R_2 + R_2/R_e)\tan N\theta_e}. \tag{2.23}$$

Figure 2.15 shows the calculated reflection coefficient $|\Gamma|$ with L/p as a parameter. In the calculation, $\Delta = 0.01\pi$, $\psi = 0$, and $\theta_1 = \theta_2$. The horizontal axis is the normalized frequency $(\theta_1 + \theta_2)/\pi - 1$. When L/p is small, Γ exhibits frequency dependence of the form $\sin x/x$. With an increase in L/p, the peak at $\theta_1 + \theta_2 = \pi$ becomes flat, and its height and width tend to saturate.

To design SAW resonators, it is convenient to define an equivalent mirror. That is, near a specific frequency ω_s, the grating reflector may be regarded as a mirror of reflection coefficient $\Gamma|_{\omega=\omega_s}$ located at a distance L_e from the end of the grating. Since the time delay due to reflection is given by $-\partial(\angle\Gamma)/\partial\omega$ (see Sect. 1.1.7), L_e is given by

$$L_e = -\frac{V_S}{2}\frac{\partial(\angle\Gamma)}{\partial\omega} = -\frac{1}{2}\frac{\partial(\angle\Gamma)}{\partial\beta}, \tag{2.24}$$

where V_S is the effective SAW velocity in the grating structure.

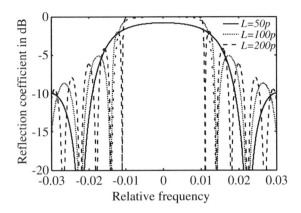

Fig. 2.15. Reflection coefficient of grating with $\Delta = 0.01\pi$ and $\psi = 0$

It should be noted that L_e is the distance from the left end of the unit cell at the incident side, and not from the middle of the unit cell.

2.3.2 Dependence of Reflection Characteristics on Parameters

Let us discuss how the reflective characteristics are dependent upon the parameters appearing in the equivalent circuit.

From (2.20), the Bragg condition is given by

$$\theta_1 + \theta_2 + \psi = m\pi. \tag{2.25}$$

When we set $\theta_i = \omega/V_i$, the Bragg frequency ω_c is given by

$$\omega_c = \frac{m\pi - \psi}{(p - w)/V_1 + w/V_2}. \tag{2.26}$$

Equation (2.20) suggests that the relation

$$(1 + T^2/4)\cos(\theta_1 + \theta_2 + \psi) - (T^2/4)\cos(\theta_1 - \theta_2 - 2\eta) = (-1)^m$$

holds at the edges of the stopband. Then the frequencies ω_+ and ω_- giving its upper and lower edges are given by

$$\omega_\pm \cong \begin{cases} \dfrac{m\pi - \psi \pm |T\cos\{(\theta_1 - \theta_2)/2 - \eta\}|}{(p - w)/V_1 + w/V_2} & (m : \text{odd}) \\[2ex] \dfrac{m\pi - \psi \pm |T\sin\{(\theta_1 - \theta_2)/2 - \eta\}|}{(p - w)/V_1 + w/V_2} & (m : \text{even}). \end{cases} \tag{2.27}$$

From (2.26) and (2.27), we obtain the following relation:

$$\frac{\omega_+ - \omega_-}{\omega_c} \cong \begin{cases} \dfrac{2|T\cos\{(\theta_1 - \theta_2)/2 - \eta\}|}{m\pi - \psi} & (m : \text{odd}) \\[4mm] \dfrac{2|T\sin\{(\theta_1 - \theta_2)/2 - \eta\}|}{m\pi - \psi} & (m : \text{even}). \end{cases} \tag{2.28}$$

Next, (2.20) also gives the attenuation coefficient $\alpha = \Im(\theta_e)/p$ of the grating mode at the Bragg frequency:

$$\alpha p = \cosh^{-1}[1 + T^2/4 - (-1)^m T^2 \cos(\theta_1 - \theta_2 - 2\eta)/4]$$

$$\cong \begin{cases} |T\cos\{(\theta_1 - \theta_2)/2 - \eta\}| & (m : \text{odd}) \\[3mm] |T\sin\{(\theta_1 - \theta_2)/2 - \eta\}| & (m : \text{even}). \end{cases} \tag{2.29}$$

This indicates that the energy storing effect is responsible not only for the resonance frequency but also the amplitude of the reflection. Comparison of (2.29) with (2.28) gives the following relation:

$$\frac{\omega_+ - \omega_-}{\omega_c} \cong \frac{2\alpha p}{m\pi - \psi}. \tag{2.30}$$

Since $R_e = \pm j R_2$ at the Bragg frequency, substitution into (2.23) gives the reflection coefficient Γ for the finite grating:

$$\Gamma = \pm j \tanh(\alpha N p), \tag{2.31}$$

where the sign is determined by that of $R_1/R_2 - 1$. This indicates that $\pm j\alpha p$ corresponds to the reflection coefficient Γ_e per strip.

It is clear from (2.26)–(2.31) that reflection in the grating structure is fully characterized by η. That is, if $\theta_1 - \theta_2 \cong 0$, namely $w \cong p/2$,

- If $\psi \ll |\Delta|$ (if the effect of the difference in the acoustic impedances is dominant), the stopbands appear for odd-order resonances with odd m. The stopband width is determined by the difference in the acoustic impedance given by $|\Delta|$.
- If $\psi \gg |\Delta|$ (if the energy storing effect is dominant), the stopbands appear for even-order resonances with even m. The stopband width is determined by ψ, and is responsible for the energy storing effect.
- In both cases, the Bragg frequency decreases with an increase in ψ due to the energy storing effect.
- The stopband width is proportional to the attenuation coefficient of the grating mode.

Note that, as seen in Fig. 2.15, the stopband for gratings with finite length is wider than that for a semi-infinite grating structure. This must be taken into account for the experimental characterization. Since $\tan(N\theta_e) = 0$ at frequencies $\omega_{0\pm}$ giving the first nulls, application of the procedure employed to derive (2.28) gives

$$\frac{\omega_{0+} - \omega_{0-}}{\omega_+ - \omega_-} \cong \begin{cases} \sqrt{1 + [\pi/NT \cos\{(\theta_1 - \theta_2)/2 - \eta\}]^2} & (m : \text{odd}) \\ \sqrt{1 + [\pi/NT \sin\{(\theta_1 - \theta_2)/2 - \eta\}]^2} & (m : \text{even}). \end{cases} \tag{2.32}$$

Then by substituting (2.30), we obtain the following relation:

$$\frac{\omega_{0+} - \omega_{0-}}{\omega_+ - \omega_-} \cong \sqrt{1 + (\pi/N\alpha p)^2}. \tag{2.33}$$

By the way, (2.23) can be rewritten as

$$\Gamma = \Gamma_0 \frac{1 - \exp(-2jN\theta_e)}{1 - \Gamma_0^2 \exp(-2jN\theta_e)} \tag{2.34}$$

$$= \Gamma_0 - \Gamma_0(1 - \Gamma_0^2) \exp(-2jN\theta_e) \sum_{n=0}^{+\infty} \left\{ \Gamma_0^2 \exp(-2jN\theta_e) \right\}^n. \tag{2.35}$$

Note that the first term Γ_0 is due to reflection at the edge for the incident side given by

$$\Gamma_0 = \frac{R_e - R_2}{R_e + R_2}, \tag{2.36}$$

and also corresponds to the reflection coefficient for a semi-infinite grating as $N \to \infty$. On the other hand, the second term is due to multiple reflection at the grating.

Note that R_e is positive and real in the passband, and converges to R_2 when the driving frequency is far from the stopband edges. On the other hand, R_e is purely imaginary in the stopband. So Γ_0 exhibits the frequency dependence shown in Fig. 2.16.

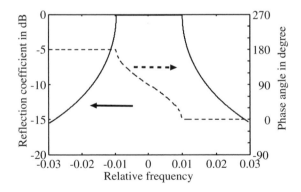

Fig. 2.16. Reflection characteristics of a semi-infinite grating with $\Delta = 0.01\pi$. (The phase shifts by π when $\Delta < 0$)

Next, let us discuss a grating structure with finite length. Equation (2.34) suggests that $|\Gamma| \cong 1$ in the stopband provided $2N\Im(\theta_e) \gg 1$. This is because, in the stopband, the field is evanescent and strips far from the incident edge scarcely contribute to the reflection coefficient (see Fig. 2.17a).

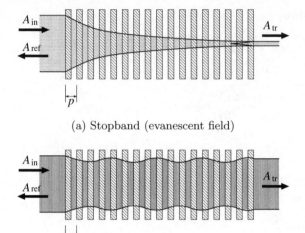

(a) Stopband (evanescent field)

(b) Passband (standing wave field due to reflection at edges)

Fig. 2.17. Field distribution in grating

On the other hand, in the passband, periodic ripples appear. This is due to the standing wave arising from the interference between the reflections at the left and right edges of the grating as shown in Fig. 2.17b. The periodicity is determined by the grating length and the effective SAW velocity. Reflection characteristics of the finite grating are shown in Fig. 2.18.

Finally, let us calculate the effective mirror distance L_e given by (2.24). Assuming $|\Delta\beta| \ll 1$ where $\Delta\beta = \beta - m\pi/p$ and $\beta p = \theta_1 + \theta_2 + \psi$, we can approximate (2.21) as

$$
\begin{aligned}
(R_e/R_2)^2 &\cong \frac{(-1)^m \Delta\beta p + T\sin(\theta_1 + \psi/2 - \eta)}{(-1)^m \Delta\beta p - T\sin(\theta_1 + \psi/2 - \eta)} \\
&\cong -[1 + (-1)^m \Delta\beta/T\sin(\theta_1 + \psi/2 - \eta)]^2.
\end{aligned}
\tag{2.37}
$$

If the grating is sufficiently long, its characteristics near the Bragg frequency may be approximated by those of the semi-infinite grating. Then substitution of (2.37) into (2.36) gives

$$
\begin{aligned}
\Gamma_0 &\cong -\frac{1 \mp j \mp j\Delta\beta/|T\sin(\theta_1 + \psi/2 - \eta)|}{1 \pm j \pm j\Delta\beta/|T\sin(\theta_1 + \psi/2 - \eta)|} \\
&= \pm j\frac{2 - j\Delta\beta/|T\sin(\theta_1 + \psi/2 - \eta)|}{2 + j\Delta\beta/|T\sin(\theta_1 + \psi/2 - \eta)|} \\
&\cong \exp[\pm j\pi/2 - j\Delta\beta/|T\sin(\theta_1 + \psi/2 - \eta)|].
\end{aligned}
\tag{2.38}
$$

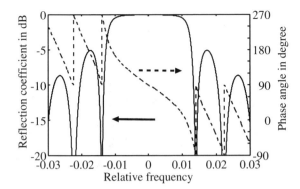

Fig. 2.18. Reflection characteristics of a finite grating with $\Delta = 0.01\pi$ and $L/p = 100$. (The phase shifts by π when $\Delta < 0$)

Since $\theta_1 + \psi/2 - \eta - m\pi = (\theta_1 - \theta_2)/2 - \eta - m\pi/2$ at $\beta p = m\pi$, this substitution into (2.24) gives

$$L_e \cong - \frac{1}{2} \frac{\partial(\angle\Gamma_0)}{\partial\beta}\bigg|_{\beta p = m\pi}$$

$$\cong \begin{cases} \dfrac{p}{2|T\cos\{(\theta_1 - \theta_2)/2 - \eta\}|} & (m : \text{odd}) \\[4mm] \dfrac{p}{2|T\sin\{(\theta_1 - \theta_2)/2 - \eta\}|} & (m : \text{even}). \end{cases} \tag{2.39}$$

Then comparison with (2.29) gives the following relation:

$$L_e = (2\alpha)^{-1}. \tag{2.40}$$

2.4 Metallic Grating

2.4.1 Fundamental Characteristics

Since most SAW devices employ piezoelectric substrates, the use of metallic gratings causes electrical reflection in addition to mechanical reflection. Namely, since the electric field associated with SAW propagation is short-circuited by the conductivity of the film, it causes a perturbation to the SAW propagation. In addition, the charge induced on the metal strips regenerates the SAWs. This effect is called electric regeneration, and is equivalent to reflection. Note that, due to this, the reflection characteristics of the structure are very dependent upon the electrical connection among the strips.

Figure 2.19 shows typical metallic gratings. (a) shows an open-circuited (OC) grating where each strip is electrically isolated. (b) shows a short-circuited (SC) grating where every strip is connected electrically in parallel.

(c) is their combination and is called a positive and negative reflection (PNR) type grating [8]. This configuration offers a larger reflection coefficient than (a) or (b).

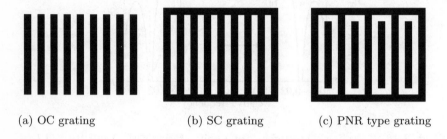

(a) OC grating   (b) SC grating   (c) PNR type grating

Fig. 2.19. Metallic grating

2.4.2 SAW Dispersion Characteristics

Since strip widths are comparable to the SAW wavelength, SAW propagation in the structure is inherently frequency dispersive.

Bløtekjær et al. proposed an efficient method to calculate the SAW dispersion relation in an infinitely long metallic grating structure where effects of the strip thickness and resistivity are ignored [1, 2]. Note that conventional numerical methods are usually ineffective for our purpose because the effects of charge concentration at strip edges are significant [9].

References [1, 2] suggest that the current and voltage induced in each strip by the SAW are not independent, and their ratio is related by the strip admittance $Y(s, \omega)$ (see Sect. B.2.1):

$$Y(s, \omega) = 2j\omega \sin(s\pi)\epsilon(\infty) \frac{\sum_{m=M_1}^{M_2} A_m P_{m+s-1}(\cos \Delta)}{\sum_{m=M_1}^{M_2} (-)^m A_m P_{m+s-1}(-\cos \Delta)}, \tag{2.41}$$

where $s = \beta_0 p/2\pi$, β_0 is the SAW wavenumber, p the grating period, $\Delta = w/p$, w the strip width, and $P_m(\theta)$ is the Legendre function of m-th order. A_m are coefficients whose ratio is determined by the following linear equations:

$$\sum_{m=M_1}^{M_2} A_m \left\{ S_{n-m} - S_n \frac{\epsilon(\infty)}{\epsilon(\beta_n/\omega)} \right\} P_{n-m}(\cos \Delta) = 0 \tag{2.42}$$

for $n = [M_1, M_2 - 1]$ where

$$S_n = \begin{cases} 1 & (n \geq 0) \\ -1 & (n < 0), \end{cases} \tag{2.43}$$

$\beta_n = \beta_0 + 2\pi n/p$ and $\epsilon(S)$ is the effective permittivity which will be described in Sect. 6.2.

For short-circuited gratings, the SAW dispersion relation is given by $Y(s,\omega)^{-1} = 0$ since the induced voltage is zero. On the other hand, for open-circuited gratings, the SAW dispersion relation is given by $Y(s,\omega) = 0$ since the induced current is zero.

Figure 2.20 shows, as a function of M_2, the error in SAW resonance frequencies ω_{o+} and ω_{o-} at the stopband edges for the open-circuited grating and those ω_{s+} and ω_{s-} for the short-circuited grating when $s=1/2$ and $M_1=-M_2$. As a substrate, 128°YX-LiNbO$_3$ was chosen. It is seen that the accuracy increases rapidly with an increase in M_2, and $M_2=1$ is sufficient for most cases.

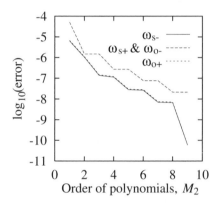

Fig. 2.20. M_2 dependence of calculation error

Figure 2.21a shows the dispersion of the SAW phase velocity calculated for 128°YX-LiNbO$_3$. The horizontal axis is the normalized frequency fp/V_B where $V_B =4025$ m/sec is the slow-shear SSBW velocity. The velocity changes irregularly at $fp/V_B \cong 0.5$. This frequency region corresponds to the stopband due to Bragg reflection. Except for this region, the SAW velocity increases monotonically with an increase in f for the short-circuited grating whereas it decreases for the open-circuited grating.

Figure 2.21b shows the attenuation with propagation. A steep peak at $f \cong 0.5V_B/p$ is due to SAW Bragg reflection, and the other loss is due to back-scattering of the incident SAW to the BAWs described in Sect. 2.2.1. The scattering does not occur at $f < 0.5V_B/p$ due to its cut-off nature. When $f > 0.62V_B/p$, the attenuation increases very rapidly by the coupling with the longitudinal SSBW. The attenuation is relatively small at $0.5 < fp/V_B < 0.62$ because the electromechanical coupling for the slow-shear SSBW is very small for 128°YX-LiNbO$_3$. Note that the attenuation characteristics for the short-circuited grating are mostly the same as those for the open-circuited grating.

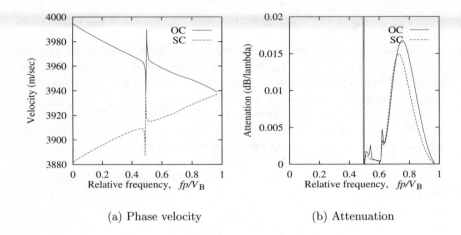

(a) Phase velocity   (b) Attenuation

Fig. 2.21. Phase velocity and attenuation of SAW in 128°YX-LiNbO$_3$

2.4.3 Approximated Dispersion Characteristics

It was shown previously that sufficient accuracy is achievable even when $M_2 = -M_1 = 1$. In this case, (2.42) gives

$$r_1 = A_1/A_0 = \frac{R_{-1} + \cos \Delta}{R_0 R_{-1} - \cos^2 \Delta}, \tag{2.44}$$

$$r_{-1} = A_{-1}/A_0 = \frac{R_0 + \cos \Delta}{R_0 R_{-1} - \cos^2 \Delta}, \tag{2.45}$$

where

$$R_n = \frac{\epsilon(\beta_n/\omega) + \epsilon(\infty)}{\epsilon(\beta_n/\omega) - \epsilon(\infty)}. \tag{2.46}$$

By substituting (2.44) and (2.45) into (2.41), the dispersion relation for the open-circuited grating is given by

$$1 + r_1 F_s(\cos \Delta) + r_{-1} F_{1-s}(\cos \Delta) = 0, \tag{2.47}$$

whereas that for the short-circuited grating is given by

$$1 - r_1 F_s(-\cos \Delta) - r_{-1} F_{1-s}(-\cos \Delta) = 0, \tag{2.48}$$

where

$$F_s(x) = \frac{P_s(x)}{P_{s-1}(x)}. \tag{2.49}$$

Figure 2.22 shows $F_s(x)$ as a function of s with x as a parameter. When x is not too small, $F_s(x)$ is mostly linear against s and x, that is, the approximation $F_s(x) \cong s(x-1) + 1$ holds.

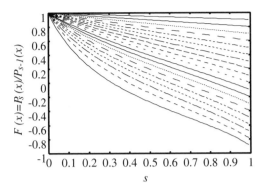

Fig. 2.22. $F_s(x) = P_s(x)/P_{s-1}(x)$. From the top, $x = 0.95, 0.9, \cdots - 0.95$

Note $\epsilon(S)$ has a null at the SAW velocity $V = V_f \equiv S_f^{-1}$ for the free surface whereas it has a pole at the SAW velocity $V = V_m \equiv S_m^{-1}$ for the metallized surface. Thus, as will be discussed in Sect. 6.2, $\epsilon(S)$ can be approximated as

$$\epsilon(S) \cong \epsilon(\infty)\frac{S^2 - S_f^2}{S^2 - S_m^2}. \tag{2.50}$$

Substitution into (2.46) gives

$$R_0 = 1 + 2\frac{(\beta/\omega)^2 - S_m^2}{S_m^2 - S_f^2}. \tag{2.51}$$

When the frequency is far from the stopband, $1/R_{-1} \cong 0$, $r_{-1} \to 0$, and $r_1 \cong 1/R_0$. Thus the dispersion relation for the open-circuited grating is given by

$$R_0 = F_s(\cos \Delta), \tag{2.52}$$

whereas that for the short-circuited grating is given by

$$R_0 = -F_s(-\cos \Delta). \tag{2.53}$$

By the substitution of (2.52) and (2.53) into (2.51), the phase velocity V_p on the open-circuited grating is given by

$$V_p = V_m/\sqrt{1 - \frac{K_V^2}{2}\{1 + F_s(\cos \Delta)\}} \cong V_m\left[1 + \frac{K_V^2}{4}\{1 + F_s(\cos \Delta)\}\right]$$
$$\cong V_f\left[1 - \frac{K_V^2}{4}\{1 - F_s(\cos \Delta)\}\right], \tag{2.54}$$

and that on the short-circuited grating is given by

$$V_p = V_m/\sqrt{1 - \frac{K_V^2}{2}\{1 - F_s(-\cos \Delta)\}}$$
$$\cong V_m\left[1 + \frac{K_V^2}{4}\{1 - F_s(-\cos \Delta)\}\right], \tag{2.55}$$

where $K_V^2 = (V_f^2 - V_m^2)/V_f^2$ is what is often employed as the electromechanical coupling factor for the SAW (see Sect. 3.2.2).

Figure 2.23 shows the change in the frequency dispersion calculated by using (2.52) and (2.53). In the calculation, $K_V^2 = 0.05$.

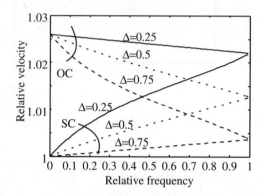

Fig. 2.23. Frequency dispersion in a metallic grating when reflection is ignored

Since $F_0(x) = 1$, the SAW velocity V_p on the open-circuited grating converges to V_f when $s \to 0$ whereas that on the short-circuited grating does to V_m. This gives negligible phase shift in each period. That is, as shown in Fig. 2.24, the effect of the short-circuited electric field is negligible for the open-circuited grating whereas variation of the surface voltage within the gaps is negligible for the short-circuited grating.

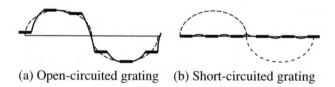

(a) Open-circuited grating　(b) Short-circuited grating

Fig. 2.24. Behavior of grating mode at $s \ll 1$

On the other hand, since $F_1(x) = x$, the SAW velocity V_p on the open-circuited grating coincides with that on the short-circuited grating at the stopband edges at $s = 1$. This is because, at $s = 1$, the voltage for each strip is equal even in the open-circuited grating, and no current flows among strips independent of the mutual electrical connection. V_p in this case is given by

$$V_p = \sqrt{\frac{2}{S_f^2 + S_m^2 - \cos \Delta (S^2 - S_f^2)}} \cong V_m \left\{ 1 + \frac{K_V^2}{4}(1 + \cos \Delta) \right\}. \quad (2.56)$$

Next, let us consider the stopband edges at $s = 1/2$. Since $R_{-1} = R_0$, (2.44) and (2.45) give

$$r_1 = r_{-1} = (R_0 - \cos \Delta)^{-1}. \tag{2.57}$$

Hence the SAW dispersion relation for the open-circuited grating is given as solutions of

$$R_0 = \cos \Delta - 2F_{1/2}(\cos \Delta) \tag{2.58}$$

or

$$R_0 = - \cos \Delta. \tag{2.59}$$

On the other hand, that for the short-circuited grating is given as solutions of

$$R_0 = \cos \Delta + 2F_{1/2}(- \cos \Delta) \tag{2.60}$$

or (2.59).

Equations (2.58)–(2.60) suggest that the SAW velocities for the open-circuited and short-circuited gratings coincide with each other at either one of two frequencies giving the stopband edges. This can be explained as follows. At the stopband edges ($s = 0.5$), the standing wave field occurs in the grating because the wavenumber is purely real. Since the system considered here is geometrically symmetrical, the allowable resonance patterns can be categorized into two types as shown in Fig. 2.25; one is symmetric and the other is antisymmetric with respect to the geometrical center for each period. For the antisymmetric resonance shown in (b), the charge distribution induced on the strip is also antisymmetric, and thus the total induced charge is zero. This means that, for antisymmetrical resonance, no current flows among strips even in the short-circuited grating. On the other hand, the symmetric case shown in (a) has nonzero total charge, and SAW propagation is influenced by the mutual electrical connection.

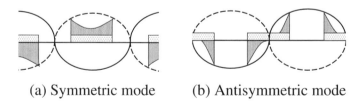

(a) Symmetric mode   (b) Antisymmetric mode

Fig. 2.25. Field distribution at $s = 0.5$

It should be noted that if none of the stopband edges for short-circuited gratings coincide with those for open-circuited gratings, it means that there exist structural and/or crystallographic asymmetries.

By substituting (2.58)–(2.60) into (2.51), we obtain the SAW velocity at the upper edge of the stopband as

$$V_p \cong V_m \left[1 + \frac{K_V^2}{4} \{ 1 - \cos \Delta + 2F_{1/2}(\cos \Delta) \} \right], \tag{2.61}$$

whereas that at the lower edge of the stopband:

$$V_p \cong V_m \left[1 + \frac{K_V^2}{4} (1 + \cos \Delta) \right], \tag{2.62}$$

which coincides with that at the upper edge of the stopband for the short-circuited grating, where the SAW velocity at the lower edge is given by

$$V_p \cong V_m \left[1 + \frac{K_V^2}{4} \{ 1 - \cos \Delta - 2F_{1/2}(-\cos \Delta) \} \right]. \tag{2.63}$$

References

1. K. Bløtekjær, K.A. Ingebrigtsen and H. Skeie: A Method for Analysing Waves in Structures Consisting of Metallic Strips on Dispersive Media, IEEE Trans. Electron. Devices, **ED-20** (1973) pp. 1133–1138.
2. K. Bløtekjær, K.A. Ingebrigtsen and H. Skeie: Acoustic Surface Waves in Piezo-electric Materials with Periodic Metallic Strips on the Substrate, IEEE Trans. Electron. Devices, **ED-20** (1973) pp. 1139–1146.
3. P. Hartemann: Ion Implanted Acoustic-Surface-Wave Resonators, Appl. Phys. Lett., **28** (1976) pp. 73–75.
4. R.V. Schmidt: Acoustic Surface Wave Velocity Perturbation in LiNbO₃ by Diffusion of Metals, Appl. Phys. Lett., **27** (1975) pp. 8–10.
5. R.C.M. Li, J.A. Alusow and R.C. Williamson: Experimental Exploration of the Limits of Achievable Q of Grooved Surface-Wave Resonators, Proc. IEEE Ultrason. Symp. (1975) pp. 279–283.
6. R.E. Collin: Field Theory of Guided Waves, Chap. 9, McGraw-Hill, New York (1960) pp. 368–408.
7. R.C.M. Li and J. Melngailis: The Influence of Stored Energy at Step Discontinuities on the Behavior of Surface-Wave Gratings , IEEE Trans. Sonics and Ultrason., **SU-22** (1975) pp. 189–198.
8. M. Takeuchi and K. Yamanouchi: New type of SAW Reflectors and Resonators Consisting of Reflecting Elements with Positive and Negative Reflection Coefficients, IEEE Trans. Ultrason., Ferroelec. and Freq. Contr., **UFFC-33** (1986) pp. 369.
9. K. Hashimoto and M. Yamaguchi: SAW Device Simulation Using Boundary Element Method, Jpn. J. Appl. Phys., **29**, Suppl. 29-1 (1990) pp. 122–124.
10. K. Hashimoto and M. Yamaguchi: Analysis of Excitation and Propagation of Acoustic Waves under Periodic Metallic-Grating Structure for SAW Device Modeling, Proc. IEEE Ultrason. Symp. (1993) pp. 143–148.

3. Interdigital Transducers

This chapter describes the theory and modeling of IDTs employed for SAW excitation and detection. First, the delta-function model is detailed. Then the effects of peripheral circuits are discussed. Unidirectional transducers (UDTs) are also included in the discussion, and the p matrix method [1] is introduced as an effective tool for the characterization of UDTs.

3.1 Fundamentals

3.1.1 Bidirectional IDTs

Interdigital transducers (IDTs) shown in Fig. 3.1 are widely used for the excitation and detection of surface acoustic waves (SAWs) [2]. Each period of an IDT consists of multiple strips aligned and connected to the bus-bars periodically. The configuration shown in Fig. 3.1a, which consists of two strips each of period p_I, is called a single-electrode-type or solid-electrode-type IDT. On the other hand, that shown in Fig. 3.1b, which consists of four strips per period is called a double-electrode-type or split-electrode-type IDT. A combination of two strips with opposite sign is called a finger-pair.

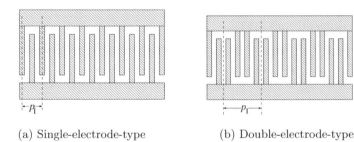

(a) Single-electrode-type   (b) Double-electrode-type

Fig. 3.1. Interdigital transducers, IDT

The single-electrode-type IDT is widely used because of its structural simplicity and relatively wide strip width ($\cong \lambda/4$) which relaxes the required definition for photolithography. Note that, for mechanical reflection, this structure is equivalent to a grating with periodicity $p = p_I/2$, and Bragg reflection occurs when $\lambda \cong p_I$ corresponding to the resonance condition for SAW excitation. This makes the IDT characteristics complex when the number of IDT finger-pairs is large.

On the other hand, although the double-electrode-type IDT has a relatively narrow strip width ($\cong \lambda/8$) and requires higher definition, Bragg reflection can be suppressed at the SAW resonance frequency because the periodicity is $p_I/4$ for mechanical reflection. For this reason, this type of IDT is frequently used when the frequency response must be controlled precisely.

Due to the symmetry in the finger geometry, the excited SAW field from each period is also symmetric with respect to the middle. Thus for IDT modeling, the SAW excitation can be regarded to occur at the middle. The effective excitation point is called the excitation center.

When piezoelectric materials are chosen as a substrate and an RF voltage is applied between two IDT bus-bars, periodic strain is generated by the electric field. Then the strain propagates as the SAW and is radiated from the IDT ends. Although SAWs excited from each period may be weak, they will be added to each other and grow, provided that the IDT period p_I corresponds to an integer multiple of the SAW wavelength.

Since SAW propagation on a piezoelectric substrate is accompanied by an electric field, the SAW induces charge if the metal strips are placed on the propagation surface.

IDT characteristics are mostly determined by (1) the finger geometry in the period, (2) the number of finger-pairs and (3) the substrate material. Of these, the substrate material dependence is a very complex function of the electric field distribution and substrate orientation. However, since detailed acoustic characteristics themselves are generally not of interest for SAW device simulation and design, SAW properties should be estimated as equivalent electrical values, and we may not be required to know how the SAW field is excited or detected by the IDT.

3.1.2 Unidirectional IDTs

The IDTs described above are symmetric with respect to the middle, and they excite SAWs toward both left and right directions with equal amplitude. We refer to this characteristic as bidirectionality. If IDTs can excite the SAW predominantly toward one of these directions, we call this type of IDT a unidirectional IDT (UDT).

Various UDTs have been proposed, and they are categorized into two types. The first one employs multi-phase electrical inputs shown in Fig. 3.2. The UDT shown in Fig. 3.2a has a periodic finger geometry with equal distance among fingers. The unidirectionality is realized by applying three-phase inputs whose phases are 120° out of phase with each other [3]. On the other hand, in the UDT shown in Fig. 3.2b, the unidirectionality is achieved by applying two-phase inputs whose phases are 90° out of phase with each other [4].

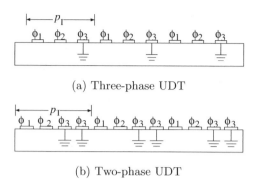

(a) Three-phase UDT

(b) Two-phase UDT

Fig. 3.2. Multi-phase UDT

The most serious problem for these configurations is the interconnection among fingers. All the fingers themselves can be realized by single-step photolithography. However, to interconnect third fingers, their series connection or an air-gap [3] structure must be given. The former results in an increased ohmic loss whereas the latter requires complicated fabrication processes.

To circumvent this problem, the two-phase UDT shown in Fig. 3.3 was proposed, and is called a group-type UDT [5]. Several finger-pairs with equal periodicity compose a group. Then two kinds of groups whose periodicity is displaced by $p_\mathrm{I}/4$ are placed alternately, and their common grounds are interconnected in series. This group structure enables us to reduce the ohmic loss.

Fig. 3.3. Group-type UDT

On the other hand, the UDTs shown in Figs. 3.4–3.6 employ asymmetric geometries, and unidirectionality appears even though the conventional single-phase signal is applied. This type of UDT is called a single-phase unidirectional transducer (SPUDT) [6]. Figure 3.4 shows, as an example, the electrode-width-controlled SPUDT (EWC/SPUDT) [7], which consists of three or four electrodes per IDT period p_I. Configuration (a) corresponds to a short-circuited grating, and no mechanical reflection occurs at the UDT resonance frequency since the grating pitch p coincides with $p_I/4$. Although configuration (b) is mechanically equivalent to a short-circuited grating, it can excite acoustic waves electrically. On the other hand, although configuration (c) is electrically inactive, it causes mechanical reflection due to the unequal electrode widths. Due to the structural symmetry, the reflection center locates in the middle of the wider electrode. Configuration (d) causes both excitation and reflection. By a combination of these structures, the spatial distributions of excitation and reflection sources can be controlled independently, so as to achieve desired UDT characteristics.

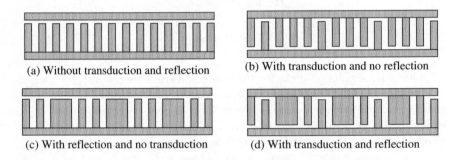

(a) Without transduction and reflection

(b) With transduction and no reflection

(c) With reflection and no transduction

(d) With transduction and reflection

Fig. 3.4. EWC/SPUDT

If we assume configuration (b) is equivalent to (d) for electrical excitation, SAWs are regarded as being excited from the center of the electrode connected to the lower bus-bar. The equivalent excitation point is called the excitation center. Thus, for configuration (d), the excitation center is displaced by $3p_I/8$ from the reflection center to the left.

Figure 3.5 shows the other type of UDT called a floating-electrode-type UDT (FEUDT) [8]. At the SPUDT resonance frequency, two electrodes mutually connected behave like a short-circuited grating in addition to two electrodes connected to the bus-bars. On the other hand, two isolated electrodes behave like an open-circuited grating. So, the reflection center locates in the middle of the isolated electrodes.

Fig. 3.5. UDT with floating electrodes (FEUDT)

For SAW excitation, the two electrodes connected to the bus-bars and the mutually connected ones are mainly responsible. So, the excitation center locates in the middle of these two electrodes. Thus the displacement between the excitation and reflection centers is $p_I/12$.

By the way, even though the conventional single-electrode-type IDT possesses geometrical symmetry, unidirectionality may occur. That is, if the substrate material possesses crystallographic asymmetry, the reflection center will be displaced from the geometrical center of the electrode where the excitation center locates. This phenomenon is called natural unidirectionality [9]. Since the single-electrode-type IDT is the simplest, UDTs employing this phenomenon are paid much attention to develop low-loss filters operating in the high frequency range.

Let us consider the SPUDT structure shown in Fig. 3.6 [10]. For electrical excitation, this structure is equivalent to a conventional double-electrode-type IDT, and the excitation center is expected to be at the point C_e in the figure. On the other hand, the thicker electrodes may mainly contribute to mechanical reflection and they are equivalent to a grating reflector with periodicity $p_I/2$. Thus the reflection center is expected to locate at the point C_r in the figure. Then it is clear that C_e is displaced by $p_I/8$ from C_r.

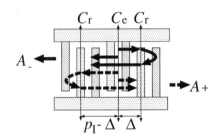

Fig. 3.6. Double-electrode-type IDT with two different film thicknesses

The distance between the excitation and reflection centers is denoted by Δ. The wave excited toward $+x$ and reflected at the reflection center returns to the excitation center. Throughout the propagation, the wave experiences a phase shift of $\angle\Gamma - 2\beta\Delta$, where Γ is the reflectivity, and it interferes with the wave excited directly toward $-x$. Similarly, the wave excited toward $-x$ and reflected at the reflection center returns to the excitation center. Its phase shift is given by $\angle\Gamma - 2\beta(p_I - \Delta)$.

As described in Sect. 2.1.2, on the usual substrates without natural unidirectionality, the phase of Γ is $\pm\pi/2$. Thus if $\beta\Delta = \pm\pi/4$, two waves propagating toward $\pm x$ are in phase whereas those toward $\mp x$ are 180° out of phase. On the other hand, if $\beta\Delta = \pm 3\pi/4$, two waves propagating toward $\pm x$ are 180° out of phase whereas those toward $\mp x$ are in phase.

Due to the periodicity of the structure, the effects of all reflected waves are summed. Thus, complete unidirectionality can be achieved when the condition $d = \pm\lambda(n/4 + 1/8)$ is satisfied. When this condition is not fulfilled, achievable performances are limited. This dependence will be discussed in detail in Sect. 3.5.3.

Note that the unidirectionality can also be realized by placing one reflector at either side of the bidirectional IDT. Although the structure is very simple, the achievable bandwidth becomes much narrower than that of the IDT itself.

3.2 Static Characteristics

3.2.1 Charge Distribution

Here, the electrostatic behavior of IDTs will be described. The term "electrostatic" means that the effect of wave excitation and/or propagation is ignored and/or negligible.

When the IDT is infinitely long and all electrode widths and spaces are equal, the static charge distribution $q(x_1)$ can be given in analytical form [11, 12]. The charge distribution $q(x_1)$ for the single-electrode-type IDT is given by [11]

$$q(x_1) = V \frac{C_s}{\sqrt{\cos(4\pi x_1/p_\mathrm{I}) - \cos(2\pi w/p_\mathrm{I})}}$$
$$\times \frac{2\sqrt{2}}{p_\mathrm{I} \cdot P_{-1/2}\{\cos(2\pi w/p_\mathrm{I})\}}, \tag{3.1}$$

and that on the double-electrode-type IDT is given by [12]

$$q(x_1) = V \frac{C_s \cos(2\pi x_1/p_\mathrm{I})}{\sqrt{-\cos(8\pi x_1/p_\mathrm{I}) - \cos(4\pi w/p_\mathrm{I})}}$$
$$\times \frac{2}{p_\mathrm{I} \cdot P_{-1/4}\{\cos(4\pi w/p_\mathrm{I})\}}, \tag{3.2}$$

where C_s is the static capacitance for the IDT per period given by

$$C_s = W\epsilon(\infty) \frac{P_{-1/2}\{\cos(2\pi w/p_\mathrm{I})\}}{P_{-1/2}\{-\cos(2\pi w/p_\mathrm{I})\}} \tag{3.3}$$

for the single-electrode-type IDT and

$$C_s = W\sqrt{2}\epsilon(\infty) \frac{P_{-1/4}\{\cos(4\pi w/p_\mathrm{I})\}}{P_{-1/4}\{-\cos(4\pi w/p_\mathrm{I})\}}, \tag{3.4}$$

for the double-electrode-type IDT, respectively. W is the aperture of the IDT, w is the electrode width, $P_v(x)$ is the Legendre function, and $\epsilon(S)$ is the effective permittivity [13] which will be detailed in Sect. 6.2. Under the low-frequency approximation, $\epsilon(\infty)$ is given by [14]

$$\epsilon(\infty) = \epsilon_0 + \sqrt{\epsilon_{11}^{T}\epsilon_{33}^{T} - \epsilon_{13}^{T2}}, \qquad\qquad (3.5)$$

where ϵ^T is the permittivity measured under the no stress condition, which is different from that measured under the zero strain condition.

Figure 3.7 shows $q(x_1)$ calculated by (3.1) and (3.2). Note that the charge concentrates at the electrode edges and diverges as $1/\sqrt{x}$. This fact makes the calculation of charge distribution on the IDT complex.

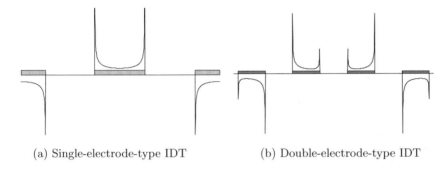

(a) Single-electrode-type IDT   (b) Double-electrode-type IDT

Fig. 3.7. Static charge distribution on an IDT

Figure 3.8 shows the w/p_I dependence of $C_s/\epsilon(\infty)W$ given by (3.3) and (3.4).

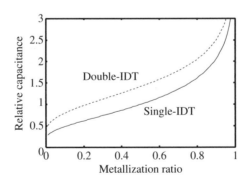

Fig. 3.8. Electrode width dependence of $C_s/\epsilon(\infty)W$

3.2.2 Electromechanical Coupling Factor

In the following analysis, a time dependence of $\exp(j\omega t)$ is assumed implicitly.

When a line charge $q(x_1) = \sigma\delta(x_1)$ is placed at $x_1 = 0$ on a flat surface, the surface electrical potential $\phi(x_1)$ associated with SAW propagation is given by

$$\phi(x_1) = j\mu_f\sigma\exp(-j\beta_{Sf}|x_1|), \tag{3.6}$$

where μ_f is a constant, β_{Sf} is the wavenumber of the SAW propagating on the free surface.

The power P supplied by the charge is given by

$$P = W\Re\left[\int_{-\infty}^{+\infty}\frac{1}{2}\phi(x_1)^*\frac{\partial q(x_1)}{\partial t}dx_1\right] = \frac{1}{2}\omega\mu_f W|\sigma|^2, \tag{3.7}$$

or, by using the relation $\sigma = C_s V/W$,

$$P = \omega C_s K_{Sf}^2|V|^2, \tag{3.8}$$

where $K_{Sf}^2 = \mu_f C_s/2W$ is the electromechanical coupling factor for the free surface.

Since the SAW is excited equally toward the $+x_1$ and $-x_1$ directions, the power $P_\pm$ for the SAW propagating toward $\pm x_1$ is given by $P/2$. Then from (3.6) and (3.8), we obtain

$$\phi(x_1) = j\sqrt{8K_{Sf}^2/\omega C_s}\exp(-j\beta_{Sf}|x_1|)\sqrt{P_\pm}. \tag{3.9}$$

Next, let us consider the case where the substrate surface is covered by a metal with an infinitesimally narrow slot at $x_1 = 0$. For the discussion, the following new variable is defined:

$$h(x_1) = \int_{-\infty}^{x_1}q(x_1)dx_1. \tag{3.10}$$

The electric field $e(x_1) = V\delta(x_1)$ is induced by the applied voltage V, and $h(x_1)$ carried by the excited SAW may be expressed as

$$h(x_1) = -j\mu_m V\exp(-j\beta_{Sm}|x_1|), \tag{3.11}$$

where μ_m is a constant, and β_{Sm} is the wavenumber for the metallized surface. By using the relation $q(x_1) = \partial h(x_1)/\partial x_1$, the power P supplied by $q(x_1)$ is given by

$$\begin{aligned}
P &= W\Re\left[\int_{-\infty}^{+\infty}\frac{1}{2}\phi(x_1)^*\frac{\partial q(x_1)}{\partial t}dx_1\right] \\
&= W\Re\left[\int_{-\infty}^{+\infty}\frac{1}{2}e^*(x_1)\frac{\partial h(x_1)}{\partial t}dx_1\right] = \omega C_s K_{Sm}^2|V|^2,
\end{aligned} \tag{3.12}$$

where $K_{Sm}^2 = \mu_m W/2C_s$ is the electromechanical coupling factor for the metallized surface.

Since the SAW is excited equally toward the $+x_1$ and $-x_1$ directions, the power $P_\pm$ for the SAW propagating toward $\pm x_1$ is given by $P/2$. Then from (3.11) and (3.12), we obtain

$$h(x_1) = -jW^{-1}\sqrt{8C_sK_{Sm}^2/\omega}\exp(-j\beta_m|x_1|)\sqrt{P_\pm}. \tag{3.13}$$

Note that, for Rayleigh-type SAWs, K_{Sf}^2 almost coincides with K_{Sm}^2, and they are approximately given by

$$K^2 = \frac{V_{Sf}^2 - V_{Sm}^2}{V_{Sf}^2} \cong 2\frac{V_{Sf} - V_{Sm}}{V_{Sf}} \equiv 2\Delta V/V \tag{3.14}$$

with sufficient accuracy [15, 16]. In this equation, V_{Sf} and V_{Sm} are the SAW velocities for the free and metallized surfaces, respectively.

In this chapter, K_{Sf}^2 and K_{Sm}^2 are not distinguished and will be simply denoted by K^2.

3.2.3 Element Factor

Let us consider the excitation efficiency for each period of the IDT. The maximum efficiency can be achieved when all charge is concentrated at a point. For this situation, the SAW excitation characteristics are given by (3.6)–(3.13). The spatial charge distribution on the IDT fingers results in a reduction of the SAW excitation efficiency. We define the element factor η as the ratio of the excitation efficiency relative to the best one [17].

From the law of superposition, the SAW amplitude radiated from the IDT can be expressed as

$$\phi(x_1) = -j\mu\int_{-\infty}^{+\infty} q(x_1')\exp(-j\beta|x_1 - x_1'|)dx_1' \tag{3.15}$$

from (3.6), where $q(x_1)$ is the charge distribution on the IDT. When x_1 locates outside the IDT, (3.15) is simplified to

$$\phi(x_1) = -2\pi j\mu Q(s\beta)\exp(-j\beta|x_1|), \tag{3.16}$$

where $s = \text{sgn}(x_1 - x_1')$ and $Q(\beta)$ is the Fourier transform of $q(x_1)$ given by

$$Q(\beta) = \frac{1}{2\pi}\int_{-\infty}^{+\infty} q(x_1)\exp(j\beta x_1)dx_1. \tag{3.17}$$

Assuming the charge distributions for each IDT period are equal to each other, (3.15) is simplified to

$$Q(\beta) = \frac{1}{2\pi}\sum_{n=0}^{N-1}\int_{(n-1/2)p_I}^{(n+1/2)p_I} q(x_1)\exp(j\beta x_1)dx_1 = Q_e(\beta)Q_a(\beta), \tag{3.18}$$

where N is the number of IDT fingers, and

$$Q_a(\beta) = \sum_{n=0}^{N-1}\exp(j\beta p_I) = \frac{\exp(j\beta N p_I) - 1}{\exp(j\beta p_I) - 1}$$

$$= \exp\{j\beta p_I(N-1)/2\}\frac{\sin(\beta N p_I/2)}{\sin(\beta p_I/2)} \tag{3.19}$$

gives the frequency dependence of the excitation efficiency in the periodic structure, and is called the array factor [17]. On the other hand,

$$Q_e(\beta) = \frac{1}{2\pi} \int_{-p_I/2}^{p_I/2} q(x_1) \exp(j\beta x_1) dx_1 \qquad (3.20)$$

gives the excitation efficiency per IDT period. Designating the static capacitance per period as C_s, maximum efficiency is realized when the charges $+C_sV$ and $-C_sV$ are placed alternately with separation of a half wavelength. For the M-th order resonance, since $\beta p_I = 2\pi M$ ($p_I = M\lambda$), the element factor η is given by

$$\eta = \frac{\pi W Q(2\pi M/p_I)}{C_s V}. \qquad (3.21)$$

Substituting (3.1) into (3.21) gives η for the single-electrode-type IDT as

$$\eta = \frac{P_{(M-1)/2}\{\cos(2\pi w/p_I)\}}{P_{-1/2}\{\cos(2\pi w/p_I)\}}. \qquad (3.22)$$

Substituting (3.2) into (3.21) gives η for the double-electrode-type IDT as

$$\eta = (-)^I \frac{P_{I-1/2}\{\cos(4\pi w/p_I)\}}{\sqrt{2}P_{-1/4}\{\cos(4\pi w/p_I)\}}. \qquad (3.23)$$

where $I = \text{int}\{(M+1)/4\}$. Note that η for even M is zero due to the symmetry in the IDT geometries.

If η is complex, the excitation center does not coincide with the geometrical center. In this case, η can be made real by redefining the geometrical center appropriately.

Figure 3.9 shows η calculated from (3.22) and (3.23) as a function of the metallization ratio $2w/p_I$, where η_0 is the factor for the single-electrode-type IDT with $w/p_I = 0.25$ given by

$$\eta_0 = \frac{P_0(0)}{P_{-1/2}(0)} = 0.84722. \qquad (3.24)$$

Although the term "element factor" is often used for η/η_0 [17], we will call η itself the element factor.

Note that, in Smith's equivalent circuit model described later, η is set to $\sqrt{2/\pi}$, which agrees well with $\eta_0 = 0.84722 = 1.062\sqrt{2/\pi}$, although the model is based upon the analogy to the BAW resonator.

3.2.4 Complex Electrode Geometries

Some types of IDTs have more complex electrode geometries with different finger widths and/or gaps, where analytical techniques will not be applicable to calculate the static capacitance and/or element factor.

For their numerical calculation, one may apply traditional numerical techniques such as the finite element method (FEM). However, most of these

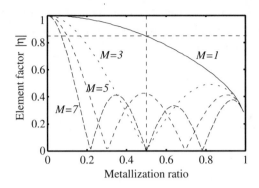

(a) Single-electrode-type IDT

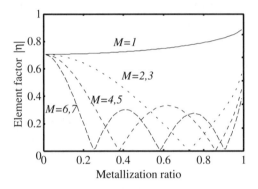

(b) Double-electrode-type IDT

Fig. 3.9. Element factor η as a function of w/p_{I}

assume continuity of the field variables and their application will result in ill convergence of the calculation due to the charge divergence at electrode edges [18].

Here, a very simple but effective method is introduced for the characterization of IDTs with arbitrary finger geometries.

Let us express the charge distribution $q_m(x)$ of the m-th electrode of the IDT as a polynomial with weighting function $1/\sqrt{1 - \{2(x - x_m)/w_m\}^2}$ which expresses the divergence at the edges. That is,

$$q_m(x) = \sum_{n=0}^{N-1} \frac{A_{mn}T_n\{2(x - x_m)/w_m\}}{\sqrt{1 - \{2(x - x_m)/w_m\}^2}}, \tag{3.25}$$

where x_m is the center position of the electrode, w_m is its width, A_{mn} is the unknown coefficient, and $T_n(r)$ is the Chebyshev polynomial given by

$$T_n(r) = \cos(n\cos^{-1} r). \tag{3.26}$$

In the electrostatic case, the surface electrical potential created by a line charge Q is given by $G_e(L) = -Q/\{\pi\epsilon(\infty)\} \cdot \log L$ where L is the distance [13]. Thus, the potential $\phi(x)$ created by the charge on the IDT is given by

$$\phi(x) = \int_{-\infty}^{+\infty} G_e(x - x')q(x')dx$$
$$= \sum_{m=1}^{M} \sum_{n=1}^{N} H_n\{2(x - x_m)/w_m\}A_{mn}(w_m/2), \tag{3.27}$$

where $H_n(r)$ is the function given by

$$H_n(r) = \begin{cases} \epsilon(\infty)^{-1} \log\left(|r| - \sqrt{r^2 - 1}\right) & (n = 0) \\ \epsilon(\infty)^{-1}\mathrm{sgn}(r)^n n^{-1}\left(|r| - \sqrt{r^2 - 1}\right)^n & (n \neq 0) \end{cases} \tag{3.28}$$

when $|r| > 1$, and

$$H_n(r) = \begin{cases} 0 & (n = 0) \\ n^{-1}\epsilon(\infty)^{-1}T_n(r) & (n \neq 0) \end{cases} \tag{3.29}$$

when $|r| < 1$.

From (3.27), we obtain $M \times N$-th order simultaneous equations with respect to A_{mn} by sampling $\phi(x)$ for N points on each electrode. From the solved A_m, the total charge Q_m on the m-th electrode is given by

$$Q_m = W \int_{-w_m/2}^{+w_m/2} q_m(x)dx = \pi W(w_m/2)A_{m0}, \tag{3.30}$$

where W is the aperture. From this, the static capacitance can be calculated.

From the A_m determined, we also obtain

$$Q(\beta) = \frac{1}{2\pi} \sum_{m=1}^{M'} \sum_{n=0}^{N} A_{mn} \int_{x_m-w_m/2}^{x_m+w_m/2} \frac{T_n\{2(x - x_m)/w_m\}}{\sqrt{1 - \{2(x - x_m)/w_m\}^2}}$$
$$\times \exp(j\beta x)dx$$
$$= \sum_{m=1}^{M'} \sum_{n=0}^{N} A_{mn}j^n(w_m/4)\exp(j\beta x_m)J_n(\beta w_m/2), \tag{3.31}$$

where M' is the number of electrodes for each period, and $J_n(x)$ is the Bessel function. The element factor can be calculated from (3.21) by substituting the static capacitance obtained from (3.30) and $Q(\beta)$ calculated from (3.31).

Figure 3.10 shows the calculated static capacitance per period of a single-electrode-type IDT with $w/p_I = 0.25$ as a function of $1/N$ [18]. For comparison, that calculated by using an unweighted polynomial instead of (3.25) is also shown in the figure.

It is seen that $N = 3$ is sufficient to achieve enough accuracy when the weighted polynomial is employed. On the other hand, when the unweighted

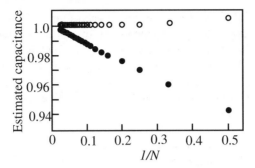

Fig. 3.10. Calculated static capacitance per period for a solid-electrode-type IDT as a function of $1/N$. ○: weighted polynomial employed, and ●: unweighted polynomial employed

polynomial is employed, the calculation converges with a $1/N$ dependence. As was mentioned above, this is due to the charge divergence at the electrode edges in the form of $1/\sqrt{x}$, and many terms are required to express this discontinuity as a sum of continuous functions.

Figure 3.11 shows the error in the calculated static capacitance as a function of $2w/p_I$ with N as a parameter [18]. It is seen that, for usual w/p_I, $N = 5$ is enough to obtain an error of less than 10^{-4}. When the gap becomes very narrow, although the error increases rapidly, sufficient accuracy can be achieved by slightly increasing N.

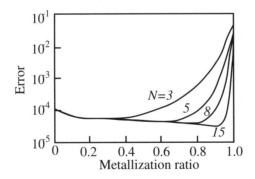

Fig. 3.11. Electrode width dependence of calculation error

3.2.5 Effect of IDT Ends

The behavior of SAW excitation at the IDT ends is somewhat different from that in its center region, the so-called end effect [19].

Figure 3.12 shows the total charge on each finger of the single-electrode-type IDT with 20 finger-pairs calculated by using the method described above. In the calculation, the region outside the IDT is assumed to be electrically free. Due to the end effect, the charge on the electrodes close to the ends changes rapidly, and variation of the order of a percent remains at the center of the IDT. This is not negligible in controlling the frequency response of SAW devices precisely.

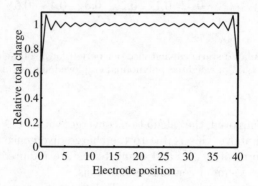

Fig. 3.12. Total charge distribution on single-electrode-type IDT with 20 finger-pairs

Figure 3.13 shows the charge distribution when unit voltage is applied to only one electrode whereas the others are zero. In this case, the applied voltage influences two or three fingers in the vicinity, and scarcely affects the others.

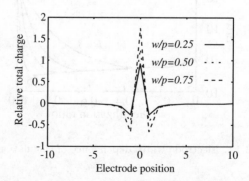

Fig. 3.13. Total charge distribution when unit voltage is applied to only one electrode

This result implies that the end effect can be reduced by placing a short-circuited grating instead of a free surface next to the IDT. The electrodes for this purpose are sometimes called dummy electrodes. If the number of dummy electrodes is sufficient, the behavior of the finger-pairs close to the IDT edges will be close to that of the finger-pair at the center.

This suggests that the charge distribution on the IDT including the dummy electrodes can be regarded as a superposition of the charge distribution when unit voltage is applied to only one electrode as shown in Fig. 3.13 [17]; and the coefficients for the summation are given by the applied voltage for each finger. Note that the redefined element factor coincides with that described in Sect. 3.2.3 at the IDT resonance frequency.

3.3 IDT Modeling

3.3.1 Delta-Function Model

The delta-function model [20] was proposed as the simplest model for an IDT, and has been widely used for device design. This model is based on the idea that the IDT is regarded as the superposition of periodic wave sources.

Divide the IDT into individual finger-pairs, and consider the case where the voltage ϕ_n is applied toward the n-th finger-pair whereas the others are kept zero. If the system is linear, the total charge q_m induced on the m-th finger-pair is proportional to ϕ_n and is expressed as $h_{mn}\phi_n$ where h_{mn} is a factor which will be discussed later. When all ϕ_n are applied simultaneously, q_m can be expressed in the form

$$q_m = \sum_n h_{mn}\phi_n, \tag{3.32}$$

because of the law of superposition.

The voltage applied to the IDT excites not only the SAW but also an electromagnetic wave and BAWs. Then h_{mn} can be expressed as a sum of contributions of the electrostatic coupling h_{mn}^{e}, that of the BAW radiation h_{mn}^{B}, and that of the SAW radiation h_{mn}^{S};

$$h_{mn} = h_{mn}^{\mathrm{e}} + h_{mn}^{\mathrm{B}} + h_{mn}^{\mathrm{S}}. \tag{3.33}$$

Usually the number of IDT finger-pairs is relatively large, and h_{mn}^{S} is assumed to be dependent only on the distance L_{mn} between the m-th and n-th finger-pairs. Then h_{mn}^{S} can be expressed as

$$h_{mn}^{\mathrm{S}} = -jh_0 \exp(-j\beta L_{mn}) \tag{3.34}$$

where β is the SAW wavenumber, and h_0 is a constant.

In Sects. 3.2.3 and 3.2.4, it was shown that $Wh(x_1)$, the charge of the SAW induced by a unit surface electric field, is given by $2\eta C_{\mathrm{s}}K^2$, where η is the element factor responsible for the SAW excitation efficiency. The charge distribution on the IDT reduces the efficiency for SAW detection as well as excitation, and reciprocity [21] suggests that the ratio is again given by the element factor η. So we get the relation $h_0 = 2\eta^2 C_s K^2$.

Let us consider the case where the voltage V_1 is applied to the input IDT with N finger-pairs and the excited SAW is detected by the output IDT with one finger-pair (see Fig. 3.14).

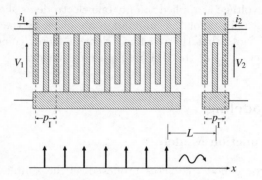

Fig. 3.14. Input and output IDTs

When the space between the IDTs is totally covered by short-circuited grating, (3.32) and (3.34) give the current I_2 flowing into the output IDT as

$$I_2/V_1|_{V_2=0} = \omega h_0 \sum_{n=0}^{N-1} \exp\{-j\beta(L + np_\mathrm{I})\}$$

$$= \omega h_0 \exp(-j\beta L) \frac{\exp(-j\beta Np_\mathrm{I}) - 1}{\exp(-j\beta p_\mathrm{I}) - 1}$$

$$= \omega h_0 \exp[-j\beta\{L + (N-1)p_\mathrm{I}/2\}] \frac{\sin(\beta Np_\mathrm{I}/2)}{\sin(\beta p_\mathrm{I}/2)}, \qquad (3.35)$$

where $L_{mn} = L + np_\mathrm{I}$. In this equation, $I_2/V_1|_{V_2=0}$ is called the transfer admittance, which characterizes the behavior of the IDT.

Figure 3.15 shows $|\sin(N\beta p_\mathrm{I}/2)/N\sin(\beta p_\mathrm{I}/2)|$. It is seen that this function takes a maximum when $p_\mathrm{I} = N\lambda$ where $\beta = 2\pi/\lambda$. Note βp_I corresponds to the normalized frequency $2\pi(f/f_\mathrm{r})$ where f_r is the IDT resonance frequency given by $V_\mathrm{S}/p_\mathrm{I}$ and V_S is the SAW velocity.

We often call the resonance peak the mainlobe whereas smaller ones are called sidelobes, and their difference is called the sidelobe level.

The IDT exhibits a frequency response with a $\mathrm{sinc}x(= \sin x/x)$ dependence because

$$\sin(\beta Np_\mathrm{I}/2)/\sin(\beta p_\mathrm{I}/2) = (-1)^{M(N-1)} \sin(\pi N\delta)/\sin(\pi\delta)$$

$$\cong N(-1)^{M(N-1)} \mathrm{sinc}(\pi N\delta), \qquad (3.36)$$

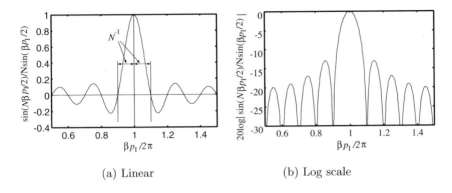

(a) Linear

(b) Log scale

Fig. 3.15. $\sin(N\beta p_{\mathrm{I}}/2)/N\sin(\beta p_{\mathrm{I}}/2)$ ($N = 10$)

where $\beta p_{\mathrm{I}} = 2\pi(M + \delta)$. It is clear that the peak height is proportional to N and the width is inversely proportional to N. Note the fractional -3 dB bandwidth is given by $1/2.257MN$.

Let us extend this result to this case where the input and output IDTs have multiple finger-pairs of N_1 and N_2 (see Fig. 3.16). In the case, by setting

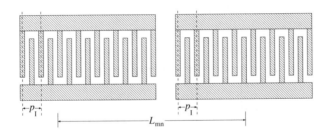

Fig. 3.16. Input and output IDTs with multiple finger-pairs

$L_{mn} = L + (m - n)p_{\mathrm{I}}$, the transfer admittance Y_{12}^{S} for SAW propagation is given by

$$
\begin{aligned}
Y_{12}^{\mathrm{S}} &= \omega h_0 \sum_{m=0}^{N_1-1} \sum_{n=0}^{N_2-1} \exp\{-j\beta(L + (m - n)p_{\mathrm{I}}\} \\
&= \omega h_0 \exp(-j\beta L) \sum_{m=0}^{N_1-1} \exp(-j\beta m p_{\mathrm{I}}) \sum_{n=0}^{N_2-1} \exp(j\beta n p_{\mathrm{I}}) \\
&= \omega h_0 \exp(-j\beta L) \frac{\sin(\beta N_1 p_{\mathrm{I}}/2)}{\sin(\beta p_{\mathrm{I}}/2)} \frac{\sin(\beta N_2 p_{\mathrm{I}}/2)}{\sin(\beta p_{\mathrm{I}}/2)}.
\end{aligned}
\tag{3.37}
$$

It is seen that Y_{12}^{S} in this case is given by the product of the transfer admittances for the input and output IDTs.

In a similar way, the input admittance $Y_{11} = I_1/V_1|_{V_2=0}$ of the IDT is given by

$$Y_{11} = j\omega \sum_{m=0}^{N_1-1} \sum_{n=0}^{N_1-1} h_{mn}. \tag{3.38}$$

By setting $L_{mn} = |m - n|p_I$, the SAW contribution Y_{11}^S, i.e., the radiation admittance, is given by

$$Y_{11}^S = \omega h_0 \left[\sum_{m=0}^{N-1} \sum_{n=0}^{N-1} \exp\{-j\beta|m - n|p_I\} \right]$$

$$= \omega h_0 \left[\sum_{n=1}^{N-1} 2(N - n) \exp(-j\beta n p_I) + N \right]$$

$$= \omega h_0 \left[2\exp(-j\beta N p_I) \sum_{\ell=1}^{N-1} \ell \exp(j\beta \ell p_I) + N \right]$$

$$= \omega h_0 \left[\frac{N \exp(-j\beta p_I) - (N - 1) - \exp(-j\beta N p_I)}{2\sin^2(\beta p_I/2)} + N \right]$$

$$= \omega h_0 \frac{2\sin^2(N\beta p_I/2) - jN\sin(\beta p_I) + j\sin(\beta N p_I)}{2\sin^2(\beta p_I/2)}. \tag{3.39}$$

Figure 3.17 shows the radiation admittance calculated from (3.39).

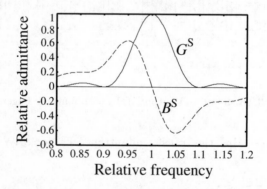

Fig. 3.17. Input admittance of IDT for SAW radiation ($N = 10$)

Since the SAW velocity is usually somewhat lower than the BAW velocity, the contribution h_{mn}^B of the BAW radiation is negligible for the discussion near the SAW resonance frequency.

The influence of the electrostatic coupling can be expressed by the static capacitance of the IDT. Thus the input admittance Y_{11} for the IDT is given by

$$Y_{11} \cong j\omega C_{\mathrm{s}} N + Y_{11}^{\mathrm{S}} \tag{3.40}$$

with reasonable accuracy.

Note that from (3.37) and (3.39), the transfer admittance Y_{12}^{S} is related to the radiation admittance Y_{ii}^{S} for the i-th IDT as

$$Y_{12}^{\mathrm{S}} = \sqrt{\Re(Y_{11}^{\mathrm{S}})\Re(Y_{22}^{\mathrm{S}})} \exp(-j\beta L). \tag{3.41}$$

3.3.2 Equivalent Circuit Model

Although the IDT characteristics can be evaluated reasonably by using the delta-function model, it cannot include the effects of internal reflection within the IDT which sometimes influence the device performance significantly. In single-phase UDTs and resonator structures which will be described in Chaps. 4 and 5, internal reflection is essential, and the model cannot be applicable to their simulation.

As an alternative, equivalent circuit models have been widely used. Figure 3.18 shows two kinds of equivalent circuits proposed by Smith et al. [22, 23]. Model (a) is based on the cascade connection of BAW resonators with thickness vibration, and is called an in-line field model, whereas model (b) is based on the cascade connection of BAW resonators with lateral vibration, and is called a crossed field model.

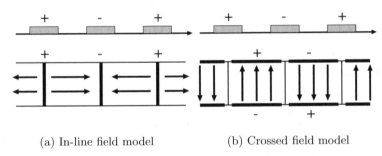

(a) In-line field model   (b) Crossed field model

Fig. 3.18. Physical model for the equivalent circuit

The equivalent circuit for the BAW resonators proposed by Mason [24] is widely used to derive the final equivalent circuit for the physical models shown in Fig. 3.18. Figure 3.19 shows the equivalent circuit expression for the in-line field model based on Mason's equivalent circuit. In the figure, $\theta = \omega/V_{\mathrm{S}}$, V_{S} is the SAW velocity, $R = 2\pi/\omega C_{\mathrm{s}} K^2$ is the effective acoustic impedance, and C_{s} is the static capacitance. The equivalent circuit for the crossed field model is simply given by removing the negative capacitance in the figure.

For the analysis of the IDT, the acoustic ports are cascade-connected whereas the electrical ports are parallel-connected as shown in Fig. 3.20.

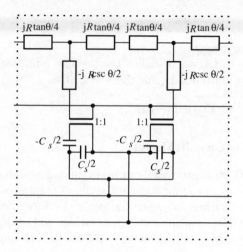

Fig. 3.19. Equivalent circuit based on an in-line field model for a unit finger-pair

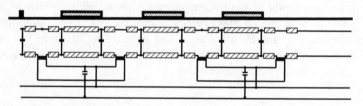

Fig. 3.20. Cascade-connection of equivalent circuit for the IDT

The crossed field model is widely used rather than the in-line field model because of its fairly good agreement with experimental results [23]. Note that the results of the crossed field model coincide with those of the delta-function model described above by setting $\eta = \sqrt{2/\pi}$.

Various models were proposed to include various second-order effects, such as mechanical reflection, the energy storing effect [25], and so on. Figure 3.21 shows an example [26]. Mechanical reflection is taken into account by giving different values for the acoustic impedances R_1 and R_2, and the energy storing effect is expressed by C_e. The change in the velocity is also considered by giving different values for the wavenumbers β_1 and β_2.

How to manipulate the equivalent circuit model and to simulate SAW devices can be found in many articles such as Ref. [27].

3.3.3 Other Models

Recently a few methods for SAW device simulation have been proposed where the unit section of the IDT is treated as a black box as shown in Fig. 3.22 and is characterized by using mathematical techniques.

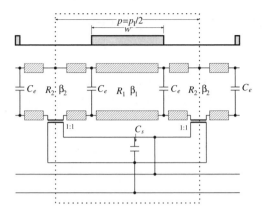

Fig. 3.21. Equivalent circuit mode for the IDT

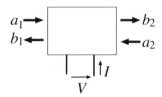

Fig. 3.22. Three-port linear circuit

One is called the p matrix method [1], where the black box is characterized by a 3-by-3 matrix where the amplitudes of the incident and reflected waves at the acoustic ports and the voltage and current (peak-to-peak) at the electrical port are chosen as variables (see Fig. 3.22), that is,

$$\begin{pmatrix} b_1 \\ b_2 \\ I \end{pmatrix} = \begin{pmatrix} p_{11} & p_{12} & p_{13} \\ p_{12} & p_{22} & p_{23} \\ -4p_{13} & -4p_{23} & p_{33} \end{pmatrix} \begin{pmatrix} a_1 \\ a_2 \\ V \end{pmatrix}. \tag{3.42}$$

It is clear that p_{11} and p_{22} correspond to the reflection coefficients at the acoustic ports, p_{12} to the transmission coefficients, p_{13} and p_{23} to the excitation efficiency, and p_{33} to the input admittance. If we can evaluate them by using numerical methods, for example the perturbation method [28], the device characteristics can be simulated by cascade-connecting the black box [29]. Usage of the p matrix will be discussed in Sect. 3.5.

The other method is called coupling-of-modes (COM) theory [30] where the black box in Fig. 3.22 is characterized by simultaneous linear differential equations. The COM analysis will be fully discussed in Chap. 7.

3.4 Influence of Peripheral Circuit

3.4.1 Summary

For RF circuit analysis, the scattering coefficient is widely used. The network analyzer commonly used for SAW device characterization measures the scattering coefficient when the device is connected to transmission lines with a specific impedance of 50 Ω.

Consider the case shown in Fig. 3.23 where the signal a_i is incident to and b_i is reflected from the circuit. Note a_i and b_i are normalized so that their square corresponds to their power. For their units, dBm and dBμ are used; 0 dBm corresponds to $|a_i|^2$ of 1 mW.

Fig. 3.23. Incidence to and reflection from a linear circuit

If the circuit is linear, b_i can be expressed as a linear combination of a_i, that is,

$$\begin{pmatrix} b_1 \\ b_2 \end{pmatrix} = \begin{pmatrix} S_{11} & S_{12} \\ S_{21} & S_{22} \end{pmatrix} \begin{pmatrix} a_1 \\ a_2 \end{pmatrix}, \tag{3.43}$$

where the matrix element S_{ij} is called the scattering coefficient.

SAW devices exhibit reciprocity which means the device characteristics are unchanged when the input and output ports are exchanged [21]. Then the relation $S_{12} = S_{21}$ holds. If the circuit is symmetrical and the port "a" is equivalent to the port "b", the relation $S_{11} = S_{22}$ holds.

Although every physical system includes nonzero power dissipation, assuming the system to have zero power dissipation sometimes makes the analysis very simple. This ideal zero dissipation condition is called the unitary condition. If the system is unitary, since $|a_1|^2 + |a_2|^2 = |b_1|^2 + |b_2|^2$ for arbitrary a_i and b_i, we obtain

$$\begin{pmatrix} S_{11} & S_{12} \\ S_{21} & S_{22} \end{pmatrix}^{-1} = \begin{pmatrix} S_{11} & S_{12} \\ S_{21} & S_{22} \end{pmatrix}^{\dagger}, \tag{3.44}$$

where $\dagger$ indicates the transpose and complex conjugate operations.

Among the matrix components, S_{11} and S_{22} are often referred to as the reflection coefficient, and are sometimes expressed by the symbol "Γ". On the other hand, S_{12} and S_{21} are referred to as the transmission coefficient, and are sometimes expressed by the symbol "T". Note $-20 \log |S_{11}|$ and $-20 \log |S_{22}|$

are called return losses whereas $-20 \log |S_{12}|$ and $-20 \log |S_{21}|$ are called insertion losses.

Let us consider the system shown in Fig. 3.24 where the device under test (DUT) is connected to the transmission line with specific impedance $R_0 = G_0^{-1}$.

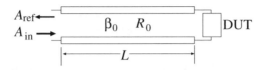

Fig. 3.24. Measurement of reflection coefficient

By taking the phase delay due to propagation into account, Γ measured at $x = L$ is given by

$$\Gamma = \Gamma_0 \exp(-2j\beta_0 L), \tag{3.45}$$

where β_0 is the wavenumber of the electromagnetic wave in the transmission line, and Γ_0 is the reflection coefficient measured at $x = 0$, which is given by

$$\Gamma_0 = \frac{G_0 - Y}{G_0 + Y} = \frac{Z - R_0}{Z + R_0}. \tag{3.46}$$

Note β_0 in (3.45) is a function of frequency. By setting $Y = G + jB$,

$$|\Gamma|^2 = \left| \frac{G - G_0 + jB}{G + G_0 + jB} \right|^2 = 1 - \frac{4GG_0}{(G + G_0)^2 + B^2}. \tag{3.47}$$

Thus, assuming $G_0 \gg G, |B|$, we obtain

$$-20 \log |\Gamma| = -10 \log \left| 1 - \frac{4GG_0}{(G + G_0)^2 + B^2} \right|$$
$$\cong -10 \log |1 - 4G/G_0| \cong 40G/G_0. \tag{3.48}$$

This means that the frequency response of the return loss directly reflects that of the radiation conductance of the connected IDT.

Figure 3.25 shows S_{11} of the IDT schematically. In the figure, G_R, which is almost independent of frequency, is due to the electrode resistivity. On the other hand, G_S which exhibits a strong peak is due to SAW radiation. Since G_B exhibits a cut-off characteristic, it is due to BAW radiation (see Sect. 3.6).

Since $|\Gamma_0| \leq 1$, the plot for Γ_0 on the complex plane with frequency as a parameter draws a contour within a unit circle as shown in Fig. 3.26. The curve for Γ can be obtained from the curve for Γ_0 by rotating it clockwise by the angle $2\beta_0 L$.

At off-resonance, IDTs are capacitive and $Y \propto j\omega$. Then from (3.46), $\Gamma = 0$ at $\omega = 1$, and Γ rotates clockwise on the rim of the unit circle with

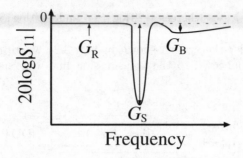

Fig. 3.25. Schematic return loss characteristics of the IDT

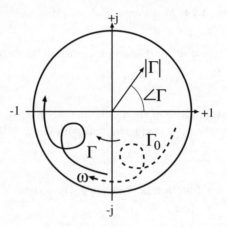

Fig. 3.26. Scattering coefficient in complex plane

an increase in ω; and at $\omega \to \infty$, $\Gamma = -1$. If nonzero conductance exists, Γ approaches the center of the circle, and if there exists resonance, the contour is a circle.

3.4.2 Smith Chart and Impedance Matching

Let us consider a plot of the normalized admittance $Y/G_0 = (G + jB)/G_0$. When B/G_0 is plotted for the range $[-\infty, +\infty]$ with G/G_0 as a parameter, the contours of Γ_0 in the complex plane are circles which do not intersect each other except at $\Gamma_0 = -1$ where $|B/G_0| \to \infty$. Similarly, when G/G_0 is plotted through the range $[0, +\infty]$ with B/G_0 as a parameter, the contours of Γ_0 in the complex plane are arcs which do not intersect each other except at $\Gamma_0 = -1$ where $|G/G_0| \to \infty$. The chart composed of these contours (see Fig. 3.27) is called a Smith chart.

Let us discuss the electrical matching of the IDT by using the Smith chart. As a first case, we try to realize the matching ($\Gamma_0 = 0$) by parallel-connecting

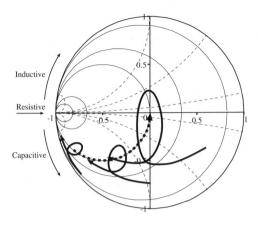

Fig. 3.27. Smith chart (admittance chart). Thin broken line: constant-B curve, thin solid line: constant-G curve, and thick broken lines: when L_p is increased

the inductance L_p with an IDT as shown in Fig. 3.28. By this connection, B changes to $B - 1/\omega L_p$. With an increase in L_p, Γ_0 at the given frequency moves counterclockwise on the constant-G curve in the Smith chart as shown in Fig. 3.27. Hence, in the configuration, electrical matching, where $|\Gamma_0| = 0$, can be realized by giving are an appropriate L_p only if the initial Γ_0 locates on the constant-G curve passing through the origin.

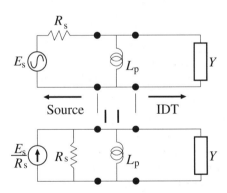

Fig. 3.28. Electrical matching by parallel connection

Next let us try to realize the matching ($\Gamma_0 = 0$) by series-connecting the inductance L_s with the IDT as shown in Fig. 3.29. In the case, it is convenient to plot the normalized impedance $Z/R_0 = (R + jX)/R_0$ where R/R_0 and X/R_0 are chosen as parameters. This is shown in Fig. 3.30, and is called an

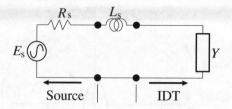

Fig. 3.29. Electrical matching by series connection

impedance chart. It is interesting to note that the impedance chart shown in
Fig. 3.30 corresponds to the mirror image of the admittance chart shown in
Fig. 3.27.

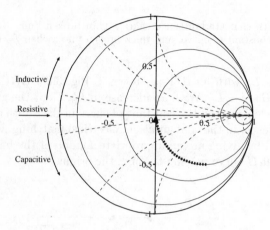

Fig. 3.30. Smith chart (impedance chart). Thin broken line: constant-X curve, thin
solid line: constant-R curve, and thick broken line: contour when L_s is increased

With this connection, X changes to $X + 1/\omega L_s$. Thus, with an increase
in L_s, Γ_0 at a given frequency moves clockwise on the constant-R curve in
the Smith chart as shown in Fig. 3.30. Hence, in this configuration, electrical
matching, where $\Gamma_0 = 0$, can be realized by giving an appropriate L_s only if
the initial Γ_0 locates on the constant-R curve passing through the origin.

For any values of Z or Y, matching ($\Gamma_0 = 0$) can be realized by using
a combination of two elements as shown in Fig. 3.31. In this case, the immittance chart shown in Fig. 3.32 is convenient where the admittance and
impedance charts are superimposed.

In method 1, firstly the parallel-connected admittance B_p is adjusted so
that Γ_0 moves to the intersection point with the constant R-curve passing
through the origin. By this adjustment, the synthesized resistance becomes

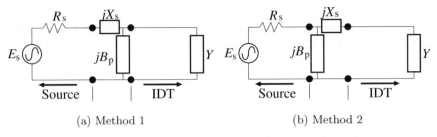

(a) Method 1   (b) Method 2

Fig. 3.31. Electrical matching by the combination of two elements

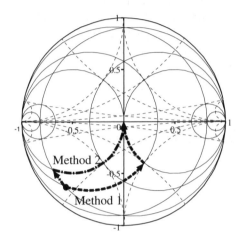

Fig. 3.32. Immittance chart

R_0. Then the series-connected impedance X_s is adjusted so that Γ_0 moves to the origin, and impedance matching is achieved.

In method 2, firstly the series-connected impedance X_s is adjusted so that Γ_0 moves to the intersection point with the constant G-curve passing through the origin. By this adjustment, the synthesized conductance becomes G_0. Then the parallel-connected susceptance B_s is adjusted so that Γ_0 moves to the origin, and impedance matching is achieved.

Which method is better is dependent on the initial Γ_0 of the IDT. Usually, the method giving shorter contour length is better.

3.4.3 Achievable Bandwidth

To add an inductive element to cancel the susceptance of the IDT is equivalent to composing an electrical resonance circuit whose resonance frequency is identical with that of the IDT. If the Q of the resonance circuit is large enough, it may limit the overall bandwidth instead of that of the IDT.

Let us consider the case when the susceptance B is parallel-connected to the IDT with the input admittance Y_{11}. Under impedance matching, namely, $\Re(Y_{11}) = G_s$ and $\Im(Y_{11}) = -B$, the resonance Q of the circuit $Y_{11} + G_s + jB$ is given by

$$Q_c = -\frac{B}{2G_s} = \frac{\Im(Y_{11})}{2\Re(Y_{11})}, \tag{3.49}$$

which is called the circuit Q. As is clear from the definition, $1/Q_c$ limits the achievable fractional -3 dB bandwidth.

Let us apply the delta-function model for further discussion. For IDTs without weighting, substitution of (3.39) and (3.40) into (3.49) gives $Q_c^{-1} = \eta^2 K^2 N/2$ at the resonance frequency. From its $\sin x/x$ dependence, $\Re(Y_{11}^S)$ has fractional -3 dB bandwidth of $1/2.257N$. Thus the achievable maximum fractional -3 dB bandwidth is given by

$$\sqrt{\eta^2 K^2/(2 \times 2.257)}$$

when $N = \sqrt{2/2.257\eta^2 K^2}$. This clearly suggests that use of substrates with large K^2 is essential to build wideband devices with low insertion loss.

3.5 p Matrix

3.5.1 Summary

Usually, the net power dissipation in the IDT is negligible, and the frequency characteristics of the IDT are determined by those of electrical mismatching with the peripheral circuit, such as signal source or load. Thus, for IDT characterization, direct use of the scattering coefficient is not so convenient because the matrix is also influenced by the peripheral circuit. To circumvent this difficulty, the p matrix method was developed [1].

The variables shown in Fig. 3.33 are related to each other by the p matrix as

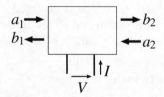

Fig. 3.33. Input and output to linear circuit

$$\begin{pmatrix} b_1 \\ b_2 \\ I \end{pmatrix} = \begin{pmatrix} p_{11} & p_{12} & p_{13} \\ p_{12} & p_{22} & p_{23} \\ -4p_{13} & -4p_{23} & p_{33} \end{pmatrix} \begin{pmatrix} a_1 \\ a_2 \\ V \end{pmatrix},$$ (3.50)

where V and I are their peak-to-peak values. Note that when V and I are also given by their effective values, the factors 4 appeared in Eq. (3.50) must be replaced by 2.

They are related to the scattering coefficient S_{ij} described in Sect. 3.1 as

$$\begin{aligned} p_{11} &= S_{11} - S_{13}^2/(S_{33}+1) \\ p_{22} &= S_{22} - S_{23}^2/(S_{33}+1) \\ p_{12} &= S_{12} - S_{13}S_{23}/(S_{33}+1) \\ p_{13} &= S_{13}/\sqrt{2R_{\mathrm{s}}}(S_{33}+1) \\ p_{23} &= S_{23}/\sqrt{2R_{\mathrm{s}}}(S_{33}+1) \\ p_{33} &= R_{\mathrm{s}}^{-1}(1-S_{33})/(1+S_{33}), \end{aligned}$$ (3.51)

where R_{s} is the characteristic impedance of the measurement system. Then p_{ij} can be evaluated experimentally from S_{ij} measured by the network analyzer.

If the system is unitary, S_{ij} satisfies the following conditions;

$$\begin{aligned} |p_{11}|^2 + |p_{12}|^2 &= 1 \\ |p_{12}|^2 + |p_{22}|^2 &= 1 \\ p_{11}p_{13}^* + p_{12}p_{23}^* + p_{13} &= 0 \\ p_{12}p_{13}^* + p_{22}p_{23}^* + p_{23} &= 0 \\ |p_{13}|^2 + |p_{23}|^2 - \tfrac{1}{2}\Re(p_{33}) &= 0. \end{aligned}$$ (3.52)

When two elements having p matrices of p^A and p^B are cascade-connected acoustically and parallel-connected electrically, the overall p matrix is given by [31]

$$p_{11} = p_{11}^A + p_{11}^B \left(\frac{p_{21}^A p_{12}^A}{1 - p_{11}^B p_{22}^A} \right),$$ (3.53)

$$p_{12} = \frac{p_{12}^A p_{12}^B}{1 - p_{11}^B p_{22}^A},$$ (3.54)

$$p_{13} = p_{13}^A + p_{12}^A \left(\frac{p_{13}^B + p_{11}^B p_{23}^A}{1 - p_{11}^B p_{22}^A} \right),$$ (3.55)

$$p_{22} = p_{22}^B + p_{22}^A \left(\frac{p_{12}^B p_{21}^B}{1 - p_{11}^B p_{22}^A} \right),$$ (3.56)

$$p_{23} = p_{23}^B + p_{21}^B \left(\frac{p_{23}^A + p_{22}^A p_{13}^B}{1 - p_{11}^B p_{22}^A} \right),$$ (3.57)

$$p_{33} = p_{33}^A + p_{33}^B + p_{32}^A \left(\frac{p_{13}^B + p_{11}^B p_{23}^A}{1 - p_{11}^B p_{22}^A} \right) + p_{31}^B \left(\frac{p_{23}^A + p_{22}^A p_{13}^B}{1 - p_{11}^B p_{22}^A} \right).$$ (3.58)

This relation is used recursively for the analysis of complex device structures.

3.5.2 IDT Characterization by Using p Matrix

Let us consider the electrical characteristics of the three-port circuit with the parallel-connected susceptance B shown in Fig. 3.34.

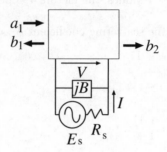

Fig. 3.34. Linear circuit with voltage source and matching circuit

When the IDT is driven by the source with internal resistance $R_s (= G_s^{-1})$ and open-circuited voltage E_s, (3.50) gives

$$\begin{pmatrix} b_1 \\ b_2 \\ I \end{pmatrix} = (E_s - R_s I) \begin{pmatrix} p_{13} \\ p_{23} \\ p_{33} + jB \end{pmatrix}, \tag{3.59}$$

since $V = E_s - R_s I$ and $a_1 = a_2 = 0$. Solving this equation, we obtain

$$I = E_s / \{R_s + (p_{33} + jB)^{-1}\}. \tag{3.60}$$

From (3.59), the directivity $D (= |b_1/b_2|)$, defined as the ratio between the acoustic outputs, is given by

$$D = |p_{13}/p_{23}|. \tag{3.61}$$

From (3.52) based on the unitary condition, one may obtain the relation

$$(1 + D^{-2})|p_{13}|^2 = \frac{1}{2} \Re(p_{33}) \tag{3.62}$$

after mathematical manipulation.

Since the maximum available power from the source is $(G_s/8)|E_s|^2$ and $|S_{13}|^2$ corresponds to the ratio of the output power $|b_1|^2$ to the maximum available power, $|S_{13}|^2$ is given by

$$|S_{13}|^2 = \frac{|b_1|^2}{(G_s/8)|E_s|^2} = \frac{|p_{13}(E_s - R_s I)|^2}{(G_s/8)|E_s|^2} = \frac{8G_s|p_{13}|^2}{|p_{33} + G_s + jB|^2}. \tag{3.63}$$

As an example, let us consider the conventional bidirectional IDT which excites SAWs with equal amplitude toward both sides. Since $D = 1$, substitution of (3.62) into (3.63) gives

$$|S_{13}|^2 = \frac{2G_s \Re(p_{33})}{|p_{33} + G_s + jB|^2}. \qquad (3.64)$$

This suggests that $|S_{13}|$ takes a maximum when $\Re(p_{33}) = G_s$ and $\Im(p_{33}) = -B$. Since p_{33} corresponds to the input admittance of the IDT, the condition is equivalent to the impedance matching condition.

Note the insertion loss $-20 \log |S_{13}|$ in this situation is 3 dB. This is due to the fact that the IDT radiates SAWs toward both sides, namely, bidirectionality. This contribution is called bidirectional loss.

Next, let us consider SAW incidence from port 1 with $E_s = 0$. Since $I = -(G_s + jB)V$ and $a_2 = 0$, (3.50) gives

$$\begin{pmatrix} b_1 \\ b_2 \\ 0 \end{pmatrix} = \begin{pmatrix} p_{11} & p_{13} \\ p_{12} & p_{23} \\ -4p_{13} & p_{33} + G_s + jB \end{pmatrix} \begin{pmatrix} a_1 \\ V \end{pmatrix}. \qquad (3.65)$$

Due to reciprocity, calculation of $S_{31} = G_s V^2 / 2|a_1|^2$ gives the same result as that given in (3.63). This suggests that bidirectional loss also occurs for detection as well as excitation. This is because the SAW is regenerated (reflected) electrically through the voltage induced by the current flowing into the load.

The reflection coefficient S_{11} is given by

$$S_{11} = \frac{b_1}{a_1} = p_{11} + \frac{4p_{13}^2}{p_{33} + G_s + jB}. \qquad (3.66)$$

Let us consider the case where mechanical reflection is negligible, i.e., $p_{11} \cong 0$ and $p_{22} \cong 0$. Substitution into (3.66) gives

$$|S_{11}|^2 = \frac{16|p_{13}|^4}{|p_{33} + G_s + jB|^2}, \qquad (3.67)$$

corresponding to reflection due to electrical regeneration. Comparison with (3.63) gives

$$|S_{11}|/|S_{13}| = \sqrt{\frac{\Re(p_{33})}{G_s(1 + D^{-2})}}. \qquad (3.68)$$

This equation suggests that if one can design G_s to be close to $\Re(p_{33})$ of the IDT to minimize the insertion loss, SAW reflection $|S_{11}|$ through electrical regeneration becomes large. This reflection is significant for practical SAW device design as will be discussed in the next chapter.

3.5.3 Discussion on Unidirectional IDTs

Let us define a parameter θ by

$$\exp(j\theta) = \frac{p_{11}}{|p_{11}|} \frac{p_{13}^*}{p_{13}}, \qquad (3.69)$$

which corresponds to the SAW phase shift between the reflection center and the excitation center for the IDT.

Applying the unitary condition of (3.52) to this equation, we obtain

$$1 - |p_{11}|^2 = D^2(1 + |p_{11}|^2 + 2|p_{11}|\cos\theta). \tag{3.70}$$

Since p_{33} corresponds to the input admittance of the IDT, impedance matching, where $|S_{33}| = 0$, is achieved when

$$G_\mathrm{s} + jB = p_{33}^*. \tag{3.71}$$

Under this condition, the scattering coefficients are reduced to

$$\begin{pmatrix} |S_{11}| & |S_{12}| & |S_{13}| \\ |S_{12}| & |S_{22}| & |S_{23}| \\ |S_{13}| & |S_{23}| & |S_{33}| \end{pmatrix} = \frac{1}{1+D^2} \begin{pmatrix} 1 & D & D^2 \\ D & D^2 & 1 \\ D^2 & 1 & 0 \end{pmatrix}, \tag{3.72}$$

from (3.61), (3.62), (3.69) and (3.70). It is interesting to note that all elements are expressed only in terms of the directivity D.

On the other hand, the zero-reflection condition where $S_{11} = 0$ is given by

$$G_\mathrm{s} + jB = -\frac{4p_{13}^2}{p_{11}} - p_{33} = -2\exp(-j\theta)\frac{\Re(p_{33})}{(1+D^{-2})|p_{11}|} - p_{33}, \tag{3.73}$$

from (3.62), (3.66) and (3.69). Thus, the condition is fulfilled by adjusting B so that

$$G_\mathrm{s} = -\Re(p_{33})\left(\frac{2\cos\theta}{(1+D^{-2})|p_{11}|} + 1\right). \tag{3.74}$$

This suggests that zero reflection is realized if $\cos\theta < -(1+D^{-2})|p_{11}|/2$ even when $\theta \neq \pi$. Under the zero-reflection condition given by (3.74), the scattering coefficients are given by

$$\begin{pmatrix} |S_{11}| & |S_{12}| & |S_{13}| \\ |S_{12}| & |S_{22}| & |S_{23}| \\ |S_{13}| & |S_{23}| & |S_{33}| \end{pmatrix} = \begin{pmatrix} 0 & D^{-1} & 1-D^{-2} \\ D^{-1} & 1-D^{-2} & D^{-2} \\ 1-D^{-2} & D^{-2} & D^{-2}(1-D^{-2}) \end{pmatrix}, \tag{3.75}$$

and are again expressed only in terms of D.

Figure 3.35 shows the insertion loss and reflection coefficient as a function of the directivity under these optimal conditions. It is seen that the difference in the insertion losses decreases rapidly with an increase in D. Thus, unidirectional IDTs are usually designed so as to fulfill the zero-reflection condition where low insertion loss and zero reflectivity are simultaneously achievable provided that D is sufficiently large.

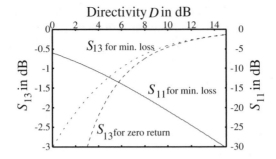

Fig. 3.35. Insertion loss and reflection coefficient under optimal conditions

3.6 BAW Radiation

3.6.1 Phase Matching Condition

Usually IDTs also act as efficient BAW transducers. In SAW devices, since the BAW excitation causes unwanted or spurious responses, materials and/or structures with low BAW radiation are desirable.

For strong excitation, BAWs radiated from each period of the IDT must be phase-matched to a certain direction. From Fig. 3.36, the phase matching condition is given by

$$p_I \cos \theta_B = n\lambda_B, \tag{3.76}$$

where λ_B is the BAW wavelength. By using the BAW wavenumber $\beta_B = 2\pi/\lambda_B$, the phase matching condition can be rewritten as

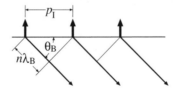

Fig. 3.36. Phase matching condition for BAW radiation

$$\beta_B \cos \theta_B = 2n\pi/p_I, \tag{3.77}$$

or by using the slowness S_B of the BAW,

$$S_B \cos \theta_B = 2n\pi/\omega p_I. \tag{3.78}$$

Note that, since piezoelectric materials employed for the SAW devices are inherently anisotropic, β_B and S_B are dependent upon the radiation direction, i.e., θ_B.

The solution of (3.78) is easily found schematically by using the slowness surface shown in Fig. 3.37a. Namely, θ_B is determined as the angle where the projection of S_B onto the S_x axis is equal to $2n\pi/\omega p_I$. It is clear that there exists a cut-off frequency ω_c given by

$$\omega_c = \frac{2\pi}{S_x|_{\theta_B=\theta_{Bc}}p_I}, \tag{3.79}$$

and the BAW is not radiated at frequencies lower than ω_c. In this equation, θ_{Bc} is the radiation angle at the right-side edge of the slowness surface, and is zero if the slowness surface is symmetrical with respect to the S_x axis.

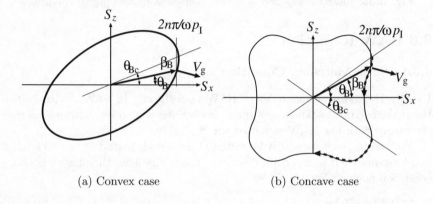

(a) Convex case   (b) Concave case

Fig. 3.37. Phase matching condition in slowness surface

For given S_x, there are an even number of solutions satisfying (3.79). Half of them correspond to the BAWs radiated from the surface toward the bulk whereas the others correspond to those coming back from the bottom surface. It should be noted that they are distinguished by the direction of the power flow instead of that of the wavevector. That is, the BAWs radiated from the top surface have a the power flow directed into the bulk whereas those reflected from the bottom surface have a the power flow in the opposite direction. So when the slowness surface is convex as shown in Fig. 3.37a, θ_B monotonically increases with frequency. On the other hand, when the slowness surface is partly concave as shown in Fig. 3.37b, two BAWs with different S_B are simultaneously excited, and with an increase in frequency, they change their radiation directions as shown by the dashed lines. It is interesting to note that for the upper branch, there exists another cut-off frequency where the power flow directs parallel to the surface, and the corresponding BAW is not radiated at frequencies above the cut-off.

3.6.2 Radiation Characteristics

As will be discussed in Sect. 6.2, the BAW power P_B radiated from the surface charge $q(x_1)$ is given by

$$P_B = \pi \omega W \int_{-\infty}^{+\infty} \frac{|Q(S\omega)|^2}{|S|} \Im[\epsilon(S)^{-1}]dS, \tag{3.80}$$

where $\epsilon(S)$ is the effective permittivity [13] and W is the aperture. This suggests that nonzero $\Im[\epsilon(S)]$ indicates radiation of BAWs whose wavevector parallel to the surface is ωS. With an applied voltage V, the conductance G_B for the BAW radiation is given by

$$G_B = \frac{2P_B}{V^2}. \tag{3.81}$$

It is known that the radiation susceptance $B(\omega)$ is not independent of the radiation conductance $G(\omega)$, and is given by [32]

$$B(\omega) = \frac{1}{\pi} \int_{-\infty}^{+\infty} \frac{G(\xi)}{\xi - \omega}d\xi. \tag{3.82}$$

This relation is called the Hilbert transform.

For the evaluation of $G_B(\omega)$ and $B_B(\omega)$, we use the static charge distribution given by (3.1) or (3.2).

Figure 3.38 shows the calculated return loss of the IDT on 45°YZ-Li$_2$B$_4$O$_7$ [33]. In the calculation, $Q(\beta)$ given by (3.18) was employed. A relatively large return loss exists for frequencies above the SAW resonance frequency of 40 MHz, and is due to the BAW radiation. The result agrees with experiment, and the effectiveness of this method is demonstrated. It is seen that the BAW radiation exhibits a cut-off nature, and the BAW is radiated for a wider frequency range than the SAW.

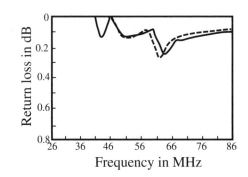

Fig. 3.38. Return loss characteristics of IDT on 45°YZ-Li$_4$B$_4$O$_7$. Solid line: experiment, and dashed line: theory

To distinguish the contributions of each BAW component, we add the subscript m in the effective permittivity $\epsilon_m(S)$. Then the radiated power P_m associated with the m-th component is given by

$$P_m = \pi\omega W \int_{-S_{\mathrm{Bcm}}}^{+S_{\mathrm{Bcm}}} \frac{|Q(\omega S)|^2}{|S|} \Im[\epsilon_m(S)^{-1}]dS, \tag{3.83}$$

where S_{Bcm} is the slowness of the m-th component at the cut-off.

Next, let us designate the slowness in the saggital plane ($x_1 - x_3$ plane) of the m-th component as $S = S_{\mathrm{Bm}}(\theta)\cos\theta$. Then, (3.83) can be rewritten as

$$P_m = \pi\omega W \int_0^\pi |Q(\omega S_{\mathrm{Bcm}}(\theta)\cos\theta)|^2 F_m(\theta,\omega)d\theta, \tag{3.84}$$

where

$$F_m(\theta,\omega) = \Im[\epsilon_m(S_{\mathrm{Bcm}}(\theta)\cos\theta,\omega)^{-1}]$$
$$\times \left(S_{\mathrm{Bm}}^{-1} \left| \frac{dS_{\mathrm{Bcm}}(\theta)}{d\theta} \right| + |\tan\theta| \right) \tag{3.85}$$

is a function responsible for the angular dependence of the BAW radiated efficiency. In (3.85), $S_{\mathrm{Bcm}}^{-1} dS_{\mathrm{Bcm}}(\theta)/d\theta$ is due to the anisotropy of the substrate, and its value is usually small.

For the case where the angular dependence is negligible, i.e., $F_m(\theta)$ is constant, the BAW radiation conductance was calculated for the IDT with 10 finger-pairs. Figure 3.39 shows the result. The horizontal axis is the frequency normalized by the BAW cut-off frequency. The SAW radiation admittance $G_{\mathrm{S}} + jB_{\mathrm{S}}$ for the IDT was already given in Fig. 3.17. It is seen that G_{B} takes a maximum value at a frequency a little higher than the BAW cut-off frequency, and a relatively large G_{B} remains at frequencies higher than the cut-off frequency.

In the figure, the radiation susceptance B_{B} for the BAW is also shown. Comparing with the susceptance B_{S} for the SAW radiation, it is interesting to note that only the positive peak appears and the negative has mostly disappeared.

By the way, if we ignore the contribution of the anisotropy in (3.85), we obtain

$$F_m(\theta,\omega) = \Im[\epsilon_m(S_{\mathrm{Bcm}}(\theta)\cos\theta)^{-1}]|\tan\theta|. \tag{3.86}$$

In the case where the θ dependence of $F(\theta,\omega)$ is also negligible, we obtain the relation

$$\Im[\epsilon_m(S)^{-1}] \propto |\cot\theta| = 1/\sqrt{(S_{\mathrm{Bcm}}/S)^2 - 1}.$$

This means that, even when $F(\theta,\omega)$ is constant, the BAW is strongly excited at $S \cong S_{\mathrm{Bcm}}$ where its beam directs almost parallel to the surface.

This frequency response can be explained schematically by using the slowness surface shown in Fig. 3.40, where the charge distribution $Q(\omega S)$ with

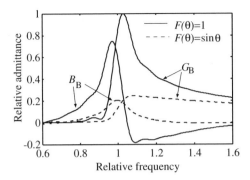

Fig. 3.39. BAW radiation characteristics. Solid line: when the effect of the surface boundary condition is weak (small η), and broken line: when the effect of the surface boundary condition is significant (large η)

a peak at $\beta p = 2\pi$ is superimposed. With an increase in the frequency, the peak moves toward the origin, and the BAW components with $\beta = \beta_B \cos \theta$ smaller than the cut-off are radiated.

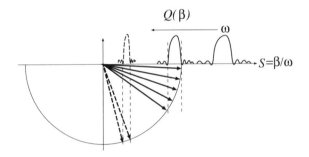

Fig. 3.40. Slowness surface and BAW radiation. $Q(\beta)$: charge distribution

As a function of S, the peak width of $Q(\omega S)$ is inversely proportional to ω. When the BAW beam directs parallel to the surface, namely $\omega \cong \omega_c$, the BAW components with a wide range of β are excited. With an increase in ω, the range of β is decreased. Hence G_B takes a maximum just above the cut-off frequency where the peak in $Q(\omega S)$ is totally involved in the slowness surface, and then G_B decreases monotonically with an increase in ω.

As described in Sect. 1.2.1, even in isotropic materials, the radiation characteristics of the SV- and L-type BAWs are dependent upon their radiation angle, and they are not radiated parallel to the surface due to the surface boundary condition. Figure 3.39 also shows, as an example, G_B for the case where $F_m(\theta, \omega) = \sin \theta$. This angular dependence significantly influences G_B

near the cut-off frequency, where the BAW beam directs almost parallel to the surface.

References

1. G. Tobolka: Mixed Matrix Representation of SAW Transducers, IEEE Trans. Sonics and Ultrason., **SU-26** (1979) pp. 426–428.
2. R.M. White and F.W. Voltmer: Direct Piezoelectric Coupling to Surface Elastic Waves, Appl. Phys. Lett., **17** (1965) pp. 314–316.
3. R.C. Rosenfeld, R.B. Brown and C.S. Hartmann: Unidirectional Acoustic Surface Wave Filters with 2 dB Insertion Loss, Proc. IEEE Ultrason. Symp. (1974) pp. 425–428.
4. D.C. Malocha: Quadrature 3-Phase Unidirectional Transducer, IEEE Trans. Sonics and Ultrason., **SU-26** (1979) pp. 313–315.
5. K. Yamanouchi, F. Nyfeller and K. Shibayama: Low Insertion Loss Acoustic Surface Wave Filter Using Group-type Unidirectional Interdigital Transducers, Proc. IEEE Ultrason. Symp. (1975) pp. 317–321.
6. T. Kodama, H. Kawabata, Y. Yasuhara and H. Sato: Design of Low-Loss SAW Filters Employing Distributed Acoustic Reflection Transducers, Proc. IEEE Ultrason. Symp. (1986) pp. 313–324.
7. C.S. Hartmann and B.P. Abott: Overview of Design Challenges for Single Phase Unidirectional SAW Filters, Proc. IEEE Ultrason. Symp. (1989) pp. 79–89.
8. K. Yamanouchi and H. Furuyashiki: New Low-Loss SAW Filter Using Internal Floating Electrode Reflection Types of Single-Phase Unidirectional Transducer, Electron. Lett., **20** (1984) pp. 989–990.
9. P.V. Wright: Natural Single-Phase Unidirectional Transducer, Proc. IEEE Ultrason. Symp. (1985) pp. 58–63.
10. C.S. Hartmann, P.V. Wright, R.J. Kansy and E.M. Garber: Analysis of SAW Interdigital Transducer with Internal Reflections and the Application to the Design of Single-Phase Unidirectional Transducers, Proc. IEEE Ultrason. Symp. (1982) pp. 40–45.
11. H. Engan: Excitation of Elastic Surface Waves by Spatial Harmonics of Interdigital Transducers, IEEE Trans. Electron. Device, **ED-16** (1969) pp. 1014–1017.
12. H. Engan: Surface Acoustic Wave Multi-Electrode Transducers, IEEE Trans. Sonics and Ultrason., **SU-22** (1975) pp. 395–401.
13. R.F. Milsom, N.H.C. Reilly and M. Redwood: Analysis of Generation and Detection of Surface and Bulk Acoustic Waves by Interdigital Transducers, IEEE Trans. Sonics and Ultrason., **SU-24** (1977) pp. 147–166.
14. D.P. Morgan: Surface-Wave Devices for Signal Processing, Chap. 3, Elsevier, Amsterdam (1985) pp. 39–55.
15. J.J. Campbell and W.R. Jones: A Method for Optimal Crystal Cuts and Propagation Directions for Excitation of Piezoelectric Surface Waves, IEEE Trans. Sonics and Ultrason., **SU-15** (1968) pp. 209–217.
16. K.A. Ingebrigtsen: Surface Waves in Piezoelectrics, J. Appl. Phys., **40** (1969) pp. 2681–2686.
17. S. Datta, B.J. Hunsinger and D.C. Malocha: A Generalized Model for Periodic Transducers with Arbitrary Voltages, IEEE Trans. Sonics and Ultrason. **SU-26**, 3 (1980) pp. 235–242.
18. K. Hashimoto Y. Koseki and M. Yamaguchi: Boundary Element Method Analysis of Interdigital Transducers Having Arbitrary Metallisation Ratio, Japan. J. Appl. Phys., **30**, Suppl. 30-1 (1991) pp. 1425–1427.

19. C.S. Hartmann and B.G. Secrest: End Effects in Interdigital Surface Wave Transducers, Proc. IEEE Ultrason. Symp. (1972) pp. 413–416.
20. R.H. Tancrell and M.G. Holland: Acoustic Surface Wave Filters, Proc. IEEE, **59** (1971) pp. 393–409.
21. D.P. Morgan: Surface-Wave Devices for Signal Processing, Chap. 3, Elsevier, Amsterdam (1985) pp. 343–353.
22. W.R. Smith, H.M. Gerard, J.H. Collins, T.M. Reeder and H.J. Show: Analysis of Interdigital Surface Wave Transducers By Use of an Equivalent Circuit Model, IEEE Trans. Microwave Theory and tech., **MTT-17** (1969) pp. 856–864.
23. W.R. Smith: Experimental Distinction Between Crossed-Field and In-Line Three-Port Circuit Models for Interdigital Transducers, IEEE Trans. Microwave Theory and tech., **MTT-17** (1974) pp. 960–964.
24. W.P. Mason (ed): Physical Acoustics, Vol. 1A, Academic Press (1964) pp. 335–416.
25. R.C.M. Li and J. Melngailis: The Influence of Stored Energy at Step Discontinuities on the Behavior of Surface-Wave Gratings, IEEE Trans. Sonics and Ultrason., **SU-22** (1975) pp. 189–198.
26. T. Kojima and K. Shibayama: An Analysis of an Equivalent Circuit Model for an Interdigital Surface-Acoustic-Wave Transducer, Jpn. J. Appl. Phys., **27**, Suppl. 27-1 (1988) pp. 163–165.
27. G.S. Kino: Acoustic Waves: Devices, Imaging, & Analog Signal Processing, Prentice-Hall, Englewood Cliffs (1987).
28. B.A. Auld: Acoustic Waves and Fields in Solids, Vol. II, Chap. 12, Wiley and Sons, New York (1973) pp. 271–332.
29. C.C.W. Ruppel, W. Ruile, G. Sholl, K.C. Wagner and O. Männer: Review of Models for Low-Loss Filter Design and Applications, Proc. IEEE Ultrason. Symp. (1994) pp. 313–324.
30. D.P. Chen and H.A. Haus: Analysis of Metal-Strip SAW Grating and Transducers, IEEE Trans. Sonics and Ultrason., **SU-26** (1985) pp. 395–408.
31. B.P. Abbott, C.S. Hartmann and D.C. Malocha: A Coupling-of-Modes Analysis of Chirped Transducers Containing Reflective Electrode Geometries, Proc. IEEE Ultrason. Symp. (1989) pp. 129–134.
32. A. Nalamwar and M. Epstein: Immittance Characterisation of Acoustic Surface-Wave Transducer, Proc. IEEE, **60** (1072) pp. 336–337.
33. K. Hashimoto: On Leaky Surface Acoustic Wave and Bulk Acoustic Wave Launched from an Interdigital Transducer, Ph D thesis, Tokyo Institute of Technology (1988) in Japanese.

4. Transversal Filters

This chapter deals with transversal filters. Starting from their fundamentals, device modeling and the design and analysis of spurious responses are discussed. In addition, various UDT-based low-loss transversal filters are detailed, which presently receive much attention for use in IF filters for mobile phones.

4.1 Basics

4.1.1 Weighting

Since the IDT possesses its own frequency dependence, filters can be constructed by placing two IDTs in parallel as shown in Fig. 4.1. This type of filter is called a transversal filter [1]. Appropriate weighting for the IDTs enables us to synthesize the desired frequency response.

For the weighting, various techniques were proposed. Among them, the apodization shown in Fig. 4.1a is most widely used. This is due to the fact that the weighting function can be controlled easily and precisely because the SAW excitation strength is proportional to the overlap length between adjacent electrodes. The withdrawal weighting [2] shown in Fig. 4.1b is also employed in cases where reduction of the diffraction and/or insertion loss is crucial. The width weighting shown in Fig. 4.1c is a variation that offers limited weighting capability. However this technique is scarcely used because the relation between the electrode width and SAW excitation strength is complicated. The series weighting [3] or dogleg weighting [4] shown in Fig. 4.1d is another variation of the withdrawal weighting. In this case, the SAW excitation strength is adjusted by the number of finger subdivision.

Let A_{in} denote the weight for the n-th finger-pair of IDT-i. Among SAWs excited by IDT-1, IDT-2 can detect only those which pass through the interdigitated region of IDT-2. Thus, as an extension of the delta-function model described in Sect. 3.3.1, the transfer admittance Y_{12}^S due to the SAW is given by

$$Y_{12}^S = \omega h_0 \sum_{m=-N_1}^{N_1} \sum_{n=-N_2}^{N_2} \mathrm{Cr}(A_{1m}, A_{2n}) \exp(-j\beta|L + (m-n)p_I|), \qquad (4.1)$$

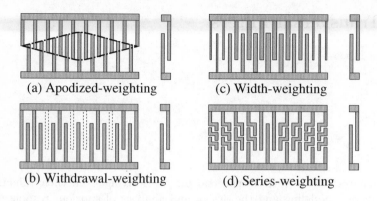

(a) Apodized-weighting   (c) Width-weighting

(b) Withdrawal-weighting   (d) Series-weighting

Fig. 4.1. Weighting techniques for IDT

where $\mathrm{Cr}(A_{1m}, A_{1n}) = \mathrm{Min}(|A_{1m}|, |A_{2n}|)\mathrm{sgn}(A_{1m}A_{2n})$, L is the distance between the IDT centers, $\mathrm{Min}(x, y)$ is a function giving the smallest among the arguments, and $\mathrm{sgn}(x)$ gives sign of its argument.

First, we consider the case shown in Fig. 4.2a where the withdrawal weighting is applied for both IDTs or the case shown in Fig. 4.2b where IDT-2 is withdrawal weighted and IDT-1 is apodized. In this case, since the

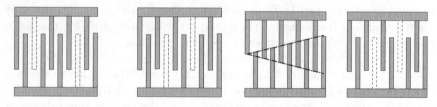

(a) Withdrawal weighted IDTs   (b) Withdrawal and apodize weighted IDTs

Fig. 4.2. Cases where products between transfer functions hold

SAW amplitude excited by IDT-2 is uniform throughout the aperture, Y_{12}^{S} is given by

$$Y_{12}^{\mathrm{S}} = \omega h_0 |A_{21}|^{-1} \sum_{m=-N_1}^{N_1} \sum_{n=-N_2}^{N_2} A_{1m} A_{2n} \exp[-j\beta\{L + (m - n)p_{\mathrm{I}}\}]$$

$$\equiv \omega h_0 |A_{21}|^{-1} \exp(-j\beta L) H_1(\omega) H_2(\omega)^*, \qquad (4.2)$$

where $H_i(\omega)$ is the transfer function of IDT-i given by

$$H_i(\omega) = \sum_{n=-N_i}^{N_i} A_{in} \exp(-jn\beta p_{\mathrm{I}}). \qquad (4.3)$$

Because $\beta = \omega/V_{\mathrm{S}}$, $H_i(\omega)$ corresponds to the frequency response of the IDT itself when the internal mechanical and electrical reflections are ignored. In (4.2), Y_{12}^{S} was subdivided into three components, i.e., $\exp(-j\beta L)$ corresponding to the phase lag due to propagation, and $H_1(\omega)$ and $H_2(\omega)$ responsible for the frequency responses of individual IDTs. In this case, the total transfer function Y_{12}^{S} is given by the product of the transfer functions of the IDTs.

Note that, when both IDTs are apodized as shown in Fig. 4.3, since the structure is equivalent to an array of SAW devices with narrow aperture, its frequency response is not given by the product of the transfer functions.

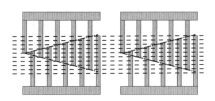

Fig. 4.3. When both IDTs are apodized

It is known that the impulse response is given by the inverse Fourier transform of the frequency response. So, the impulse response $h(t)$ associated with $H(\omega S_{\mathrm{S}})$ is given by

$$h(t) = \frac{1}{2\pi} \int_{-\infty}^{+\infty} H(\omega S_{\mathrm{S}}) \exp(j\omega t)d\omega = \sum_{n=-N}^{N} A_n \delta(t - n\tau), \tag{4.4}$$

where $\delta(t)$ is the delta-function, and $\tau = p_{\mathrm{I}}/V_{\mathrm{S}}$ is the time delay with SAW propagation for a period, i.e., unit finger-pair length. As is clear from (4.4), the weighting of the IDT corresponds to the waveform of the impulse response. Note that because of its finite impulse response length, this type of filter is also called a finite impulse response (FIR) filter.

The fact that the Fourier transform of the weighting function corresponds to the frequency response suggests that the existence of sidelobes in the frequency response is due to a nonzero gradient in the weighting function. For example, the large sidelobes appearing in unapodized IDTs is due to a large discontinuity at the IDT ends. This means that giving a smooth weighting function is essential in order to suppress the sidelobes.

When the weighting for the IDT is symmetrical with respect to the center, i.e., $A_n = A_{-n}$, as shown in Fig. 4.4, (4.3) can be written as

$$H(\omega) = A_0 + 2\sum_{n=1}^{N} A_n \cos(n\omega p_{\mathrm{I}}/V_{\mathrm{S}}). \tag{4.5}$$

Thus $H(\omega)$ is always real. From this, if both the input and output IDTs are symmetrical, (4.2) can be rewritten as

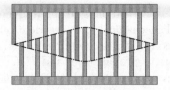

Fig. 4.4. Symmetrical weighting

$$Y_{12}^S = \omega h_0 |A_{21}|^{-1} \exp(-j\beta L) H_1(\omega) H_2(\omega). \tag{4.6}$$

This suggests that Y_{12}^S has linear phase response where the group delay is independent of ω and only the amplitude is dependent on ω. This feature is desirable for use in communication systems because undesirable signals can be rejected without affecting the object signal.

By the way, very long weighting function must be given to control the frequency response precisely. In this case, the device size becomes large and the effects of diffraction, which will be discussed in Sect. 4.3.1, may be significant.

If linear phase characteristics are not required, there is an alternative approach called minimum phase design [5]. That is, for a given $|H(\omega)|$, the design offers minimized $-\angle H(\omega)$ for arbitrary ω.

Designing the minimum phase weighting function is relatively simple, and the weighting function can be derived uniquely from that with linear phase [5], which can be designed by using various techniques described in Sect. 4.2.

Figure 4.5 shows, as an example, weighting functions having the same $|H(\omega)|$ under the linear phase and minimum phase designs. It is seen that the weighting function by the minimum phase design concentrates on the left-hand side more than that of the linear phase design. Thus, the effective distance between the receiving IDT can be shortened, and the effects of diffraction will be reduced.

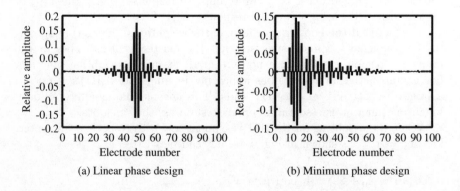

(a) Linear phase design   (b) Minimum phase design

Fig. 4.5. Linear phase and minimum phase designed weighting functions

4.1.2 Basic Properties of Weighted IDTs

The delta-function or equivalent circuit models described in the previous chapter are applicable for the simulation of the transversal filter employing apodized IDTs by subdividing it into narrow tracks as shown in Fig. 4.6. Each track is equivalent to a transversal filter without weighting and its admittance matrix is given by the analysis based on the delta-function or equivalent circuit models. Then the overall admittance matrix is simply given by the sum of the admittance matrices for all tracks [6].

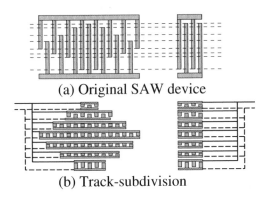

(a) Original SAW device

(b) Track-subdivision

Fig. 4.6. Subdivision of SAW device

Let us consider the case where every SAW excited by the apodized IDT-1 intersects the unapodized IDT-2.

The delta-function model analysis described in Sect. 3.3.1 suggests that the radiation admittance Y_{22}^S is given by

$$\frac{Y_{22}^S}{\omega h_0} = \sum_{m=-N_2}^{N_2} \sum_{n=-N_2}^{N_2} |A_{21}| \mathrm{sgn}(A_{2m}) \mathrm{sgn}(A_{2n}) \exp(-j\beta|(m-n)p_I|)$$

(4.7)

for the unweighted or withdrawal-weighted IDT-2. From this, its real part, i.e., the radiation conductance $G_{22} = \Re(Y_{22}^S)$, is given by

$$\frac{G_{22}^S}{\omega h_0} = \Re \left[\sum_{m=-N_2}^{N_2} \sum_{n=-N_2}^{N_2} |A_{21}| \mathrm{sgn}(A_{2m}) \mathrm{sgn}(A_{2n}) \exp(-j\beta|(m-n)p_I|) \right]$$

$$= \Re \left[\sum_{m=-N_2}^{N_2} \sum_{n=-N_2}^{N_2} |A_{21}| \mathrm{sgn}(A_{2m}) \mathrm{sgn}(A_{2n}) \exp\{-j\beta(m-n)p_I\} \right]$$

$$= |A_{21}|^{-1} |H_2(\omega)|^2.$$

(4.8)

On the other hand, the radiation admittance Y_{11}^S of the apodized IDT-1 is given by

$$Y_{11}^S = \omega h_0 \sum_{m=-N_1}^{N_1} \sum_{n=-N_1}^{N_1} \mathrm{Cr}(A_{1m}, A_{1n}) \exp(-j\beta|(m-n)p_I|). \tag{4.9}$$

In this case, due to the term $\mathrm{Cr}(A_{1m}, A_{1n})$, the radiation conductance $G_{11} = \Re(Y_{11}^S)$ is not proportional to $|H_1(\omega)|^2$.

Since the transfer function Y_{12}^S has already been derived in (4.1), all components required for SAW device simulation have been obtained.

For the case where both IDTs are unapodized, we derived the following relation as (3.41):

$$|Y_{12}^S|^2 = \Re(Y_{11}^S)\Re(Y_{22}^S).$$

On the other hand, when either IDT is apodized, (4.1), (4.8) and (4.9) suggest that this relation does not hold. So we define

$$\mu = \frac{|Y_{12}^S|}{\sqrt{\Re(Y_{11}^S)\Re(Y_{22}^S)}}. \tag{4.10}$$

From (4.6)–(4.9), $\mu \leq 1$. We refer to $-20\log\mu$ as the apodization loss. This loss originates from the fact that the SAW field excited by the apodized IDT is not uniform within its aperture, and the SAW amplitude detectable by the unapodized IDT is limited by its average over the aperture. For withdrawal-weighted IDTs, since the excited SAW field is uniform throughout the aperture, the apodization loss does not occur.

4.1.3 Effects of Peripheral Circuits

Analysis. The characteristics of the transversal filter are quite sensitive to the peripheral circuits connected to the IDTs. To discuss their effects, let us express the SAW device by the admittance matrix Y as shown in Fig. 4.7, where B_1 and B_2 are the susceptances for impedance matching, and R_{in} and R_{out} are the input and output resistances of the peripheral circuits. Note that, although $Y_{12} = Y_{21}$ due to the reciprocity relation, we distinguish them for further discussion.

The circuit analysis for Fig. 4.7b gives

$$\begin{pmatrix} i_1 \\ i_2 \end{pmatrix} = \begin{pmatrix} Y_{11} + jB_1 & Y_{12} \\ Y_{21} & Y_{22} + jB_2 \end{pmatrix} \begin{pmatrix} e_1 \\ e_2 \end{pmatrix}. \tag{4.11}$$

By substituting $i_1 = G_{in}(E_s - e_1)$ and $i_2 = -G_{out}e_2$, we obtain

$$\begin{pmatrix} G_{in}E_s \\ 0 \end{pmatrix} = \begin{pmatrix} Y_{11} + jB_1 + G_{in} & Y_{12} \\ Y_{21} & Y_{22} + jB_2 + G_{out} \end{pmatrix} \begin{pmatrix} e_1 \\ e_2 \end{pmatrix}. \tag{4.12}$$

Then, the output voltage $E_{out} = e_2$ is given by

$$E_{out} = \frac{-Y_{21}G_{in}E_s}{(G_{in} + jB_1 + Y_{11})(G_{out} + jB_2 + Y_{22}) - Y_{12}Y_{21}}. \tag{4.13}$$

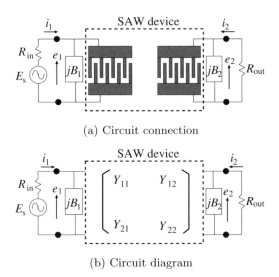

(a) Circuit connection

(b) Circuit diagram

Fig. 4.7. Transversal filter with peripheral circuits

By using the scattering coefficient S_{12}, the insertion loss IL is expressed as $-20 \log_{10} |S_{12}|$, corresponding to the output power relative to the available maximum power from the source. If $R_{\text{in}} = R_{\text{out}}$, the maximum power is achieved when the input and output ports are directly connected. On the other hand, if $R_{\text{in}} \neq R_{\text{out}}$, the maximum power is achieved when the ideal transformer of $n = \sqrt{G_{\text{in}}/G_{\text{out}}}$ is inserted as shown in Fig. 4.8. In the case,

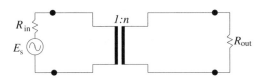

Fig. 4.8. Maximum power transfer

the output voltage E_{out} is $nE_{\text{s}}/2 \equiv E_{\text{max}}$. So S_{12} is given by

$$S_{12} = \frac{E_{\text{out}}}{E_{\text{max}}} = \frac{-2Y_{21}\sqrt{G_{\text{in}}G_{\text{out}}}}{(G_{\text{in}} + jB_1 + Y_{11})(G_{\text{out}} + jB_2 + Y_{22}) - Y_{12}Y_{21}}. \tag{4.14}$$

Then the frequency response of the SAW filter including the peripheral circuits can be evaluated by substituting Y_{ij} estimated by the delta-function or equivalent circuit model into (4.14).

Origin of Insertion Loss. Since Y_{12} corresponds to the current induced in IDT-1 by the voltage across IDT-2, the term $Y_{12}Y_{21}$ in the denominator of

(4.14) is due to electrical regeneration by the IDTs. The effects of this term will be discussed later.

Let S_{12}^0 denote S_{12} when the term $Y_{12}Y_{21}$ is removed, i.e.,

$$S_{12}^0 = \frac{-2Y_{21}\sqrt{G_{\text{in}}G_{\text{out}}}}{(G_{\text{in}} + jB_1 + Y_{11})(G_{\text{out}} + jB_2 + Y_{22})}. \tag{4.15}$$

Then the insertion loss of the SAW device IL $= -20\log|S_{12}^0|$ is subdivided into

$$\text{IL} = \text{IL}a + \text{IL}b + \text{IL1} + \text{IL2}, \tag{4.16}$$

where

$$\text{IL1} = -10\log\left|\frac{4G_{11}G_{\text{in}}}{(G_{\text{in}} + G_{11})^2 + (B_1 + B_{11})^2}\right|, \tag{4.17}$$

$$\text{IL2} = -10\log\left|\frac{4G_{22}G_{\text{out}}}{(G_{\text{out}} + G_{11})^2 + (B_2 + B_{22})^2}\right|, \tag{4.18}$$

are the mismatching losses for IDT-1 and IDT-2, respectively, with the peripheral circuits, and

$$\text{IL}a = -20\log\left|\frac{Y_{21}}{\sqrt{G_{11}G_{22}}}\right| = -20\log\mu \tag{4.19}$$

is the apodization loss defined in (4.10). The SAW propagation loss is included in it if it exists; and

$$\text{IL}b = 20\log 2 \tag{4.20}$$

is the bidirectional loss of the IDTs.

TTE. Here we discuss the impulse response of the SAW device. When an impulse is applied to IDT-1, the response due to the electrostatic coupling appears in IDT-2 with negligible time delay. This response is called electrical feedthrough. Then the SAW response appears in IDT-2. Since Y_{12} corresponds to the effect of the voltage on IDT-2 to IDT-1, a nonzero Y_{12} causes SAW regeneration or electrical reflection by the voltage induced on IDT-2 through the current flow to the peripheral circuit. SAWs reflected by IDT-2 are again reflected by IDT-1, and are finally detected by IDT-2 (see Fig. 4.9). This response is called a triple transit echo (TTE).

Figure 4.10 shows the impulse response of the SAW device including electrical feedthrough and TTE. Added to these, multiple reflected echoes and BAW responses are often observed. Recent network analyzers offer the ability to evaluate the impulse response numerically from the measured frequency response.

Figure 4.11 shows the influence of the TTE to the frequency response of a SAW transversal filter. Due to the difference in the time delay 2τ between the main response and the TTE, a ripple with short period $1/2\tau$ occurs due to their interference. As will be discussed below, this ripple becomes significant at frequencies where $|S_{12}|$ is large.

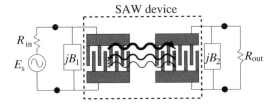

Fig. 4.9. TTE

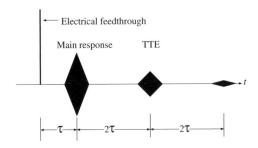

Fig. 4.10. Impulse response including electrical feedthrough and TTE

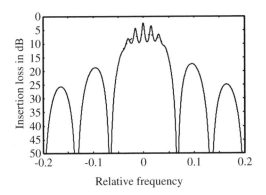

Fig. 4.11. Influence of TTE on frequency response. Solid line: S_{12}, and dashed line: S_{12}^0

Modification of (4.14) gives

$$|S_{12}| = \mu T_{\text{in}} T_{\text{out}} \left| \sum_{n=0}^{+\infty} \{\mu^2 \Gamma_{\text{in}} \Gamma_{\text{out}} \exp(-2j\beta L)\}^n \right| \tag{4.21}$$

from the relation $Y_{12} = Y_{21} = \mu\sqrt{G_{11}G_{22}} \exp(-j\beta L)$. In this equation, T_{in} and T_{out} are the transmission coefficients between electrical and acoustic ports given by (3.64), and Γ_{in} and Γ_{out} are the reflection coefficients given by (3.67) for each IDT when mechanical reflection is ignored, i.e., $p_{11} = 0$.

That is,

$$T_{in} = \frac{\sqrt{2G_{in}G_{11}}}{|G_{in} + jB_1 + Y_{11}|} \tag{4.22}$$

$$T_{out} = \frac{\sqrt{2G_{out}G_{22}}}{|G_{out} + jB_2 + Y_{22}|} \tag{4.23}$$

$$\Gamma_{in} = \frac{G_{11}}{|G_{in} + jB_1 + Y_{11}|} \tag{4.24}$$

$$\Gamma_{out} = \frac{G_{22}}{|G_{out} + jB_1 + Y_{11}|}. \tag{4.25}$$

By using these expressions, (4.15) gives

$$|S_{12}^0| = \mu T_{in} T_{out}. \tag{4.26}$$

Each component in the summation of (4.21), i.e.,

$$U_n = \mu T_{in} T_{out} \{\mu^2 \Gamma_{in} \Gamma_{out} \exp(-2j\beta L)\}^n \tag{4.27}$$

corresponds to the amplitude of the n-th echo. Thus, the TTE can be suppressed by reducing $\mu^2 \Gamma_{in} \Gamma_{out}$.

On the other hand, (4.22)–(4.25) suggest

$$\frac{\Gamma_{in}}{T_{in}} = \sqrt{\frac{G_{11}}{2G_{in}}}, \tag{4.28}$$

$$\frac{\Gamma_{out}}{T_{out}} = \sqrt{\frac{G_{22}}{2G_{out}}}. \tag{4.29}$$

Thus, to reduce Γ_{in} and Γ_{out}, we must reduce either G_{11}/G_{in} and G_{22}/G_{out} or T_{in} and T_{out}. However, both choices will result in increased insertion loss. Namely, the TTE cannot be suppressed without increasing the insertion loss.

From (4.27)–(4.29), we get

$$\left|\frac{U_1}{U_0}\right| = \mu |S_{12}^0| \sqrt{\frac{G_{11}G_{22}}{4G_{in}G_{out}}}. \tag{4.30}$$

This suggests that giving B_i is not effective for TTE suppression because it increases the insertion loss by the same amount. In other words, the TTE should be suppressed by giving smaller G_{ii}.

Let us define the TTE ripple r as the deviation from $|S_{12}^0|$. From (4.14), it is given by

$$r = \frac{1 + \mu^2 \Gamma_{in} \Gamma_{out}}{1 - \mu^2 \Gamma_{in} \Gamma_{out}}. \tag{4.31}$$

Figure 4.12 shows the relationship between the insertion loss and the TTE ripple when $G_{11} = G_{22}$, $G_{in} = G_{out}$, $\Im(Y_{11}) = -B_1$, and $\Im(Y_{22}) = -B_2$. In the figure, $\alpha = -20 \log \mu$. It is seen that the insertion loss of 20 dB must be accepted so as to suppress the TTE ripple by up to 0.1 dB.

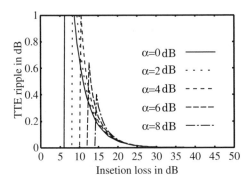

Fig. 4.12. Insertion loss and TTE ripple when $G_{11} = G_{22}$ and $G_{in} = G_{out}$

4.2 Design of Transversal Filters

4.2.1 Fourier Transforms

Let us consider the transfer function $H(\omega)$ of the IDT:

$$H(\omega) = \sum_{n=-N}^{N} A_n \exp(-2\pi jn\omega/\omega_S), \qquad (4.32)$$

where $\omega_S = 2\pi V_S/p_I$ is the SAW resonance frequency.

In the case where $N \to \infty$, A_n corresponds to the Fourier expansion coefficient of $H(\omega)$. Thus, if the desired transfer function $H_d(\omega)$ is specified, the optimal choice of A_n is given by

$$A_n \propto \frac{1}{\omega_S} \int_{-\omega_S/2}^{+\omega_S/2} H_d(\omega) \exp(2\pi jn\omega/\omega_S)d\omega \qquad (4.33)$$

in the sense of the least square error. Note that, since A_n is real, the relation $H_d(-\omega) = H_d(\omega)^*$ must hold.

As a specification, let us consider the ideal flat passband shape:

$$H_d(\omega) = \begin{cases} 1 & (|\omega| < \omega_w) \\ 0 & (|\omega| > \omega_w). \end{cases} \qquad (4.34)$$

Then, substitution into (4.33) gives

$$A_m \propto \frac{\sin(2\pi m\omega_w/\omega_S)}{\pi m} = (2\omega_w/\omega_S)\text{sinc}(2\pi m\omega_w/\omega_S). \qquad (4.35)$$

To understand the resultant weighting function, let us discuss the building-blocks of unapodized IDTs. So as to compensate round passband edges (see Fig. 4.13a) of the unapodized IDT, we must add another term which increases the amplitudes at the shoulders. The term is also required to reduce the peak amplitude. Then the frequency response of the term must be sign inverted and possess a narrower passband than the original one (see Fig. 4.13b), and the corresponding weighting function must be longer and be sign inverted. Then the total weighting function will have sign-inverted regions as shown in Fig. 4.13c.

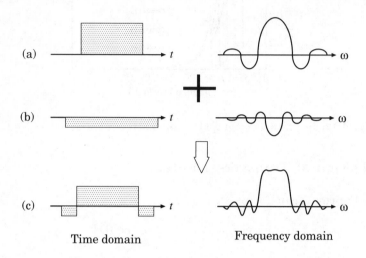

(a)

(b)

(c)

Time domain   Frequency domain

Fig. 4.13. Effect of sign-inverted region in weighting function

It is clear that a further longer weighting function must be added to fill the dips appearing in the resultant frequency response and to achieve a more flattened passband. In addition, the total weighting function must be smoothed so as to suppress the sidelobes.

For the case where $\omega_w = 0.1\omega_S$, $H(\omega)$ was calculated from (4.32) by using A_m determined by (4.35). Figure 4.14 shows the calculated $H(\omega)$. In the calculation, the number of IDT finger-pairs was limited to 51. Although a rectangular passband shape is realized, a very large ripple occurs within and close to the passband. This is called Gibb's phenomenon, which arises from the difficulty of approximating discontinuous functions by a sum of continuous functions with finite gradient.

The window function technique is widely used to suppress this phenomenon. The window function $w(t)$ is a function to extract a portion of the original signal $f(t)$ with infinite extent for a finite time interval $t = [-t_c, +t_c]$, and the resultant function is given by $f(t)w(t)$. For example, the function for extraction without weighting, i.e.,

$$w(t) = 1 \qquad (|t| < t_c) \tag{4.36}$$

is called a rectangular window. The Fourier transform of $f(t)w(t)$ corresponding to a transfer function $H(\omega)$ is given by

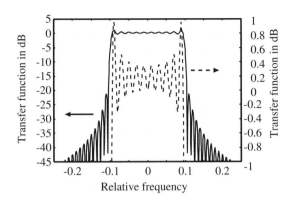

Fig. 4.14. Frequency response of device designed by using Fourier expansion

$$H(\omega) = \frac{1}{2\pi} \int_{-\infty}^{+\infty} W(\omega - \xi)F(\xi)d\xi. \tag{4.37}$$

This means that $H(\omega)$ is given by the convolution of the Fourier transforms $F(\omega)$ and $W(\omega)$ of $f(t)$ and $w(t)$, respectively.

For a rectangular window, $W(\omega) = 2t_\mathrm{c}\mathrm{sinc}(\omega t_\mathrm{c})$. This function has a relatively large ripple or sidelobes symmetrically with respect to the peak at $\omega = 0$. Thus, by the convolution, a large ripple will occur at frequencies where $F(\omega)$ changes rapidly. This ripple can be avoided by the use of window functions with a small sidelobe level. A narrower peak width of $W(\omega)$ is also desired so as not to deform the original $H(\omega)$.

The followings are typical window functions:

$$w(t) = 0.54 + 0.46 \cos r$$

Hamming window

$$w(t) = 0.42 + 0.50 \cos r + 0.08 \cos(2r)$$

Blackmann window

$$w(t) = 0.35875 + 0.48829 \cos r + 0.14128 \cos(2r) + 0.01168 \cos(3r)$$

Blackmann–Harris window

where $r = \pi t/t_\mathrm{c}$. Figure 4.15 shows their Fourier transforms. Large sidelobe suppression can be achieved in Blackmann–Harris, Blackmann, Hamming and rectangular windows. However, the peak is widened by the same order. Thus, an appropriate window function should be chosen under this trade-off relation.

Figure 4.16 shows the frequency response of an IDT designed for the flat passband shape with the Hamming window. Comparing with the result shown in Fig. 4.14, it is seen that a flat passband with small ripple is realized under the trade-off of widened transition bandwidth.

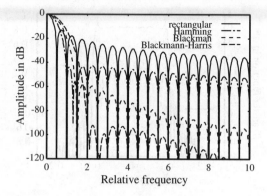

Fig. 4.15. Fourier transform of window functions

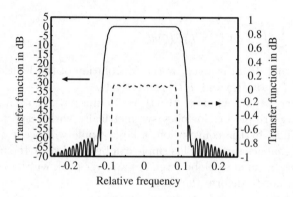

Fig. 4.16. Frequency response of device designed by using Fourier expansion with Hamming window

Note that the window function itself is often used as the weighting function for the IDT when the flat passband shape is not required.

4.2.2 Remez Exchange Method

The design technique based on minimization of the maximum error between the designed function $H(\omega)$ and the desired function $H_d(\omega)$ is called the mini-max method. Usually the device performance is evaluated from the maximum error, so the mini-max method gives optimal solutions in this sense.

Let us discuss the following transfer function $H(\omega)$;

$$H(\omega) = \sum_{n=-N}^{N-1} A_n \exp(-2\pi j n\omega/\omega_S). \tag{4.38}$$

For each frequency range $[\omega_{li}, \omega_{hi}]$, the desired frequency response and the maximum error are specified as H_i and Err_i, respectively. Then the object function in the sense of the mini-max is given by

$$\mathrm{Max}(w_i \mathrm{Err}_i) \rightarrow \mathrm{Min}. \tag{4.39}$$

where w_i is the weight to specify the ripple level for each frequency range.

The method to perform this optimization and determine A_m is called the Remez exchange method. Fortunately, its source code was published in the original paper and other books [7–9], and this technique is widely used for designing practical SAW devices.

Figure 4.17 shows the frequency response of a SAW device with a flat passband designed by the Remez exchange method. In the calculation, the number of IDT finger-pairs is 212.5, and the specification is as follows: attenuation of -50 dB in the rejection bands $f = [0, 0.78]$ and $f = [1.22, 2]$ and ripple level of 0.05 dB in the passband $f = [0.8, 1.2]$. It is seen that the desired frequency response with equi-ripple is realized.

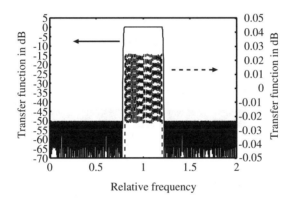

Fig. 4.17. Frequency response of SAW device designed by Remez exchange method

The Remez exchange method can also synthesize asymmetric frequency responses. A designed example is shown in Fig. 4.18. In the calculation, the number of IDT finger-pairs is 105, and the specification is as follows: attenuation of -40 dB in the rejection bands $f = [0, 0.75]$ and $f = [1.25, 2]$, ripple of 0.05 dB in the passband $f = [1, 1.2]$ and -20 dB down at $f = [0.8, 0.95]$.

By specifying continuous functions for H_i and Err_i, an arbitrary curved passband shape can be synthesized by using this method. Figure 4.19 shows, as an example, the frequency response of the Nyquist filter [10].

4.2.3 Linear Programming

The method of finding a solution giving the maximum (or minimum) object function under restrictions given by simultaneous inequalities is called linear

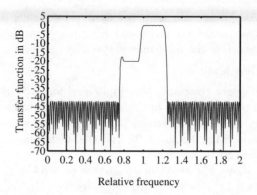

Fig. 4.18. Frequency response of SAW filter with asymmetric passband shape designed by Remez exchange method

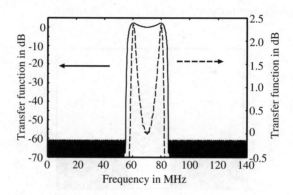

Fig. 4.19. Frequency response of Nyquist SAW filter designed by Remez exchange method

programming. This method allows us to perform optimal design for a wide variety of transversal filters, for example, those with nonlinear phase response [11].

The design specification can be written as follows:

$$w_i\{H(\omega) - H_d(\omega_i)\} \leq +\delta, \tag{4.40}$$

$$w_i\{H(\omega) - H_d(\omega_i)\} \geq -\delta, \tag{4.41}$$

where w_i is the weight and δ is the deviation from the desired value. Then the object function is specified by

$$\delta \rightarrow \text{Min.} \tag{4.42}$$

Since software for linear programming is commercially available, optimal design can be performed simply by preparing the linear inequalities of (4.40) and (4.41) and the object function of (4.42) as their input data. Note that applicable problems are limited because this method consumes much computer power and resources.

4.3 Spurious Responses

4.3.1 Diffraction

Influence on Device Performance. In the analyses discussed before, SAWs excited by an IDT are assumed to be plane waves. However as was shown in Sect. 1.1.4, the plane wave spreads with propagation due to diffraction, and finally forms a circular wave (see Fig. 4.20). The influence of diffraction becomes significant when the aperture is small and the propagation distance is large.

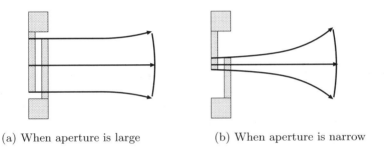

(a) When aperture is large   (b) When aperture is narrow

Fig. 4.20. Relation between aperture and diffraction

Since transversal filters sometimes include small aperture regions due to apodization, their characteristics are often significantly affected by diffraction. Although this problem may be circumvented by employing withdrawal weighting or dogleg weighting, its employment loses design flexibility, and achievable performance is limited.

The effects of diffracted SAWs are small in the passband where the SAW fields excited by each finger are added in phase. On the other hand, diffracted SAWs significantly influence device characteristics in the rejection band where the SAW fields cancel each other and their strength becomes comparable to that of the diffracted ones. Then, as shown in Fig. 4.21, periodic nulls may disappear and out-of-band rejection may deteriorate. In addition, diffraction may widen the transition band.

The effect of diffraction must be compensated for the design of high-performance SAW transversal filters. Figure 4.22 shows one technique for compensation where the aperture of the receiving IDT is increased to capture diffracted SAWs.

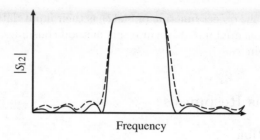

$$|S_{12}|$$

Frequency

Fig. 4.21. Influence of diffraction of frequency response

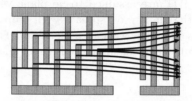

Fig. 4.22. Increasing aperture for capturing diffracted SAWs

Analysis Based on the Free Surface Approximation. This subsection describes a numerical technique for diffraction analysis based on the free surface approximation [12, 13]. Figure 4.23 shows the coordinate system employed for the analysis. For simplicity, the power flow angle is assumed to be zero.

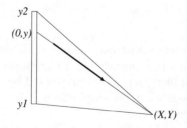

Fig. 4.23. Employed coordinate system

When a point source is placed at the origin, the surface electrical potential $G(X,Y)$ of the excited SAW may be expressed in the Fourier integral form of

$$G(X,Y) = \frac{1}{2\pi} \int_{-\infty}^{+\infty} F(\beta_y) \exp[-j\{\beta_x(\beta_y)X + \beta_y Y\}] d\beta_y, \qquad (4.43)$$

where (β_x, β_y) is the SAW wavevector and $F(\beta_y)$ is the excitation efficiency for the field with β_y.

From the law of superposition, the total SAW field excited by the charge $q(x, y)$ is given by the convolution between $q(x, y)$ and $G(X, Y)$ in (4.43), i.e.,

$$\phi(X, Y) = \int_{-\infty}^{+\infty} \int_{-\infty}^{+\infty} G(X - x, Y - y)q(x, y)dxdy. \tag{4.44}$$

Since our concern is SAW excitation by the IDT, it is expected that the contribution of SAW fields with large $|\beta_y|$ is negligible. So we expand the β_y dependence of β_x as

$$\beta_x = \beta_{x0} - \xi\beta_y^2/\beta_{x0}. \tag{4.45}$$

This is called the parabolic approximation. In addition, the angular dependence of $F(\beta_y)$ is ignored and $F(\beta_y)$ is set constant. Then $G(X, Y)$ in (4.43) is analytically given by

$$G(X, Y) \cong \frac{F}{2\pi} \exp(-j\beta_{x0}|X| - jY^2/4\zeta|X|)$$
$$\times \int_{-\infty}^{+\infty} \exp\{j\zeta X(\beta_y + Y/2\zeta X)^2\}d\beta_y$$
$$= \frac{F}{\sqrt{2\pi|\zeta X|}} \exp(-j\beta_{x0}|X| - jY^2/4\zeta|X|), \tag{4.46}$$

where $\zeta = \xi/\beta_{x0}$.

The charge distribution on each finger is ignored for the input IDT, and is approximated by a line source. Then $\phi(X, Y)$ in (4.44) is given by

$$\phi(X, Y) = \sum_{n=1}^{N} \frac{F\exp\{-j\beta_{x0}|X - x_n|\}}{\sqrt{2\pi|\zeta(X - x_n)|}}$$
$$\times \int_{y_n-w_n/2}^{y_n+w_n/2} \exp\{+j(Y - y)^2/4\zeta|X - x_n|\}dy$$
$$= \sum_{n=1}^{N} F\exp\{-j\beta_{x0}|X - x_n|\}$$
$$\times \left[\mathrm{Ei}\left(\frac{|Y - y_n - w_n/2|}{\sqrt{4|\zeta(X - x_n)|}}\right) - \mathrm{Ei}\left(\frac{|Y - y_n + w_n/2|}{\sqrt{4|\zeta(X - x_n)|}}\right) \right], \tag{4.47}$$

where N is the number of IDT fingers, w_n is the aperture length of the n-th electrode, (x_n, y_n) is the location of its center, and

$$\mathrm{Ei}(x) = \frac{1}{\sqrt{2\pi}} \int_0^x \frac{\exp(-sjt)}{\sqrt{t}}dt, \tag{4.48}$$

where $s = \mathrm{sgn}(\zeta)$.

When the charge distribution on each finger is also ignored for the output IDT, (4.47) gives the total charge Q induced on the output IDT as

$$
Q = \sum_{m=1}^{M} \int_{Y_m - W_m/2}^{Y_m + W_m/2} \phi(X_m, Y) dY
$$

$$
= \sum_{m=1}^{M} \sum_{n=1}^{N} \sqrt{4|\zeta(X_m - x_n)|} \exp\{-j\beta_{x0}|X_m - x_n|\}
$$

$$
\times (\mathrm{Eii}_{++} - \mathrm{Eii}_{-+} - \mathrm{Eii}_{+-} + \mathrm{Eii}_{--}), \tag{4.49}
$$

where

$$
\mathrm{Eii}_{pq} = \mathrm{Eii}\left(\frac{|Y_m + pW_m/2 - y_n - qw_n/2|}{\sqrt{4|\zeta(X_m - x_n)|}}\right). \tag{4.50}
$$

The function

$$
\mathrm{Eii}(x) = \int_0^x \mathrm{Ei}(x) dx \tag{4.51}
$$

can be evaluated numerically as follows. When $x \leq 4$,

$$
\mathrm{Eii}(x) = \sum_{k=0}^{6} c_k t^{2k+1} - sj \sum_{k=0}^{5} d_k t^{2k+2}, \tag{4.52}
$$

where $t = x^2$, and the coefficients c_k and d_k are given by:

k	c_k	d_k
0	7.9788456×10^{-1}	1.3298076×10^{-1}
1	$-2.6596152 \times 10^{-2}$	$-4.7493129 \times 10^{-3}$
2	7.3878200×10^{-4}	1.0074300×10^{-4}
3	$-1.2177725 \times 10^{-5}$	$-1.3192536 \times 10^{-6}$
4	1.2933859×10^{-7}	1.1572400×10^{-8}
5	$-9.5184240 \times 10^{-10}$	$-7.2422792 \times 10^{-11}$
6	$5.1253053 \times 10^{-12}$	

On the other hand, when $x > 4$,

$$
\mathrm{Eii}(x) = \sqrt{2t} \exp(-sj\pi/4) + \sum_{k=0}^{8} (b_k - sja_k)(4/t)^k \exp(-sjt)
$$

$$
+ sj\sqrt{2/\pi}, \tag{4.53}
$$

where the coefficients a_k and b_k are given by:

k	a_k	b_k
0	0	$-1.7776364 \times 10^{-8}$
1	4.8319936×10^{-6}	$-9.9732860 \times 10^{-2}$
2	3.7259642×10^{-2}	$-6.4257124 \times 10^{-5}$
3	1.6108580×10^{-3}	2.3888603×10^{-2}
4	$-2.9712984 \times 10^{-2}$	$-.1.2381365 \times 10^{-3}$
5	2.9086760×10^{-2}	$-2.7171204 \times 10^{-2}$
6	$-1.3605636 \times 10^{-2}$	3.1883772×10^{-2}
7	2.6535702×10^{-3}	$-1.6677158 \times 10^{-2}$
8	0	3.5073032×10^{-3}

For the calculation, we must determine ξ from the slowness surface in advance. The free software VCAL [14] allows us to determine the SAW velocity V_S as a function of the offset angle θ from the X-axis. We expand $V_S(\theta)^{-1}$ as

$$V_S(\theta)^{-1} \cong V_0^{-1}(1 - \gamma\theta^2/2). \tag{4.54}$$

The relations $\beta_x = \omega/V_S(\theta)\cos\theta$ and $\beta_y = \omega/V_S(\theta)\sin\theta$ give

$$\beta_x/\omega \cong V_0^{-1}(1 - \theta^2/2)(1 - \gamma\theta^2/2) \cong V_0^{-1}\{1 - (\beta_y/\beta_x)^2(1 + \gamma)/2\}$$
$$\cong V_0^{-1} - V_0(\beta_y/\omega)^2(1 + \gamma)/2. \tag{4.55}$$

Thus, comparison with (4.45) gives the relation

$$\xi = (1 + \gamma)/2. \tag{4.56}$$

This method was applied for the simulation of a transversal filter composed of unapodized and Hamming-weighted IDTs. In the calculation, γ was chosen to be 0.378, corresponding to that for ST-cut quartz. Figure 4.24 shows the result. The solid line is the result obtained by using the free surface approximation whereas the broken line is that obtained by the delta-function model analysis where diffraction is not taken into account. It is seen that, due to diffraction, the response at frequencies higher than the passband is considerably deteriorated. On the other hand, the effect at low frequencies is relatively small. Although not shown, the simulation agrees fairly well with experiment.

4.3.2 Bulk Waves

Effects of Bulk Waves. In SAW devices, responses due to acoustic waves except SAWs are called spurious BAW responses.

Figure 4.25 shows a typical frequency response of a SAW transversal filter [15], where the substrate is X-112°Y LiTaO3, and both input and output IDTs are not apodized. In the figure, the solid line is the response when the bottom surface was kept as cut whereas the broken line is that when the bottom surface was sand-blasted. It is seen that there exist two types of spurious BAW responses; one is suppressed by sand-blasting whereas the other is not.

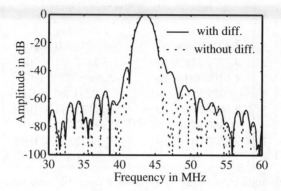

Fig. 4.24. Influence of diffraction on SAW device characteristics. Solid line: simulation by free surface approximation, and broken line: simulation by the delta-function model

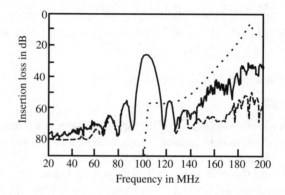

Fig. 4.25. Influence of sand-blasting of bottom surface on SAW device characteristics. Solid line: as-cut, broken line: sand-blasted, and dotted line: simulation by using effective permittivity

In fact, there are several types of spurious BAW response shown in Figure 4.26. Type (a) arrives at the receiving IDT after reflection at the bottom surface. This type of BAW, which is called a deep bulk acoustic wave (DBAW) [16], can be suppressed by sand-blasting the bottom surface. Since the DBAW is faster than the SAW and its radiation angle increases with an increase in frequency, the response appears at frequencies higher than the SAW resonance frequency. In Fig. 4.25, the spurious response at 130–200 MHz is mainly due to the DBAW. On the other hand, type (b) arrives at the receiving IDT without reflection at the bottom surface, and is called a SSBW [17] or SBAW [18] described in Sect. 1.2.5. Physically the SSBW is equivalent to the DBAW, and the SSBW is detected by the receiving IDT only when the BAW is radiated parallel to the surface. Thus, the SSBW also appears at a higher frequency than the SAW resonance frequency. The SSBW exhibits a frequency response similar to that of the SAW, and is not influenced by sand-blasting of the bottom surface. In Fig. 4.25, the spurious response at 170–200 MHz is mainly due to the SSBW. Type (c) is the thickness resonance of the BAW between the IDT and the bottom surface. Its resonance appears periodically at integer times $2d/V_B$ where d is the substrate thickness. Since usually $d \gg \lambda$, this spurious response appears at frequencies lower than the SAW resonance frequency. In Fig. 4.25, this response is seen at 50 MHz. Usually, practical SAW devices employ piezoelectric substrates of 200–500 μm thickness due to its easy handling. From this, the spurious response becomes serious for devices operating at relatively low frequencies.

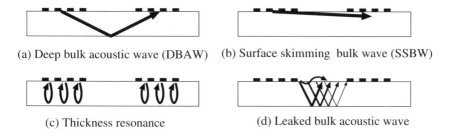

(a) Deep bulk acoustic wave (DBAW)   (b) Surface skimming bulk wave (SSBW)

(c) Thickness resonance   (d) Leaked bulk acoustic wave

Fig. 4.26. Spurious BAW responses

Note that, if a leaky SAW is employed, other types of spurious BAW response is observed. This is due to the BAW generated by the leakage from the leaky SAW and reflected at the bottom surface (see Fig. 4.26d). This wave is called a leaked BAW and will be discussed in Sect. 8.1.2.

Evaluation of BAW Response by Effective Permittivity. In Sect. 3.6.2, we showed a method of calculating the radiation conductance $G_B(\omega)$ for a BAW based on the effective permittivity. For the simulation of SAW transversal filters, we must derive the transfer function Y_{12}^B for the BAW.

For a SAW, we showed the following relation as (4.10):

$$Y_{12} = \mu \sqrt{G_{11}G_{22}} \exp(-j\beta_p L), \tag{4.57}$$

where μ is the parameter responsible for the propagation loss and the apodization loss, and $|\mu| \le 1$. As for the DBAW and SSBW responses, $|\mu| < 1$ because a portion of the BAWs excited by the IDT are diffused into the bulk and are not detected by the receiving IDT. Here we estimate Y_{12}^B by simply assuming $\mu = 1$ and applying the method described in Sect. 3.6.2. This estimation gives the worst case of the spurious level.

The dotted line in Fig. 4.25 shows the spurious BAW response estimated by using this method [15]. The estimation agrees fairly well with experiment for the case where sand-blasting was not applied. Scattering by sand-blasting results in an increase in the propagation loss for the DBAW.

It should be noted that μ can be roughly estimated as 0.5 (6 dB) from the difference between theory and experiment, and the value may be considerably smaller than expectation from the field decay in the form of $1/\sqrt{r}$ (see Sect. 1.2.1). This is because the DBAW excited by the IDT propagates as a plane wave for a while due to the finite IDT length, and diffraction into the depth with propagation is not significant in this case [19].

Delta-Function Model Analysis of SSBW Response. In the estimation described above, in addition to the contribution of the SSBW, that of the DBAW is included, although the spurious DBAW response is not significant in practice because it is suppressed by sand-blasting of the bottom surface. Here the delta-function model developed for the SAW response is applied to the analysis of the spurious SSBW response [20].

As will be shown in Sect. 6.2.1, the surface electrical potential $h_B(x_1)$ due to the SSBW excited by a line-source is given by

$$h_B(x_1) = -\frac{K_B^2}{2\pi\epsilon(\infty)} H_0^{(2)}(S_B\omega x_1)U(x_1/x_c), \qquad (4.58)$$

where K_B^2 is the electromechanical coupling factor for the SSBW and $H_0^{(2)}(\theta)$ is the Hankel function of the second kind approximated to

$$H_0^{(2)}(\theta) \cong \sqrt{2/\pi\theta}\exp(-j\theta - j\pi/4), \qquad (4.59)$$

and $U(r)$ is a function responsible for the effects of the surface boundary condition given by

$$U(r) = \frac{2}{\sqrt{\pi}}\int_0^{+\infty} t^2\exp(-t^2)/(t^2 - jr)dt \cong \begin{cases} 1 & (r \ll 1) \\ j/2r & (r \gg 1). \end{cases} \qquad (4.60)$$

It is seen that $h_B(x_1)$ is characterized by the parameter x_c: when $x_1 \ll x_c$, the SSBW is scarcely influenced by the surface boundary condition, and the SSBW amplitude decays with propagation in the form of $x_1^{-0.5}$. On the other hand, when $x_1 \gg x_c$, the SSBW is much influenced and its amplitude decays in the form of $x_1^{-1.5}$. Note that x_c is dependent upon the substrate material and boundary condition. Figure 4.27 shows $U(r)$.

The transfer admittance of IDTs for a SSBW can be derived by applying (4.58) to the delta-function model described in Sect. 3.3.1.

This method was applied for a transversal filter on X-112°Y LiTaO₃. Figure 4.28 shows the result [20]. In the figure, L is due to a longitudinal-wave-type SSBW, and it agrees well with experiment. Note that spurious responses due to a shear-wave-type SSBW are not shown in the figure because its insertion loss is larger than 100 dB.

4.3.3 Other Parasitic Effects

There are other parasitic effects which deteriorate the performance of SAW transversal filters. The capacitive coupling between input and output IDTs

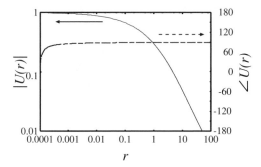

Fig. 4.27. $U(r)$

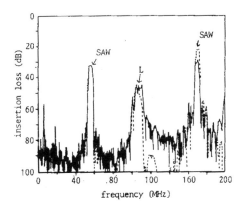

Fig. 4.28. Delta function model analysis of SSBW response. Solid line: experiment, and broken line: simulation

and/or bonding pads is representative, and is called electrical feedthrough. Usually its frequency dependence is small, and it deteriorates the out-of-band rejection level and causes ripple in the passband.

Figure 4.29 shows a typical layout of a SAW transversal filter. A guard electrode is placed between the input and output IDTs so as to shunt the static capacitance between the IDTs to ground. Suppression of electrical feedthrough becomes difficult in the GHz range.

The electrode resistance is also significant because it increases the insertion loss and deteriorates the transition band response [21].

Although increasing electrode thickness is preferable to reduce the resistance, it simultaneously increases the reflection at electrode fingers and IDT ends. Reflection within the IDT can be suppressed by employing a double-electrode-type IDT. The effect of reflection at the IDT ends can be reduced by giving dummy electrodes to both ends and cutting them obliquely as shown

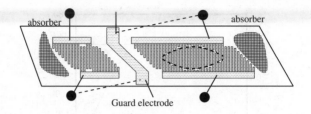

Fig. 4.29. Typical configuration of SAW transversal filter

in Fig. 4.29. Acoustic absorbers are placed at the other ends to suppress reflection from the substrate edges.

4.4 Low-Loss Transversal Filters

4.4.1 Multi-IDT Structures

The achievable performance of SAW transversal filters mentioned above is limited by the trade-off between the TTE level and insertion loss arising from the bidirectionality of the IDTs.

To circumvent this problem, several techniques have been proposed. Figure 4.30 shows a transversal filter with three IDTs [22] where two IDTs at both sides have the same pattern and the middle IDT is symmetric with respect to the center. If the center IDT is impedance-matched with the load,

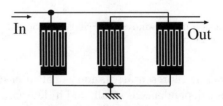

Fig. 4.30. Three-IDT type transversal filter

the IDT can detect all SAWs excited toward the inside from the side IDTs without electrical regeneration. Thus SAW reflection does not occur, and then the TTE is suppressed. Since all IDTs can be impedance-matched without deteriorating the TTE level, the low insertion loss and low TTE level can simultaneously be achievable.

Note that since SAWs excited toward the outside from the side IDTs are spoiled, the bidirectional loss of 3 dB still remains.

As an extension of this method, increasing the number of IDTs was proposed to decrease the bidirectional loss. That is, input and output IDTs are placed alternately as shown in Fig. 4.31, and the structure is called an interdigitated-interdigital transducer (IIDT) [23].

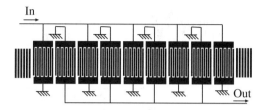

Fig. 4.31. IIDT-type transversal filter

For the same reason as with the three-IDT structure, the TTE can be suppressed when all launching IDTs are identical and the receiving IDTs are symmetrical. Note that they must be impedance-matched with the source and load. The bidirectional loss decreases as $-10\log(1 - 1/N)$ with an increase in the number N of launching IDTs. Usually N is chosen to be $5 - 7$ under trade-off with the device size, and reflectors are placed at both ends of the IDT array to reduce the remaining tiny bidirectional loss.

The distance d between the IDTs critically influences device performance. That is, although the TTE is suppressed at a frequency where the receiving IDTs are impedance-matched, considerable TTE occurs at its adjacent frequencies. Note that the TTE extends the impulse response. Thus, if we choose d as an even multiple of $\lambda/4$, the passband tends to be narrow because the TTE is in phase with the original response and the mainlobe in the impulse response becomes longer. On the other hand, if we choose d as an odd multiple of $\lambda/4$, the passband width tends to increase because the TTE becomes 180° out of phase and the mainlobe in the impulse response becomes shorter.

On the other hand, the stopband characteristics are mostly determined by those of the IDTs. Thus, they can be controlled by giving appropriate weighting to the IDTs. Withdrawal- and phase-weightings [24] are often used so that the apodization loss does not arise. Desirable out-of-band rejection may be achieved by simply weighting the number of IDT finger-pairs [25].

Figure 4.32 shows the frequency response of an IIDT filter employing a leaky SAW on 36°YX-LiTaO$_3$ [26]. This device was developed by Fujitsu Ltd. After impedance matching, the insertion loss of 2.8 dB, -3 dB bandwidth of 37 MHz, and out-of-band rejection level of 30 dB were achieved.

Device characteristics of another IIDT filter employing a longitudinal-type leaky SAW on (0, 47.3°, 90°) Li$_2$B$_4$O$_7$ will also be shown in Sect. 8.2.2.

The out-of-band rejection can be improved by the cascade-connection of two IIDT devices [27]. If the number of IDT finger-pairs is sufficiently large, the susceptance of the IDT becomes zero at a frequency a little higher than its resonance frequency. Thus, if the IDTs are mutually connected as shown

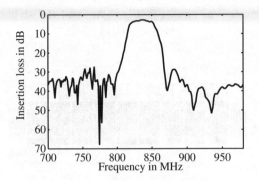

Fig. 4.32. Frequency response of IIDT filter on $36°$YX-LiTaO$_3$ [26]

in Fig. 4.33, 100% power transfer between the IDTs can be achieved because their electrical conductances are equal and the susceptances are zero. That is, the impedance matching condition is fulfilled. In the out-of-band, since the susceptance becomes much larger than the conductance, large electrical mismatching occurs. Thus, large out-of-band rejection can be achieved without increasing the insertion loss in the passband. This method is called image connection [27].

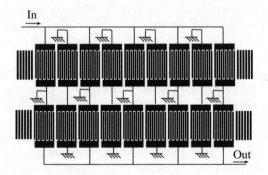

Fig. 4.33. Image connection

4.4.2 Transversal Filters Employing SPUDTs

Use of SPUDTs described in Sect. 3.1.2 is another solution to suppress the TTE without increasing the insertion loss [28].

As was shown in (3.74), the zero-reflection condition ($S_{11} = 0$) can be realized provided that

$$|p_{11}| = -\frac{2\cos\theta}{(1 + D^{-2})[1 + G_s/\Re(p_{33})]} = -\frac{2\cos\theta}{1 + D^{-2} + G_s/2|p_{13}|^2}, \quad (4.61)$$

even when the directivity D is finite. Under this situation, mechanical reflection is totally canceled by electrical regeneration. So as to fulfill the condition given by (4.61) for a wide range of frequencies, some sort of weighting must be applied to the mechanical reflection p_{11} [28, 29].

In Sect. 3.5.2, we derived the following relation as (3.63):

$$|S_{13}| = \frac{\sqrt{8G_s}|p_{13}|}{|p_{33} + G_s + jB|}, \quad (4.62)$$

and the unitary condition suggested the following relation as (3.62):

$$\Re(p_{33}) = 2(1 + D^{-2})|p_{13}|^2. \quad (4.63)$$

In addition, from the Hilbert transform relation of (3.82), the following relation holds:

$$\Im[p_{33}(\omega)] = \frac{1}{\pi}\int_{-\infty}^{+\infty}\frac{\Re[p_{33}(\xi)]}{\xi - \omega}d\xi + \omega C_0. \quad (4.64)$$

This suggests that $|S_{13}|$ is determined only by $\Re(p_{33})$. Thus, if $|S_{13}|$, ϕ and D are specified and p_{13} can be determined as the inverse problem of (4.62), $|p_{11}|$ can be determined by substituting the determined p_{13} into (4.61). Note that D and ϕ are usually frequency dispersive and are also dependent on the IDT pattern. In addition, the phase of S_{13} is usually not specified and this problem is nonlinear, so iterative optimization techniques must be applied for the design [29].

Recently, a new concept was proposed for transversal filters employing the SPUDT [30]. The idea is to give the resonance to the SPUDT itself to extend and/or control its impulse response. This can be realized by inverting the directivity partially within the SPUDT structure. This type of SPUDT is called a resonant SPUDT (RSPUDT).

The traditional SPUDTs are designed to cancel electrical regeneration by mechanical reflection. Thus the length of the impulse response is approximately determined by the physical size of the SPUDT. This type is referred to as a nonresonant SPUDT. If the internal resonance is given to the SPUDT, it extends the impulse response of the SPUDT as shown in Fig. 4.34a. So if the weighting for the reflection is properly designed, a rectangular-shaped passband might be realized. SPUDTs based on this idea are called weak-resonant SPUDTs. In this case, the passband width is mostly determined by the weighting for the SAW excitation. As an extension of this idea, by increasing the internal resonance and extending the impulse response as shown in Fig. 4.34b, a narrower and rectangular passband shape might be realized. SPUDTs based on this idea are called strong-resonant SPUDTs. Although the required device size can be reduced by giving stronger resonance, distortion in the group delay becomes obvious as a trade-off.

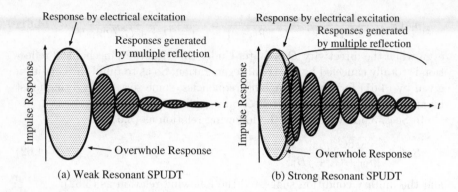

Fig. 4.34. Impulse response of resonant SPUDT

Figure 4.35 shows frequency responses of the resonant SPUDTs developed by Siemens Corporate Technology [31]. Device (a) employs weak-resonant SPUDTs on ST-cut quartz, and exhibits an insertion loss of 4 dB and shape factor (3 dB − 30 dB) of 1.55. It is interesting to compare with the result of conventional apodization, such as Fig. 4.16. The rough periodicity of the sidelobes suggests that a rectangular passband is realized by using SPUDTs with a small number of finger-pairs.

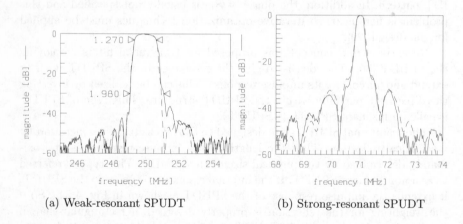

Fig. 4.35. Frequency response of resonant SPUDT [31]. Sold line: experiment, and broken line: simulation

On the other hand, device (b) employs strong-resonant SPUDTs on 37.5° Y-cut quartz, and offers an insertion loss of 7 dB. The periodicity of the sidelobes relative to the passband width indicates reduced device size.

4.4.3 Combination of SPUDTs and Reflectors

Present computer technologies enables us to simulate and design complicated device structures. Then various structures using SPUDTs were proposed recently to achieve superior frequency response and low insertion loss adding to the compact device size.

Externally Coupled Resonator Filter. It is known that a controlled passband shape can be realized by cascade-connecting multiple reflectors as shown in Fig. 4.36 [32]. This type of filter is called an externally coupled resonator filter (ECRF).

Fig. 4.36. Configuration of externally coupled resonator filter

If two reflectors are placed with a displacement of $\lambda/4$, the structure is equivalent to the Fabry–Perot resonator described in Sect. 2.2.3, and acts as a narrow bandpass filter. That is, SAWs cannot transmit through the resonators at their stopband except in a narrow passband originating from multiple reflection. The frequency response out of the stopband is mainly determined by that of the SPUDTs. So, a desirable passband shape can be synthesized by the cascade-connection of multiple reflectors whose Q are adjusted by the numbers of strips [33].

Figure 4.37 shows the frequency response of the ECRF developed by Siemens Corporate Technology [34]. This filter consists of two UDTs and three grating reflectors. An insertion loss of 3.5 dB, −3 dB bandwidth of 0.1% and group delay ripple of 1 μsec were achieved. Although no weighting was applied to the IDTs for this device, sufficient sidelobe suppression was achieved by giving the withdrawal weighting to the IDTs [34].

Z-path Filter. Figure 4.38 shows a filter consisting of two oblique reflectors and two UDTs [35], which is called a Z-path filter. Its frequency response is synthesized by giving appropriate weighting to the reflectors. Compared with the weighting to the IDTs, weighting to the reflectors requires a smaller physical length because the SAWs pass through the same path at least twice. In addition, the zig-zag path offers a smaller device size relative to conventional in-line structures.

Figure 4.39 shows the frequency response of the Z-path filter for the GSM-IF developed by EPCOS [36]. Quartz was employed for the substrate. An

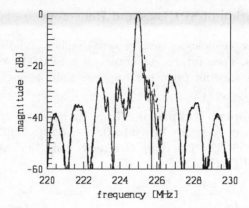

Fig. 4.37. Frequency response of externally coupled resonator filter [34]. Solid line: experiment, and broken line: simulation

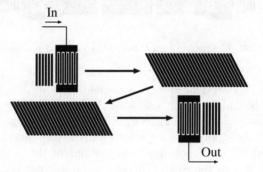

Fig. 4.38. Configuration of Z-path filter

insertion loss of 8 dB and group delay ripple of 1 μsec were obtained. The device fits into the DIP 14 or 16 package.

Multi-Track Filter. A properly designed SPUDT acts as a perfect reflector when the electrical port is short-circuited.

Let us consider parallel-connected transversal filters with a reflector sandwiched by two SPUDTs as shown in Fig. 4.40. The upper track has $2 \times \lambda/4$ shorter propagation path than the lower track. First, we discuss SAWs excited by the SPUDT and transmitted through the reflectors. Due to the difference in the propagation path length, the outputs of the detecting SPUDTs cancels each other. Then no net output signal appears, and all received SAWs are regenerated or reflected. For the same reason, SAWs arriving to the receiving SPUDTs after an odd number of reflections are reflected. On the other hand, SAWs reflected twice arrive at the receiving SPUDTs in phase, so the output signal appears. Since its frequency response is determined by the product of

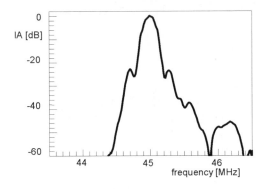

Fig. 4.39. Frequency response of Z-path filter [36]

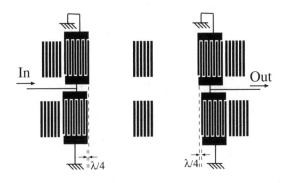

Fig. 4.40. Configuration of multi-track filter

that of the input and output SPUDTs and that of the reflector, this configuration will offer a steeper transition band with reasonable physical device
size.

By extending this principle, it is possible to construct a much steeper
passband shape by parallel-connecting N tracks where the propagation path
lengths are different from each other by $\lambda/2N$. This type of filter is called a
multi-track filter [37, 38].

It is interesting to note that the rejection band characteristics can be
improved by giving a different design to the reflectors [38].

Figure 4.41 shows the frequency response of the multi-track filter developed for the DECT-IF by Siemens Corporate Technology [38]. The device
employed EWC/SPUDTs [28] and weighted reflectors. As a substrate, X-
112°Y LiTaO$_3$ was used. An insertion loss of 8 dB (after matching), −3 dB
bandwidth of 1.1 MHz and group delay ripple of 200 nsec were achieved.

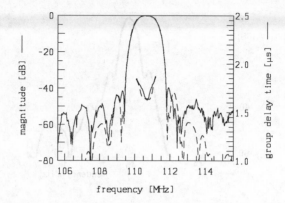

Fig. 4.41. Frequency response of multi-track filter [38]. Solid line: experiment, and broken line: simulation

References

1. H.E. Kallmann: Transverse Filters, Proc. IRE, **28** (1940) pp. 302–310.
2. C.S. Hartmann: Weighting Interdigital Surface Wave Transducers by Selective Withdrawal of Electrodes, IEEE Ultrason. Symp. (1973) pp. 423–426.
3. A. Rønnekleiv, H. Skeie and H. Hanebrekke: Design Problems in Surface Acoustic Filters, IEE International Specialist Seminor on Component Performance and Systems Applications of Surface Acoustic Wave Devices (1973) pp. 141–151.
4. K.M. Lakin, D.W.T. Mih and R.M. Tarr: A New Interdigital Electrode Transducer Geometry, IEEE Trans. Microwave Theory and Tech., **MTT-22** (1974) pp. 763.
5. M. Feldmann and J. Henaff: Design of SAW Filter with Minimum Phase Response, Proc. IEEE Ultrason. Symp. (1978) pp. 720–723.
6. R.H. Tancrell and M.G. Holland: Acoustic Surface Wave Filters, Proc. IEEE, **59** (1971) pp. 393–409.
7. J.H. McClellan, T.M. Parks and L.R. Rabiner: A Computer Program for Designing Optimum FIR Linear Phase Digital Filters, IEEE Trans. Audio and Electroacoustics, **AU-21** (1973) pp. 506–526.
8. C.K. Campbell: Surface Acoustic Wave Devices and Their Signal Processing Applications, Chap. 8, Academic Press, Boston (1989).
9. M. Feldmann and J. Henaff: Surface Acoustic Waves for Signal Processing, Altech House, Boston (1989).
10. J.C.B. Saw, T.P. Cameron and M.S. Suthers: Impact of SAW Technologies on the System Performance of High Capacity Digital Microwave Radio , Proc. IEEE Ultrason. Symp. (1993) pp. 59–65.
11. C. Ruppel, E. Ehrmann-Falkenau, H.R. Stocker and R. Veith: Optimum Design of SAW-Filters by Linear Programming, Proc. IEEE Ultrason. Symp. (1983) pp. 23–26.
12. D. Penunuri: A Numerical Technique for SAW Diffraction Simulation, IEEE Trans. Microwave Theory and Tech., **MTT-26** (1978) pp. 288.
13. E.B. Savage: Fast Computation of SAW Filter Responses Including Diffraction, Electron. Lett., **15** (1979) pp. 538–539.

14. K. Hashimoto and M. Yamaguchi: Free Software Products for Simulation and Design of Surface Acoustic Wave and Surface Transverse Wave Devices, Proc. Freq. Contr. Symp. (1996) pp. 300–307.
15. K. Hashimoto: On Leaky Surface Acoustic Wave and Bulk Acoustic Wave Launched from an Interdigital Transducer, Ph D thesis, Tokyo Institute of Technology (1988) [in Japanese].
16. C.N. Helmick, D.L. White and K.M. Lakin: Deep Bulk Acoustic Wave Devices Utilizing Interdigital Transducers, Proc. IEEE Ultrason. Symp. (1981) pp. 280–285.
17. M. Lewis: Surface Skimming Bulk Waves, SSBW, Proc. IEEE Ultrason. Symp. (1977) pp. 744–752.
18. K.F. Lau, K.H. Yen, R.S. Kagiwada and K.L. Gong: Further Investigation of Shallow Bulk Acoustic Waves Generated by Using Interdigital Transducers, Proc. IEEE Ultrason. Symp. (1977) pp. 996–1001.
19. M. Yamaguchi, K. Hashimoto, M. Tanno and H. Kogo: Effects of Diffraction on Frequency Response of Bulk-Acoustic-Wave-Beam Filters, Electron. Lett., 20, 7 (1984) pp. 275-277.
20. K. Hashimoto and M. Yamaguchi: Delta Function Model Analysis of SSBW Spurious Response in SAW Devices, Proc. IEEE Freq. Contr. Symp. (1993) pp. 639–644.
21. O. Männer and G. Visintini: Analysis and Compensation of Metal Resistivity Effects in Apodized SAW Transducers, Proc. IEEE Ultrason. Symp. (1989) pp. 1–6.
22. M.F. Lewis: Triple-Transit Echo Suppression in Surface-Acoustic-Wave Devices, Electron. Lett., 8 (1972) pp. 553–554.
23. M.F. Lewis: SAW Filters Employing Interdigitated Interdigital Transducer IIDT, Proc. IEEE Ultrason. Symp. (1982) pp. 12–17.
24. M. Hikita, Y. Kinoshita, H. Kojima and T. Tabuchi: Phase Weighting for Low Loss SAW Filters, Proc. IEEE Ultrason. Symp. (1980) pp. 308–312.
25. O. Ikata, Y. Satoh, T. Miyashita, T. Matsuda and Y. Fujiwara: Development of 800 MHz Band SAW Filters Using Weighting for the Number of Finger Pairs, Proc. IEEE Ultrason. Symp. (1990) pp. 83–86.
26. SAW device data sheet, Fujitsu Ltd., FAR-F5CB-836M50-G201.
27. M. Hikita, H. Kojima, T. Tabuchi and Y. Kinoshita: 800-MHz High-Performance SAW Filter Using New Resonant Configuration, IEEE Trans. Microwave Theory and Techn., MTT-33, 6 (1985) pp. 510–518.
28. C.S. Hartmann and B.P. Abbott: Overview of Design Challenges for Single Phase Unidirectional SAW Filters, Proc. IEEE Ultrason. Symp. (1989) pp. 79–89.
29. E.M. Garber, D.S. Yip and D.K. Henderson: Design of High Selectivity DART SPUDT Filters on Quartz and Lithium Tantalate, Proc. IEEE Ultrason. Symp. (1994) pp. 7–12.
30. P. Ventura, M. Solal, P. Dufilie, J. Desbois, M. Doisy and J.M. Hodé: Synthesis of SPUDT Filters with Simultaneous Reflection and Transduction Optimization, Proc. IEEE Ultrason. Symp. (1992) pp. 71–75.
31. C.C.W. Ruppel, R. Dill, J. Franz, S. Kurp and W. Ruile: Design of Generalized SPUDT Filters, Proc. IEEE Ultrason. Symp. (1996) pp. 165–168.
32. R.V. Shmidt and P.S. Cross: Externally Coupled Resonator-Filter (ECRF), IEEE Trans. Sonics and Ultrason., SU-26, 2 (1979) pp. 88–93.
33. P.S. Cross, R.V. Shmidt and H.A. Haus: Acoustically Cascaded ASW Resonator-Filters, Proc. IEEE Ultrason. Symp. (1976) pp. 277–286.
34. G. Scholl, R. Dill, W. Ruile and C.C.W. Ruppel: New Resonator Filter with High Sidelobe Suppression, Proc. IEEE Ultrason. Symp. (1992) pp. 117–121.

35. J. Machui and W. Ruile: Z-Path IF-Filters for Mobile Telephones, Proc. IEEE Ultrason. Symp. (1992) pp. 147–150.
36. J. Machui, J. Bauregger, G. Riha and I. Schropp: SAW Devices in Cellular and Cordless Phones, Proc. IEEE Ultrason. Symp. (1995) pp. 121–130.
37. M. Solal and J.M. Hodé: A New Compact SAW Filter for Mobile Radio, Proc. IEEE Ultrason. Symp. (1993) pp. 105–109.
38. R. Dill, J. Machui and G. Müller: A Novel SAW Filter for IF-Filtering in DECT Systems, Proc. IEEE Ultrason. Symp. (1995) pp. 51–54.

5. Resonators

This chapter deals with resonators-based devices. First, fundamentals of one-port and two-port SAW resonators are discussed. Then we deal with parasitic effects such as the transverse mode. Last, various types of resonator-based filters are reviewed.

5.1 One-Port SAW Resonators

5.1.1 Introduction

Figure 5.1a shows a typical configuration of the one-port SAW resonator where two grating reflectors are placed at both ends of the IDT [1]. Very steep resonance can be detected by the IDT when the device is designed so that the resonance frequencies of the IDT and reflectors coincide with each other. As will be discussed later, the length of the gaps between the IDT and reflector significantly influence the resonance characteristics.

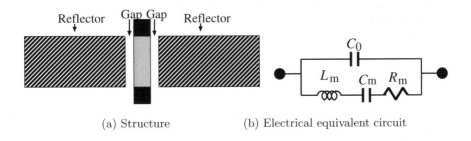

(a) Structure   (b) Electrical equivalent circuit

Fig. 5.1. One-port SAW resonator

Figure 5.1b shows its equivalent circuit near resonance, where C_m and L_m are the motional capacitance and inductance, respectively, corresponding to the contributions of elasticity and inertia. On the other hand, C_0 is the static capacitance of the IDT, and R_m is the motional resistance corresponding to the contribution of damping.

Figure 5.2 shows typical electrical resonance characteristics of the one-port SAW resonator. Just above the resonance frequency ω_r where the conductance G takes a maximum, there exists the antiresonance frequency ω_a where the resistance G^{-1} takes a maximum.

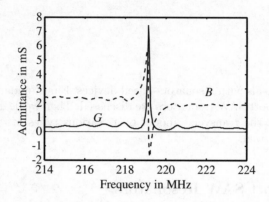

Fig. 5.2. Electrical resonance characteristics of one-port SAW resonator

From the equivalent circuit, since

$$\omega_r = \frac{1}{\sqrt{L_m C_m}}, \tag{5.1}$$

$$\omega_a = \frac{1}{\sqrt{L_m C_m C_0 / (C_m + C_0)}}, \tag{5.2}$$

we can determine L_m and C_m from the measured C_0, ω_r and ω_a. Note R_m can be determined by G^{-1} at $\omega = \omega_r$.

The capacitance ratio γ is frequently used as a measure of the resonator performance, and is given by

$$\gamma = \frac{C_0}{C_m} = \frac{1}{(\omega_a/\omega_r)^2 - 1}, \tag{5.3}$$

which corresponds to the inverse of the effective electromechanical coupling factor. The quality factor Q at the resonance frequency is called the resonance Q, and is denoted by Q_r. This is also an important measure, which is given by

$$Q_r = \omega_r L_m / R_m = 1/\omega_r C_m R_m. \tag{5.4}$$

Originally, Q is defined as the ratio of the stored energy to the dissipated energy in a half cycle, and it is known that the resonance characteristics are well expressed by

$$Y \propto \frac{j}{1 + (j\omega/\omega_r)^2 + jQ^{-1}}.$$

Figure 5.3 shows various methods to estimate Q. In method (a), Q is estimated by the inverse of the fractional width between frequencies where $G = \Re(Y)$ becomes the half height of its peak or $|Y|$ reduces to $1/\sqrt{2}$ of its peak value.

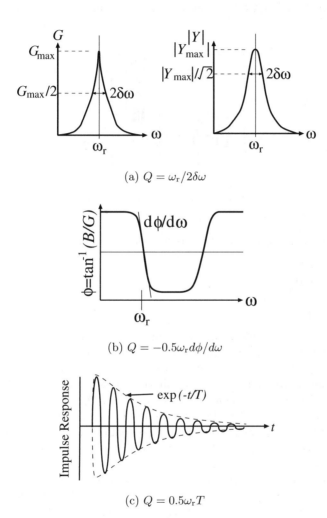

(a) $Q = \omega_r/2\delta\omega$

(b) $Q = -0.5\omega_r d\phi/d\omega$

(c) $Q = 0.5\omega_r T$

Fig. 5.3. Estimation of Q

In method (b), Q is evaluated from the gradient of $\phi = \angle Y$ at the resonance frequency by using the relation $Q = -0.5\omega_r d\phi/d\omega$. This means that if Q is sufficiently large, the frequency giving the specified ϕ does not change under variation of ϕ.

Since $-d\phi/d\omega$ corresponds to the group delay T, namely, the time constant of the impulse response, then as shown in Fig. 5.3c, Q is also estimated by the relation $Q = 0.5\omega_{\mathrm{r}}T$.

SAW oscillators are one of the most important applications of one-port SAW resonators. Figure 5.4 shows a typical circuit configuration of a SAW oscillator.

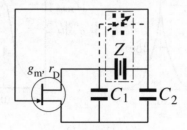

Fig. 5.4. SAW oscillator circuit

From the circuit analysis, the oscillation condition is given by

$$X = \frac{C_1 + C_2}{\omega C_1 C_2}, \tag{5.5}$$

$$\frac{R}{X} < \frac{g_{\mathrm{m}}}{\omega(C_1 + C_2)}, \tag{5.6}$$

where g_{m} is the mutual conductance of the FET, and $Z = R + jX$ is the total impedance of the resonator and capacitance. In the analysis, the drain resistance r_{D} of the FET is assumed to be sufficiently large.

Equation (5.5) suggests that oscillation occurs when $X > 0$, i.e., Z is inductive. When the gradient of X is steep, i.e., Q_{r} is large, the oscillation frequency is less affected by the variation of circuit parameters. In addition, since a larger Q_{r} means smaller power dissipation, we can use smaller gain amplifiers, which are preferable to reduce thermal noise. In addition, since large oscillation power is applicable, a large oscillation level relative to the thermal noise level is achievable. From these features, we can suppress the frequency drift due to nonlinear coupling between the thermal noise and oscillation signal. Since the drift occurs in a moment (order of msec. or less), we refer to its stability as short-term stability.

From the equivalent circuit shown in Fig. 5.1b, we obtain the following relation;

$$\frac{X}{R} = -Q_{\mathrm{r}}\gamma\frac{(\omega - \omega_{\mathrm{r}})(\omega - \omega_{\mathrm{a}})}{\omega_{\mathrm{r}}\omega} - \frac{\gamma}{Q_{\mathrm{r}}}. \tag{5.7}$$

Then, at $\omega = (\omega_{\mathrm{r}} + \omega_{\mathrm{a}})/2$, X/R takes a maximum value of

$$\frac{X}{R} \cong Q_r \gamma \left(\frac{\omega_a - \omega_r}{4\omega_r}\right)^2 - \frac{\gamma}{Q_r} \cong \frac{Q_r}{64\gamma} - \frac{\gamma}{Q_r}. \tag{5.8}$$

Equations (5.6) and (5.8) suggest that a larger $M = Q_r/\gamma$ enables us to use FETs with smaller g_m. In this sense, $M = Q_r/\gamma$ is referred to as the figure of merit of the resonator.

A smaller γ enables us to control the oscillation frequency over a wider range by a capacitance parallel-connected to the resonator. This feature is preferable for use in a voltage-controlled oscillator (VCO) where the varicap is employed as a voltage adjustable capacitance.

Drift in the oscillation frequency also arises from the SAW velocity change and thermal expansion due to variation in the environmental temperature. The drift can be reduced by choosing an appropriate substrate material. Since the time scale for the drift is in the order of seconds or minutes, we refer to its stability to as mid-term stability. In addition, drift also occurs due to the oxidization of electrode materials and/or contamination of the substrate surface. Since the time scale for the drift is in the order of days to years, we refer to its stability as long-term stability.

For SAW resonators the frequency stability is very important and has been investigated aggressively from various aspects. Formerly, the mid-term and long-term stabilities were mainly discussed. However, because modern communication systems employ complicated signal processing techniques and/or tight frequency allocations, short-term stability also becomes important. Detailed discussions on stabilities can be seen in Ref. [2].

5.1.2 Fabry–Perot Model

One-port SAW resonators are characterized by using the Fabry–Perot model shown in Fig. 5.5 [3]. In this model, the internal reflection by the IDT is

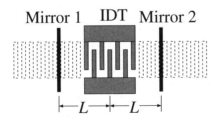

Fig. 5.5. Fabry–Perot model

ignored and the grating reflectors are replaced by mirrors located at a distance L from the IDT center. We will refer to the region between the mirrors as the (resonant) cavity and the distance $2L$ as the cavity length.

When an impulse is applied to the resonator through a constant-voltage source with zero internal impedance, the resulting current flow is schematically shown in Fig. 5.6.

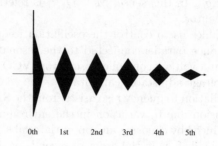

| 0th | 1st | 2nd | 3rd | 4th | 5th |

Fig. 5.6. Impulse response

When the input admittance of the IDT before placing the mirrors is $Y_S = G_S + jB_S + j\omega C_0$, the input admittance Y_{11} of the IDT after placement can be obtained by summing all echo responses as

$$Y_{11} = Y_S + 2\sum_{n=1}^{+\infty} Y_t^{(n)}, \tag{5.9}$$

where $Y_t^{(n)}$ is the transfer admittance between the IDT and its mirror image for the n-th echo. From (3.41), it is expected that, when the weighting and propagation losses are negligible, $Y_t^{(n)}$ is given by $G_S \Gamma^n \exp(-2nj\beta L)$, where Γ is the reflection coefficient of the mirrors. Substituting this relation into (5.9) gives

$$Y_{11} = Y_S + \sum_{n=1}^{+\infty} 2G_S \Gamma^n \exp(-2nj\beta L) = Y_S + 2G_S \frac{-\Gamma \exp(-2j\beta L)}{\Gamma \exp(-2j\beta L) - 1}$$

$$= j\omega C_0 + jB_S + G_S \frac{1 + \Gamma \exp(-2j\beta L)}{1 - \Gamma \exp(-2j\beta L)}, \tag{5.10}$$

where β is the wavenumber of the SAW. Equation (5.10) suggests that resonances, where the denominator becomes small, appear periodically, and the peak height of $\Re(Y_{11})$ is given by $G_S(1 + |\Gamma|)/(1 - |\Gamma|)$. Thus, typical SAW resonators exhibit multiple resonances as shown in Fig. 5.7. This type of multi-mode resonances is called longitudinal mode resonance. Note that the grating reflectors possess strong frequency dependence, and the number of resonance peaks are limited in practice.

Note that L can be determined from the gradient of $\angle \Gamma$ at the resonance frequency. For example, L is given by (2.24) from the equivalent circuit discussed in Sect. 2.3.

By setting $\Gamma = |\Gamma| \exp(j\phi)$, the resonance condition is given by

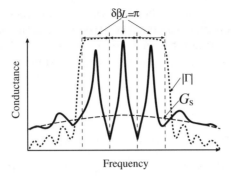

Fig. 5.7. Response of SAW resonator

$$2\beta L - \phi = 2m\pi. \tag{5.11}$$

From the discussion given in Sect. 2.1.2, the maximum SAW reflection is given at $\phi = \pm\pi/2$. Usually, SAW resonators are designed to resonate under this condition, then L should be chosen to be $(m/2 \pm 1/8)\lambda$. In this case, the radiation conductance $G_{\max}$ is given by,

$$G_{\max} = R_m^{-1} = \frac{G_S(1 + |\Gamma|)}{1 - |\Gamma|} \cong \frac{2G_S}{1 - |\Gamma|}. \tag{5.12}$$

Next, the resonance Q, Q_r is obtained by the resonance frequency ω_r relative to the frequency width $2\delta\omega$ giving half the radiation conductance of its peak value;

$$Q_r = \frac{\omega_r}{2\delta\omega} = \frac{2\pi(L/p_I)|\Gamma|}{1 - |\Gamma|} \cong \frac{2\pi(L/p_I)}{1 - |\Gamma|}. \tag{5.13}$$

This suggests that larger L/p_I and $|\Gamma|$ are preferable to achieve higher Q_r.

From this, the motional capacitance C_m is given by

$$\omega_r C_m = 1/R_m Q_r = \pi^{-1}(p_I/L)G_S, \tag{5.14}$$

and the capacitance ratio γ is given by

$$\gamma = \frac{\omega_r C_0}{\omega_r C_m} = \pi \frac{L}{p_I} \frac{\omega_r C_0}{G_S}. \tag{5.15}$$

This suggests that a larger $G_S/\omega_r C_0$ is preferable to achieve a smaller γ. Note that increasing L/p_I results in an increase in γ as a trade-off with increased Q_r.

The figure of merit M is given by

$$M = \frac{Q}{\gamma} = 2\frac{G_S}{\omega_r C_0} \frac{1}{1 - |\Gamma|}. \tag{5.16}$$

Thus, to achieve a larger M, the grating reflector should be designed to achieve a larger $|\Gamma|$, whereas the IDT should be designed to achieve a larger $G_S/\omega_r C_0$.

Usually, L is chosen below the limit where an unwanted longitudinal mode appears, and is adjusted so that the structure resonates at the frequency where $|\Gamma|$ of the grating reflector takes a maximum. On the other hand, the IDT is designed so that $G_S/\omega_r C_0$ becomes a maximum at that frequency.

5.2 Spurious Responses

5.2.1 Beam Diffraction and Transverse Modes

The diffraction of the SAW excited by the IDT affects the resonator performance. In high Q SAW resonators, since SAWs propagate back and forth within the cavity many times, their diffraction causes dissipation of SAW energy from the cavity and results in a reduced Q_r (see Fig. 5.8).

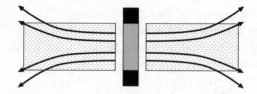

Fig. 5.8. SAW beam spreading due to diffraction

To circumvent this problem, total reflection at the boundary between the bus-bars and grating reflector is used to guide the SAW energy toward the transverse direction as shown in Fig. 5.9. This is called energy trapping for the transverse direction. Note that when the trapping is strong enough,

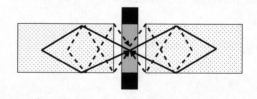

Fig. 5.9. Energy trapping in waveguide structure

higher-order resonances may appear (see Fig. 5.10). Since the higher-order modes resonate at different frequencies, they will be inharmonic spurious resonances near the main response. These are called higher-order transverse mode resonances [4].

For use in oscillators, these inharmonic spurious resonances may cause an undesirable jump in the oscillation frequency, and may limit the adjustable

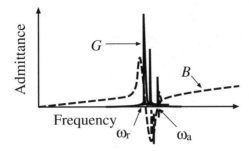

Fig. 5.10. Spurious responses due to higher-order transverse modes

frequency range for use in VCOs. For use in filters, on the other hand, their suppression is important because they cause ripples in the passband and/or undesirable peaks in the rejection band.

Since the cut-off frequency of each transverse mode is dependent upon the aperture of the grating and the bus-bar width, their adjustment makes suppression of the transverse modes possible. In addition, since the transverse modes have their own field distributions toward the transverse direction, their excitation amplitudes can be controlled by giving appropriate apodization [5] or an overlap length [6] different from the aperture to the IDT (see Fig. 5.11).

Fig. 5.11. IDT with different apertures for excitation and guidance

5.2.2 Transverse-Mode Analysis

Scalar Potential Theory. The dispersion relation of the modes propagating in the system shown in Fig. 5.12 is analyzed by using scalar potential theory [7]. In the figure, w_G and w_B are the widths of regions G and B, respectively. In the following analysis, w_B is assumed to be sufficiently large for simplicity.

Let us express the x and y components of the wavevector of the mode propagating in region G as β_x and $\beta_{Gy}^{\pm}$, respectively, where $\pm$ indicates prop-

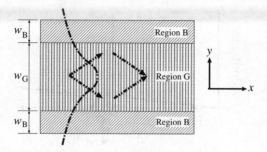

Fig. 5.12. Grating structure

agation toward the $\pm y$ direction. Here we assume that the field distribution of the mode propagating in region G is given by

$$\phi = \{\phi_G^+ \exp(-j\beta_{Gy}^+ y) + \phi_G^- \exp(+j\beta_{Gy}^- y)\} \exp(-j\beta_x x), \tag{5.17}$$

where $\phi_G^\pm$ is a constant determined by the boundary condition.

Similarly, the field distribution in region B may be expressed as

$$\phi = \begin{cases} \phi_B^+ \exp(-\alpha_{By}^+ y) \exp(-j\beta_x x) & (y > w_G/2) \\ \phi_B^- \exp(+\alpha_{By}^- y) \exp(-j\beta_x x) & (y < -w_G/2) \end{cases} \tag{5.18}$$

where $\alpha_{By}^\pm = -j\beta_{By}^\pm$, and the constant $\phi_B^\pm$ will be determined by the boundary condition.

Following conventional scalar potential theory, the continuity of ϕ and $\partial\phi/\partial y$ is employed as the boundary condition at $y = \pm w_G/2$. Then, from (5.17) and (5.18), we get

$$\frac{(j\beta_{Gy}^+ - \alpha_{By}^+)(j\beta_{Gy}^- - \alpha_{By}^-)}{(j\beta_{Gy}^- + \alpha_{By}^+)(j\beta_{Gy}^+ + \alpha_{By}^-)} = \exp\{+j(\beta_{Gy}^+ + \beta_{Gy}^-)w_G\}. \tag{5.19}$$

When the system is symmetric with respect to the x-axis, (5.19) splits into

$$\alpha_{By} = \beta_{Gy} \tan(\beta_{Gy} w_G/2) \tag{5.20}$$

and

$$\alpha_{By} = -\beta_{Gy} \cot(\beta_{Gy} w_G/2), \tag{5.21}$$

because $\beta_{Gy}^+ = \beta_{Gy}^-$ and $\alpha_{By}^+ = \alpha_{By}^-$. The solutions of (5.20) are called symmetric modes because $\phi^+ = \phi^-$ whereas those of (5.21) are called antisymmetric modes because $\phi^+ = -\phi^-$. It should be noted that, if the IDT is not weighted or is weighted symmetrically toward the transverse direction, the antisymmetric modes are not excited and detected by the IDT.

Analysis Based on the Parabolic Approximation. Here we employ the parabolic approximation discussed in Sect. 4.3.1. For simplicity, the symmetry of the structure with respect to the x-axis is assumed. Then the relation between β_x and β_{Gy} for the SAW propagating in region G is expressed as

$$\beta_x/\omega = (V_{G0})^{-1}\{1 - \xi_G(V_{G0}\beta_{Gy}/\omega)^2\}, \tag{5.22}$$

where ξ_G is a parameter responsible for the anisotropy in region G, and is 0.5 when the material is isotropic; its derivation was given in Sect. 4.3.1, and V_{G0} is the SAW velocity toward the x-direction in region G. Figure 5.13 shows the change in the slowness surface with the sign of ξ.

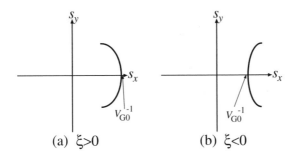

(a) $\xi > 0$ (b) $\xi < 0$

Fig. 5.13. Slowness surface

Next, following the parabolic approximation, let us approximate the relation between β_x and $\alpha_{By} = j\beta_{By}$ in region B as

$$\beta_x/\omega = (V_{B0})^{-1}\{1 + \xi_B(V_{B0}\alpha_{By}/\omega)^2\}, \tag{5.23}$$

where ξ_B is a parameter responsible for the anisotropy in region B, and V_{B0} is the SAW velocity toward the x-direction in region B. Since the influence of the film thickness on the anisotropy is small [8], we approximate ξ_B by setting it equal to ξ_G.

When a mode is trapped in region G, its SAW field composes a standing wave pattern where β_{Gy} is real, and is evanescent in region B where α_{By} is real. Then so that solutions with real β_x exist, the phase velocity $V_p = \omega/\beta_x$ must satisfy

$$V_{G0} < V_p < V_{B0} \ (\xi > 0)$$
$$V_{B0} < V_p < V_{G0} \ (\xi < 0)$$

from (5.22) and (5.23). Thus, when $\xi > 0$, the SAW velocity in the grating (region G) must be smaller than that in the bus-bars (region B). On the other hand, when $\xi < 0$, the SAW velocity in the grating (region G) must be larger than that in the bus-bars (region B).

If this condition is satisfied, since

$$\beta_{Gy} = \frac{\omega}{V_{G0}}\sqrt{(1 - V_{G0}/V_p)/\xi}$$

and

$$\alpha_{By} = \frac{\omega}{V_{B0}}\sqrt{(V_{B0}/V_p - 1)/\xi},$$

the dispersion relation for the symmetric modes is given by

$$\frac{V_{G0}}{V_{B0}}\sqrt{\frac{V_{B0} - V_p}{V_p - V_{G0}}} = \tan\left\{\frac{\pi w_G}{\lambda_p}\sqrt{\frac{V_p(V_p - V_{G0})}{\xi V_{G0}^2}}\right\} \tag{5.24}$$

from (5.20), and that for the antisymmetric mode is given by

$$\frac{V_{G0}}{V_{B0}}\sqrt{\frac{V_{B0} - V_p}{V_p - V_{G0}}} = -\cot\left\{\frac{\pi w_G}{\lambda_p}\sqrt{\frac{V_p(V_p - V_{G0})}{\xi V_{G0}^2}}\right\} \tag{5.25}$$

from (5.21), where $\lambda_p = 2\pi V_p/\omega$ is the wavelength of the mode toward the x-direction.

Assuming $|(V_{B0} - V_{G0})/V_{G0}| \ll 1$, (5.24) is simplified to

$$\sqrt{\frac{1 - \hat{V}}{1 + \hat{V}}} = \tan\left\{\pi\hat{w}_G\sqrt{\frac{1 + \hat{V}}{2}}\right\}, \tag{5.26}$$

where

$$\hat{V} = \frac{2V_p - V_{B0} - V_{G0}}{V_{B0} - V_{G0}} \tag{5.27}$$

is the normalized phase velocity, and

$$\hat{w}_G = \frac{w_G}{\lambda_p}\sqrt{\frac{V_{B0} - V_{G0}}{\xi}}, \tag{5.28}$$

is the normalized aperture. Similarly, (5.25) is simplified to

$$\sqrt{\frac{1 - \hat{V}}{1 + \hat{V}}} = -\cot\left\{\pi\hat{w}_G\sqrt{\frac{1 + \hat{V}}{2}}\right\}. \tag{5.29}$$

Figure 5.14 shows $\hat{V}$ of the transverse modes calculated from (5.26) and (5.29) as a function of $\hat{w}_G$. When $\hat{w}_G$ becomes an integer, higher symmetric modes appear, and when $\hat{w}_G$ becomes a half-integer, higher antisymmetric modes appear. Then (5.28) suggests the critical w_G becomes large if $|\xi|$ is large and/or the difference in V_{B0} and V_{G0} is small.

To suppress all higher transverse modes, $\hat{w}_G$ must be smaller than 0.5. However, when the IDT is symmetric, the limitation for $\hat{w}_G$ is relaxed to be unity since the antisymmetric modes are not excited inherently.

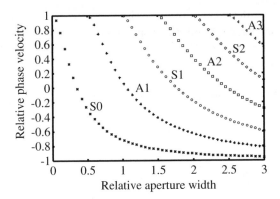

Fig. 5.14. Relation between aperture and phase velocity of transverse modes

Simulation Technique. Let us consider the equivalent circuit shown in Fig. 5.15. Individual parallel-connected series-resonance circuits correspond to transverse mode resonances.

From this equivalent circuit, the resonance frequency of the n-th mode is given by $1/\sqrt{C_m^{(n)} L_m^{(n)}}$. Thus, by using the phase velocity V_n of each mode, we obtain the following relation:

$$\frac{C_m^{(n)} L_m^{(n)}}{C_m^{(1)} L_m^{(1)}} = \left(\frac{V_1}{V_n}\right)^2. \tag{5.30}$$

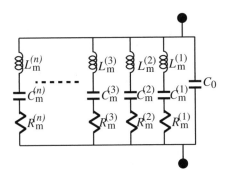

Fig. 5.15. Equivalent circuit

Let us assume the power P_k carried by the k-th mode is given by

$$P_k = \int_{-\infty}^{+\infty} |\phi_k(y)|^2 dy, \tag{5.31}$$

where $\phi_k(y)$ is the field distribution of the k-th mode. When the k-th and n-th modes are propagating simultaneously, since no interaction occurs between them, the power carried by these two modes must be given by $P_k + P_n$. Thus, for arbitrary $\phi_k(y)$ and $\phi_n(y)$, the relation

$$\int_{-\infty}^{+\infty} |\phi_k(y) + \phi_n(y)|^2 dy = \int_{-\infty}^{+\infty} |\phi_k(y)|^2 dy + \int_{-\infty}^{+\infty} |\phi_n(y)|^2 dy \qquad (5.32)$$

must hold when $k \neq n$. By rearranging, we obtain an orthogonal relation among the modes:

$$\int_{-\infty}^{+\infty} \phi_k(y)\phi_n(y)^* dy = \delta_{kn} P_k, \qquad (5.33)$$

where δ_{kn} is Kronecker's delta.

Next, the total field $\phi(y)$ is assumed to be expressed as a sum of the eigenmodes. That is,

$$\phi(y) = \sum_{k=1}^{\infty} A_k \phi_k(y) / \sqrt{P_k} \qquad (5.34)$$

where A_k is the amplitude of the k-th mode. Note that terms due to leaky modes and/or evanescent modes must be included in this expansion.

Let us assume that the SAW field excited by the IDT is uniform over its aperture. That is,

$$\phi(y) = \begin{cases} \phi_0 & (|y| < w/2) \\ 0 & (|y| > w/2). \end{cases} \qquad (5.35)$$

Multiplying (5.35) by $\phi_n(y)^*$ and integrating over y, we obtain

$$\int_{-\infty}^{+\infty} \phi(y)\phi_n(y)^* dy = \sum_{k=1}^{\infty} A_k \frac{\int_{-\infty}^{+\infty} \phi_k(y)\phi_n(y)^* dy}{\sqrt{\int_{-\infty}^{+\infty} |\phi_k(y)|^2 dy}}$$

$$= A_n \sqrt{\int_{-\infty}^{+\infty} |\phi_n(y)|^2 dy}. \qquad (5.36)$$

Then, the unknown A_n is given by

$$A_n = \frac{\phi_0 \int_{-w/2}^{+w/2} \phi_n(y)^* dy}{\sqrt{\int_{-\infty}^{+\infty} |\phi_n(y)|^2 dy}}. \qquad (5.37)$$

If we apply the one-dimensional analysis to this procedure, $A = \phi_0\sqrt{w}$ because of the uniform field distribution. Thus, if we can derive the motional capacitance $\hat{C}_m^{(0)}$ of the resonator by using another technique for the one-dimensional case, the motional capacitance $C_m^{(n)}$ of the n-th mode is given by

$$\frac{C_m^{(n)}}{\hat{C}_m^{(0)}} = \frac{1}{w} \frac{\left| \int_{-w/2}^{+w/2} \phi_n(y) dy \right|^2}{\int_{-\infty}^{+\infty} |\phi_n(y)|^2 dy}, \tag{5.38}$$

from (5.37).

Substituting the field distribution derived previously into (5.38), we obtain

$$\frac{C_m^{(n)}}{\hat{C}_m^{(0)}} = 2 \frac{\sin^2(\beta_{Gy} w_G/2)}{1 + (\beta_{Gy} w_G/2)^{-1} \cot(\beta_{Gy} w_G/2)} \tag{5.39}$$

for the symmetric modes. Of course, $C_m^{(n)} = 0$ for the antisymmetric modes.

Figure 5.16 shows $C_m^{(n)}/\hat{C}_m^{(0)}$ calculated by the substitution of (5.26)–(5.28) into (5.39). For all modes, $C_m^{(n)}$ increases monotonically with an in-

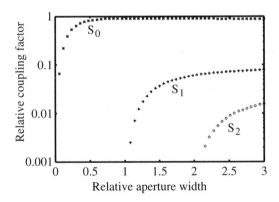

Fig. 5.16. Relation between aperture and $C_m^{(n)}/\hat{C}_m^{(0)}$ for transverse modes

crease in the normalized aperture $\hat{w}_G$. This is because small $\hat{w}_G$ results in increased penetration depth into the bus-bar region (Fig. 5.17a). In addition, $C_m^{(n)}/\hat{C}_m^{(0)}$ for the S_0 mode is much larger than those of the S_1 and S_2 modes. This is because there exist sign inversions in the field distributions of the S_1 and S_2 modes (Fig. 5.17c). The electromechanical coupling factor for the S_0 mode almost coincides with that of the one-dimensional analysis provided $\hat{w}_G > 0.6$ because most of the energy is concentrated within the grating region and its distribution is mostly uniform (Fig. 5.17b).

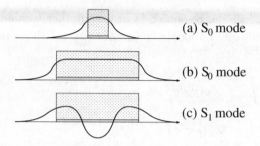

Fig. 5.17. Mode profile of S_n modes

5.2.3 Effect of BAW Radiation

Since the BAWs are not efficiently reflected by the reflectors, the spurious BAW response is not significant in resonators. However, since the energy leak due to the BAW results in a reduction of the resonance Q, its behavior must be clarified.

The cut-off frequency for the BAW radiation is considerably higher than the SAW resonance frequency. Thus, if the IDT and grating reflectors are designed to operate at their fundamental resonance, the BAW is not radiated from the IDT and the grating themselves.

However, as described in Sect. 2.2.2, this is not due to the suppression of the BAW excitation but due to the interference and cancellation of BAWs excited by the periodic structure. Then the interference results in the SAW velocity reduction, namely, the energy storing effect.

Since the lengths of the IDT and reflector gratings are finite, their periodicity disappears at their ends, where the interference among the scattered BAWs is weakened. This means that BAW excitation occurs at any frequency. In addition, a variation in the phase shift occurs at the grating edge. This phenomenon may be expressed in the equivalent circuit shown in Fig. 5.18 by replacing the capacitance responsible for the energy storing effect with a resistance.

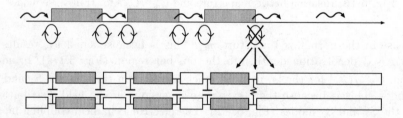

Fig. 5.18. Equivalent circuit for BAW radiation at a discontinuity

Let the distribution of the SAW field and the reflection source be denoted by $u(x_1)$ and $r(x_1)$, respectively. Then the field $u(x_1)r(x_1)$ generated by the reflection can be expressed in the Fourier integral form

$$u(x_1)r(x_1) = \int_{-\infty}^{+\infty} U_r(\beta) \exp(-j\beta x_1)d\beta, \qquad (5.40)$$

where $U_r(\beta)$ is the Fourier transform of $u(x_1)r(x_1)$ given by

$$U_r(\beta) = \frac{1}{2\pi} \int_{-\infty}^{+\infty} u(x_1)r(x_1) \exp(+j\beta x_1)dx. \qquad (5.41)$$

Let us express $u(x_1)$ as

$$u(x_1) = \hat{u}_+(x_1) \exp(-j\beta_S x_1) + \hat{u}_-(x_1) \exp(+j\beta_S x_1), \qquad (5.42)$$

and $r(x_1)$ as

$$r(x_1) = \hat{r}(x_1)\{\exp(-2\pi j x_1/p) + \exp(+2\pi j x_1/p)\}, \qquad (5.43)$$

where p is the periodicity of the gratings and β_S is the wavevector of the SAW. Since our main concern is behavior at resonance, we set $\beta_S = \pi/p$. Then (5.41) can be rewritten as

$$U_r(\beta) = U_r^+(\beta + \beta_S) + U_r^-(\beta - \beta_S) + U_r^+(\beta - 3\beta_S) + U_r^-(\beta + 3\beta_S), \qquad (5.44)$$

where

$$U_r^\pm(\beta) = \frac{1}{2\pi} \int_{-\infty}^{+\infty} \hat{u}_\pm(x_1)\hat{r}(x_1) \exp(j\beta x)dx \qquad (5.45)$$

is the Fourier transform of $\hat{u}_\pm(x_1)\hat{r}(x_1)$. Since both $\hat{u}_\pm(x_1)$ and $\hat{r}(x_1)$ are relatively smooth functions, their product $\hat{u}_\pm(x_1)\hat{r}(x_1)$ will also be smooth. Thus its Fourier transform $U_r^\pm(\beta)$ has a peak at $\beta \cong 0$ (see Fig. 5.19).

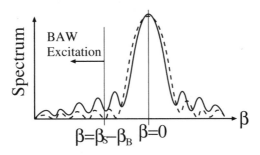

Fig. 5.19. Fourier transform of source distribution

In (5.44), the terms $U_r^\pm(\beta \mp 3\beta_S)$ at $\beta = \pm 2\beta_S$ give the contribution of the reflected SAWs. The power P_B due to the BAW radiation is approximately given by

$$P_{\mathrm{B}} = \int_{+\beta_{\mathrm{B}}}^{+\infty} F(\beta)|U_{\mathrm{r}}(\beta)|^2 d\beta + \int_{-\infty}^{-\beta_{\mathrm{B}}} F(\beta)|U_{\mathrm{r}}(\beta)|^2 d\beta$$

$$\cong \int_{+\beta_{\mathrm{B}}}^{+\infty} F(\beta)|U_{\mathrm{r}}^-(\beta - \beta_{\mathrm{S}})|^2 d\beta + \int_{-\infty}^{-\beta_{\mathrm{B}}} F(\beta)|U_{\mathrm{r}}^+(\beta + \beta_{\mathrm{S}})|^2 d\beta,$$

$$(5.46)$$

where β_{B} is the wavenumber of the BAW at its cut-off, and $F(\beta)$ is the angular dependence of the BAW radiation efficiency. In (5.46), the first term is due to the BAW scattered toward the $+x_1$ direction whereas the second term is due to that toward the $-x_1$ direction. Thus, as shown in Fig. 5.19, the BAW radiation is governed by the behavior of $U_{\mathrm{r}}^\pm(\beta)$ with β slightly apart from its peak, and a narrower $U_{\mathrm{r}}^\pm(\beta)$ is desirable for a reduction of the BAW radiation.

As an example, let us consider the field distributions shown in Fig. 5.20. In case (b), since $u(x_1)r(x_1)$ is smooth, $U_{\mathrm{r}}^\pm(\beta)$ concentrates at $\beta \cong 0$, and then P_{B} is small. On the other hand, in case (a), since $u(x_1)r(x_1)$ concentrates at the gaps between the IDT and reflectors, $U_{\mathrm{r}}^\pm(\beta)$ may spread widely. Thus, it is expected that SAW resonators with field distribution (b) will possess a higher resonance Q than those with field distribution (a).

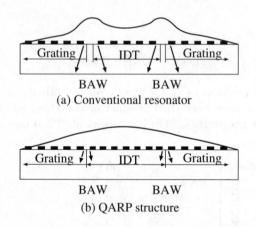

Fig. 5.20. Field distribution at resonance

Let us consider a resonator where the IDT has the same periodicity as the reflectors, and the gaps of $\pm p/8$ are given so that the device resonates at their Bragg frequency. At resonance, since the SAW fields both in the IDT and reflectors are evanescent, the field distribution is expected to be similar to the one shown in Fig. 5.20a. Next, let us consider a resonator without gaps where the IDT periodicity is adjusted so that the resonance frequency coincides with the Bragg frequency of the gratings. In this case, a standing wave pattern is generated within the IDT, and then the field distribution is expected to be similar to that shown in Fig. 5.20b. This structure is called QARP (quasi-constant acoustic reflection periodicity) [9], and is known to have a higher resonance Q than structure (a).

5.3 Two-Port SAW Resonators

5.3.1 Summary

Figure 5.21 shows a two-port SAW resonator where the grating reflectors are placed on both sides of a conventional transversal filter. When the device is designed so that the reflectors resonate at the IDT resonance frequency, the transfer admittance becomes very large at resonance, and a very narrow but low-loss passband is realized.

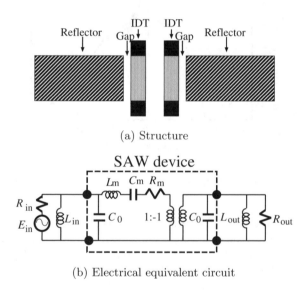

(a) Structure

(b) Electrical equivalent circuit

Fig. 5.21. Two-port SAW resonator

The oscillator can be constructed by using this device as a feedback element as shown in Fig. 5.22. This configuration is widely used for operation in the UHF range [10] because of its insensitivity to parasitic circuit elements.

Figure 5.23 shows a typical frequency response of a two-port SAW resonator. It is seen that the steep resonance peak is overlapped by the conventional $\sin x/x$-type frequency response of the transversal filter, which seems to be negligible for the discussion on resonance characteristics.

Figure 5.21b shows the equivalent circuit near the resonance frequency. Although the circuit is similar to that for the one-port SAW resonator shown

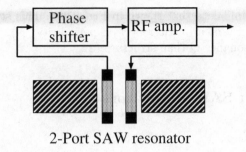

2-Port SAW resonator

Fig. 5.22. Oscillator circuit employing two-port SAW resonator

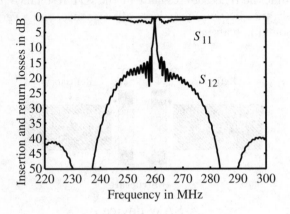

Fig. 5.23. Typical frequency response of two-port SAW resonator

in Fig. 5.1b, the resonance circuit is involved as a shunt element between two IDTs. This is because the structure is equivalent to a one-port SAW resonator if two IDTs are electrically parallel-connected. In the figure, L_{in} and L_{out} are the matching inductances to cancel the static capacitances C_0 of the IDTs.

When $R_{in} = R_{out}$, the scattering matrix element S_{21} is given by

$$S_{21} = \frac{-jQ_r^{-1}(\omega/\omega_r)(2R_{in}/R_m)}{1 - (\omega/\omega_r)^2 + jQ_r^{-1}(\omega/\omega_r)(1 + 2R_{in}/R_m)}, \tag{5.47}$$

where Q_r is the resonance Q.

From (5.47), $S_{21}|_{\omega=\omega_r}$ is given by

$$S_{21}|_{\omega=\omega_r} = \frac{-1}{1 + R_m/2R_{in}}, \tag{5.48}$$

and the fractional -3 dB bandwidth $2\delta\omega/\omega_r$ is given by

$$\frac{2\delta\omega}{\omega_r} = Q_r^{-1}(1 + 2R_{in}/R_m) = Q_r^{-1}(1 + 2R_{in}/R_m) \equiv Q_L^{-1}, \tag{5.49}$$

where Q_L is called the loaded Q. With this terminology, the resonance Q is also called the unloaded Q.

Note that the resonance characteristics due to C_0 and L_{in} or L_{out} must be taken into account. That is, the -3 dB bandwidth is also limited by the resonance Q of the matching circuit. It is referred to the circuit Q and is given by $Q_c = \omega C_0 R_{in} = \gamma Q_r^{-1}(R_{in}/R_m)$. The overall bandwidth is limited by the smaller one between Q_L^{-1} and Q_c^{-1}. Thus the maximum fractional bandwidth is achieved when $Q_L = Q_c$, and the value is given by

$$\left.\frac{2\delta\omega}{\omega_r}\right|_{max} = Q_r^{-1}\frac{\sqrt{1 + 8Q_r^2/\gamma} + 1}{2} \cong \sqrt{2/\gamma}, \tag{5.50}$$

where $\gamma = C_m/C_0$ is the capacitance ratio. Under this condition,

$$S_{21}|_{\omega=\omega_r} = -\frac{1}{1 + 2\left\{\sqrt{1 + 8Q_r^2/\gamma} - 1\right\}^{-1}} \cong -\frac{1}{1 + Q_r^{-1}\sqrt{\gamma/2}}. \tag{5.51}$$

This suggests that smaller γ and larger Q_r are desired to achieve low insertion loss.

5.3.2 Fabry–Perot model

To understand the fundamental characteristics of two-port SAW resonators, the Fabry–Perot model shown in Fig. 5.24 is employed, which has already been applied for the characterization of one-port SAW resonators in Sect. 5.1.2. In the figure, D is the distance between the IDTs.

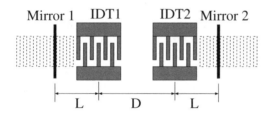

Fig. 5.24. Fabry–Perot model for two-port SAW resonator

When the input admittance of the IDT before placing the mirrors is $Y_S = G_S + jB_S + j\omega C_0$, the input admittance Y_{11} of the IDTs after placement is given by summing up all echo responses:

$$Y_{11} = Y_S + G_S \sum_{n=0}^{+\infty}[\Gamma\exp(-2j\beta L) + \Gamma\exp\{-2j\beta(L + D)\}$$
$$+2\Gamma^2\exp\{-2j\beta(2L + D)\}](\Gamma)^{2n}\exp\{-2jn\beta(2L + D)\}$$

$$= j\omega C_0 + jB_S + G_S \frac{[1 + \Gamma \exp(-2j\beta L)][1 + \Gamma \exp\{-2j\beta(L+D)\}]}{1 - \Gamma^2 \exp\{-2j\beta(2L+D)\}},$$

$$(5.52)$$

where β is the wavenumber of the SAW and Γ is the reflection coefficient of the mirrors. In a similar way, the transfer admittance Y_{12} is given by

$$Y_{12} = G_S \sum_{n=0}^{+\infty} [\exp(-j\beta D) + 2\Gamma \exp\{-j\beta(2L+D)\}$$

$$+ \Gamma^2 \exp\{-j\beta(4L+D)\}] \times (\Gamma)^{2n} \exp\{-2jn\beta(2L+D)\}$$

$$= G_S \exp(-j\beta D) \frac{\{1 + \Gamma \exp(-2j\beta L)\}^2}{1 - \Gamma^2 \exp\{-2j\beta(2L+D)\}}. \tag{5.53}$$

From (5.52) and (5.53), the resonance condition is given by $\beta(2L+D) - \phi = m\pi$ where $\phi = \angle\Gamma$. Then when m is even, namely when $\beta(2L+D) - \phi = 2M\pi$,

$$Y_{11} = j\omega C_0 + jB_S + G_S \frac{|1 + |\Gamma| \exp(j\beta D)|^2}{1 - |\Gamma|^2}, \tag{5.54}$$

$$Y_{12} = G_S \exp(-j\beta D) \frac{\{1 + |\Gamma| \exp(j\beta D)\}^2}{1 - |\Gamma|^2}. \tag{5.55}$$

On the other hand, when m is odd, namely when $\beta(2L+D) - \phi = (2M+1)\pi$,

$$Y_{11} = j\omega C_0 + jB_S + G_S \frac{|1 - |\Gamma| \exp(j\beta D)|^2}{1 - |\Gamma|^2}, \tag{5.56}$$

$$Y_{12} = G_S \exp(-j\beta D) \frac{\{1 - |\Gamma| \exp(j\beta D)\}^2}{1 - |\Gamma|^2}. \tag{5.57}$$

Thus setting $\beta D = m\pi$ and $2\beta L - \phi = 2m'\pi$ maximizes the desired resonance and suppresses adjacent resonances.

The discussion given in Sect. 2.1.2 suggests that grating reflectors offer maximum SAW reflection at $\phi = \pm\pi/2$. Since SAW resonators are usually designed to resonate under this condition, L is chosen to be $(m'/2 \pm 1/8)\lambda$.

When $\beta D = m\pi$, (5.52) and (5.53) are reduced to

$$Y_{11} = j\omega C_0 + jB_S + G_S \frac{1 + \Gamma \exp(-2j\beta L)}{1 - \Gamma \exp(-2j\beta L)} \tag{5.58}$$

$$Y_{12} = G_S \exp(-j\beta D) \frac{1 + \Gamma \exp(-2j\beta L)}{1 - \Gamma \exp(-2j\beta L)}. \tag{5.59}$$

The result shown in (5.58) coincides with that obtained for the one-port SAW resonator as in (5.10). Then the discussions on the resonance Q and capacitance ratio γ for the one-port SAW resonator given in Sect. 5.1.2 are readily applicable to the two-port SAW resonator. Since terms in Y_{11} related to G_S are proportional to Y_{12}, similar results can easily be obtained for Y_{12}.

5.3.3 Multi-Mode Resonator Filter

Among multi-mode resonator filters, the longitudinally coupled double-mode resonator filter has exactly the same structure as the conventional two-port SAW resonator discussed previously, but is designed to resonate at two different frequencies [11].

Since the two-port SAW resonator has a symmetric structure, one of the two resonant modes is symmetric with respect to the center of the structure whereas the other is antisymmetric. These symmetric and antisymmetric modes can be excited selectively by connecting two IDTs as shown in Fig. 5.25.

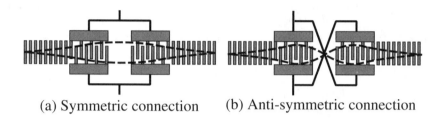

(a) Symmetric connection   (b) Anti-symmetric connection

Fig. 5.25. Double-mode resonator filter

Figure 5.26 shows the electrical equivalent circuit for the double-mode resonator filter, where Y_s and Y_a are the motional admittances for the symmetric and antisymmetric modes respectively, and ω_r^s and ω_r^a are their resonance frequencies.

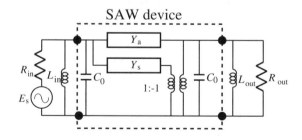

Fig. 5.26. Equivalent circuit for double-mode resonator filter

When $Y_{\text{out}} = Y_{\text{in}} = (G_{\text{in}} + 1/j\omega L_{\text{in}})^{-1}$, S_{21} is given by

$$S_{21} = \frac{2\Re(Y_{\text{in}})(Y_a - Y_s)}{(2Y_a + Y_{\text{in}})(2Y_s + Y_{\text{in}})}. \tag{5.60}$$

If the difference in the resonance frequencies is sufficiently large so that two resonant peaks do not overlap each other, we obtain

$$S_{21}|_{\omega=\omega_r^s} \cong -\frac{2\Re(Y_{in})R_m^{s-1}}{(j\omega_r^s C_0 + Y_{in})(2R_m^{s-1} + j\omega_r^s C_0 + Y_{in})}, \tag{5.61}$$

$$S_{21}|_{\omega=\omega_r^a} \cong \frac{2\Re(Y_{in})R_m^{a-1}}{(j\omega_r^a C_0 + Y_{in})(2R_m^{a-1} + j\omega_r^a C_0 + Y_{in})}. \tag{5.62}$$

Under the matching condition $\Im(Y_{in}) = -\omega C_0$, they reduce to

$$S_{21}|_{\omega=\omega_r^s} \cong \frac{-1}{1 + R_m^s G_{in}/2} \tag{5.63}$$

$$S_{21}|_{\omega=\omega_r^a} \cong \frac{1}{1 + R_m^a G_{in}/2}. \tag{5.64}$$

Thus, if $R_m^s, R_m^a \ll R_{in}$, $S_{21}|_{\omega=\omega_r^s} \cong -1$ and $S_{21}|_{\omega=\omega_r^a} \cong 1$.

Let us assume the loaded Qs for $(2Y_s + Y_{in})$ and $(2Y_a + Y_{in})$ are equal to each other. At $\omega_c = (\omega_r^a + \omega_r^s)/2$, we obtain

$$\begin{aligned}
S_{21}|_{\omega=\omega_c} &= \frac{4jG_{in}\Im(Y_a)}{|2Y_a + G_{in}|^2} \\
&= \frac{4jR_m G_{in}Q\{1 - (\omega_c/\omega_r)^2\}}{(2 + R_m G_{in})^2 + (R_m G_{in}Q)^2\{1 - (\omega_c/\omega_r)^2\}^2},
\end{aligned} \tag{5.65}$$

because $(2Y_a + Y_{in})_{\omega=\omega_c} = (2Y_s + Y_{in})_{\omega=\omega_c}^*$. Then $S_{21}|_{\omega=\omega_c}$ takes a maximum value of

$$S_{21}|_{\omega=\omega_c} = \frac{j}{1 + R_m G_{in}/2}. \tag{5.66}$$

when ω_c is chosen so that $\omega_c/\omega_r = \sqrt{1 \pm Q^{-1}(1 + 2/R_m G_{in})} \cong 1 \pm Q^{-1}(1/2 + 1/R_m G_{in})$. Then from (5.63), (5.64) and (5.66), we get,

$$|S_{21}|_{\omega=\omega_c}| = |S_{21}|_{\omega=\omega_r^s}| = |S_{21}|_{\omega=\omega_r^a}|.$$

Thus, in this situation, a flat frequency response as shown in Fig. 5.27 can be achieved.

By extending this principle, resonator filters using multiple higher modes can be synthesized, and a wider passband width may be obtained.

Figure 5.28 shows a variation [11], where the first and third resonant modes are combined and the second mode is suppressed due to the symmetry of the structure. Note that the piezoelectricity of the substrate must be strong enough to achieve low insertion loss and wide bandwidth simultaneously.

Figure 5.29 shows the frequency response of the double-mode resonator filter for GSM-Rx developed by Fujitsu Ltd. [12]. As a substrate, 42°YX-LiTaO$_3$ [13] was employed (see Sect. 8.1.3). Low insertion loss and good out-of-band rejection are achieved. The attained insertion loss and -3 dB bandwidth were 2.0 dB and 38 MHz, respectively.

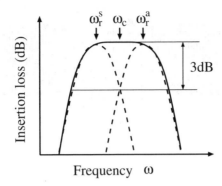

Fig. 5.27. Frequency response of double-mode resonator filter

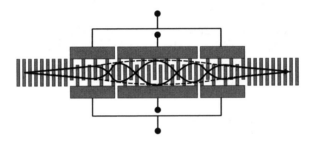

Fig. 5.28. Double-mode resonator filter employing first and third modes

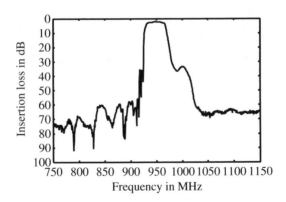

Fig. 5.29. Frequency response of double-mode resonator filter for GSM-Rx [12]

Figure 5.30 shows the configuration of a double-mode resonator filter employing lateral coupling of two one-port SAW resonators through the transverse mode. This is called a transversely coupled resonator filter [14]. Although the achievable bandwidth is narrow, better out-of-band rejection can

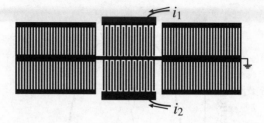

Fig. 5.30. Transversely coupled resonator filter

be achieved than the longitudinally coupled double-mode resonator filter because of the lack of a direct acoustic path between the IDTs.

Figure 5.31 shows the frequency response of the transversely coupled resonator filter developed by Fujitsu Lab. Ltd. [15]. As a substrate, ST-cut quartz was employed, and a grating was placed at the spacer region between two tracks so as to widen the bandwidth. Low spurious and superior out-of-band rejection were achieved. The insertion loss and −3 dB bandwidth were 5.5 dB and 305 kHz, respectively.

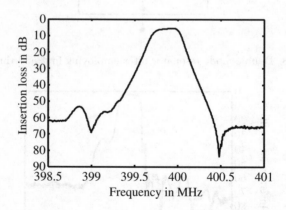

Fig. 5.31. Frequency response of transversely coupled resonator filter [15]

Recently, a multi-mode resonator filter combining multiple longitudinal and transverse modes was proposed [16], and is being aggressively studied.

5.3.4 Cascade Connection of Resonators

Let us consider the cascade connection of two two-port SAW resonators shown in Fig. 5.32, where L_{in}, L_{out} and L_c are the tuning inductors. For simplicity, we assume that the resonators are identical and $L_{in} = L_{out}$.

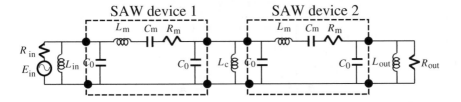

Fig. 5.32. Cascade connection of two-port SAW resonators

Under optimal design, the input impedance of the devices is $(1/R_{\rm in} + 1/j\omega L_{\rm in})^*$ from the matching condition. The input impedance looking from the center port is $(1/R_{\rm in} + 1/j\omega L_{\rm in})^*$, too. Then lossless power transfer between the two resonators can be realized by setting $L_{\rm c} = L_{\rm in}/2 = L_{\rm out}/2$.

In the parallel resonance circuit composed of $L_{\rm in}$, $L_{\rm out}$ and C_0, the resistances $R_{\rm in}$ and $R_{\rm out}$ limit their circuit Qs. On the other hand, since no resistance is included for $L_{\rm c}$, the circuit Q at the center port is high. In practice, the circuit Q is limited by the Q of the inductance.

In the rejection band, since the frequency responses of the resonators are multiplied, the total out-of-band rejection level is doubled. In addition, a large mismatching loss arises at the center port due to its high circuit Q.

Hence when N stages are cascade-connected, the out-of-band rejection level is improved by more than N whereas the insertion loss increases by N. Note that spurious resonance peaks can be removed by designing the resonators slightly different from each other.

5.4 Impedance Element Filters

Recently, filter configurations employing a SAW resonator as a circuit element have been paid much attention. This type of filter, called an impedance element filter, offers low insertion loss and high power durability, compared with acoustically coupled resonators. Although the physical size increases with an increase in the number of resonators, this problem is not significant in the GHz range.

5.4.1 π-Type Filters

Figure 5.33 shows the configuration of a π-type filter. When $G_{\rm in} = G_{\rm out}$ and $Y_{\rm p1} = Y_{\rm p2} = Y_{\rm p}$, $S_{21}(= 2E_{\rm out}/E_{\rm in})$ is given by

$$S_{21} = \frac{2G_{\rm in}}{(G_{\rm in} + Y_{\rm p})\{2 + (G_{\rm in} + Y_{\rm p})/Y_{\rm s}\}}. \tag{5.67}$$

If we assume $Y_{\rm p} = 0$, (5.67) can be simplified to

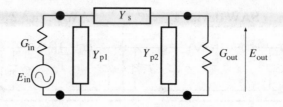

Fig. 5.33. π-type filter

$$S_{21} = \frac{1}{1 + G_{in}/2Y_s}. \tag{5.68}$$

If Y_s is a one-port SAW resonator, $S_{21} \cong 1$ at its resonance frequency whereas $S_{21} \cong 0$ at the antiresonance frequency (see Fig. 5.34a).

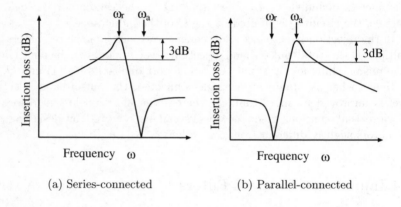

(a) Series-connected   (b) Parallel-connected

Fig. 5.34. Frequency response when a resonator is series- or parallel-connected

By using the equivalent circuit shown in Fig. 5.1b, we obtain

$$S_{21} \cong \left\{ 1 + \frac{Q_c}{j(1+\gamma^{-1})} \frac{1 - (\omega/\omega_r)^2 + jQ_r^{-1}}{1 - (\omega/\omega_a)^2 + jQ_r^{-1}} \right\}^{-1}, \tag{5.69}$$

where $\gamma = C_0/C_r$ is the capacitance ratio, $\omega_r = 1/\sqrt{C_m L_m}$ is the resonance frequency, $\omega_a = \omega_r \sqrt{1 + \gamma^{-1}}$ is the antiresonance frequency, $Q_r = \omega L_m/R_m$ is the resonance Q, and

$$Q_c = G_{in}/2\omega C_0 \tag{5.70}$$

is the circuit Q. When $\omega \cong \omega_r$ and $Q_r \gg \gamma$,

$$S_{21} \cong [1 - j\gamma Q_c \{1 - (\omega/\omega_r)^2 + jQ_r^{-1}\}]^{-1}. \tag{5.71}$$

Thus the -3 dB bandwidth $2\delta\omega/\omega_r$ is given by

$$\frac{2\delta\omega}{\omega_r} = Q_r^{-1} + (\gamma Q_c)^{-1} = (\gamma Q_c)^{-1}(1 + \gamma Q_c/Q_r) \equiv Q_L^{-1}, \qquad (5.72)$$

where Q_L is the loaded Q. Then S_{21} at the resonance frequency is given by

$$S_{21}|_{\omega=\omega_r} = (1 + \gamma Q_c/Q_r)^{-1}. \qquad (5.73)$$

Equations (5.72) and (5.73) suggest that, provided that $Q_c \ll M(= Q_r/\gamma)$, the bandwidth $2\delta\omega/\omega_r$ increases with an increase in Q_c without deteriorating S_{21}. Thus, by using resonators with large M, filters with wide bandwidth and low insertion loss can be realized.

Since $S_{21} \cong (1 - jQ_c)^{-1}$ at frequencies far from resonance, Q_c is required to be much larger than unity so as to achieve sufficient out-of-band rejection. This requirement contradicts the need for a wide bandwidth, and Q_c must be determined under this trade-off.

We can construct another type of filter by applying a one-port SAW resonator to Y_p instead of Y_s and setting $Y_s^{-1} = 0$. In the case, $S_{21} \cong 1$ at its antiresonance frequency whereas $S_{21} \cong 0$ at the resonance frequency (see Fig. 5.34b).

If both Y_p and Y_s are the resonators and the resonance frequency ω_r^s of Y_s is chosen to coincide with the antiresonance frequency ω_a^p of Y_p, then $Y_p \cong 0$ and $Y_s^{-1} \cong 0$ at $\omega = \omega_r^s = \omega_a^p$. Hence, the achievable bandwidth becomes much wider than the case where either Y_p or Y_s is the resonator. In addition, since nulls are created at both sides of the passband, the desirable passband shape, as shown in Fig. 5.35, can be achieved.

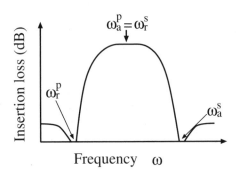

Fig. 5.35. Frequency response of π-type filter with series- and parallel-connected resonators

It is interesting to note that zero insertion loss is achieved by setting $\sqrt{2Y_sY_p + Y_p^2} = G_{in}$ provided $-1 \le Y_s/Y_p \le 0$ even though all elements are reactive. This situation will be detailed and extended in Sect. 5.4.3.

5.4.2 Lattice-Type Filters

Figure 5.36 shows the configuration of a lattice-type filter [17]. As shown in the figure, since this filter is electrically equivalent to the double-mode resonator filter described in Sect. 5.3.3, superior frequency response can be achieved.

This device possesses balanced input and output whereas the double-mode resonator filter has unbalanced input and output. This feature is promising because balanced input and output are now widely used for high-frequency mixers and amplifiers.

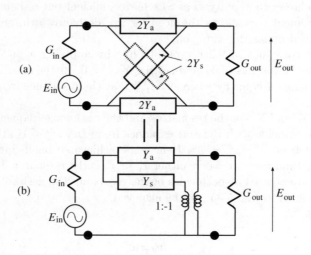

Fig. 5.36. Configuration of lattice-type filter and equivalent unbalanced filter

It should be noted that since conventional network analyzers have unbalanced input and output with common ground, they are not directly applicable for devices with balanced input and output. In fact, if the device configuration shown in Fig. 5.36 is measured by a conventional network analyzer, the device becomes the equivalent to a π-type filter because the lower series-connected element is short-circuited. Then for this measurement, balanced-to-unbalanced converters, namely baluns, should be employed.

Here we will show a technique to create nulls adjacent to the passband so as to realize sharp transition band characteristics.

Let us consider the case where a small capacitance δC is parallel-connected to Y_s and $\omega_r^a > \omega_r^s$. Application of the equivalent circuit shown in Fig. 5.1b suggests

$$Y_a - Y_s \cong j\omega(\delta C + C_0^a - C_0^s) + \frac{j\omega C_r^a}{1 - (\omega/\omega_r^a)^2} - \frac{j\omega C_r^s}{1 - (\omega/\omega_r^s)^2} \qquad (5.74)$$

within the rejection band. Then, when $C_0^{\mathrm{a}} = C_0^{\mathrm{s}}$ and $C_{\mathrm{r}}^{\mathrm{a}} = C_{\mathrm{r}}^{\mathrm{s}}$, the condition $Y_{\mathrm{a}} = Y_{\mathrm{s}}$ giving $S_{21} = 0$ is realized at frequencies $\omega_{\pm}$ satisfying

$$2\frac{\delta C}{C_{\mathrm{r}}^{\mathrm{a}}} \cong -\frac{1}{1 - (\omega_{\pm}/\omega_{\mathrm{r}}^{\mathrm{a}})} + \frac{1}{1 - (\omega_{\pm}/\omega_{\mathrm{r}}^{\mathrm{s}})}. \tag{5.75}$$

By solving (5.75), we get

$$\frac{\omega_{\pm}}{\omega_{\mathrm{c}}} \cong 1 \pm \sqrt{\left(\frac{\delta\omega}{\omega_{\mathrm{c}}}\right)^2 + \left(\frac{\delta C}{C_{\mathrm{r}}^{\mathrm{a}}}\right)^{-1}\frac{\delta\omega}{\omega_{\mathrm{c}}}}, \tag{5.76}$$

where $\delta\omega = (\omega_{\mathrm{r}}^{\mathrm{a}} - \omega_{\mathrm{r}}^{\mathrm{s}})/2$. Thus by giving a tiny δC, nulls can be generated adjacent to the passband as shown in Fig. 5.37, and sharp transition band characteristics are realized.

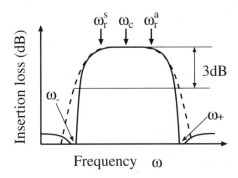

Fig. 5.37. Frequency response of lattice-type filter with δC. Solid line: with δC, and broken line: without δC

It should be noticed that as a trade-off, this technique also causes deterioration of the out-of-band rejection at $\omega < \omega_-$ and $\omega > \omega_+$. Assuming $|Y_{\mathrm{in}}| \gg |2Y_{\mathrm{a}}|$, (5.60) gives S_{21} at $\omega \ll \omega_-$ and $\omega \gg \omega_+$ as

$$S_{21} \cong 2j\omega\delta C/G_{\mathrm{in}} = 2j\frac{\delta C}{C_0}Q_{\mathrm{c}}, \tag{5.77}$$

where $Q_{\mathrm{c}} = \omega C_0/G_{\mathrm{in}}$.

This technique is also applicable to the double-mode resonator filter discussed in Sect. 5.3.3 [18].

5.4.3 Ladder-Type Filters

Figure 5.38a shows the basic configuration of the ladder-type filter consisting of cascade-connected multiple stages [19]. All stages composed of two SAW resonators are identical. Note that stages labeled with even numbers are the mirror image of those labeled with odd numbers. In practical device

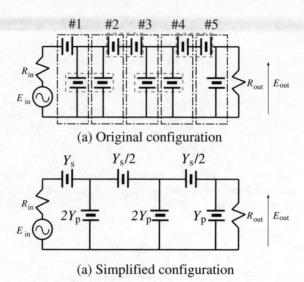

(a) Original configuration

(a) Simplified configuration

Fig. 5.38. Ladder-type filter with five stages

design, series-connected resonators and parallel-connected resonators can be combined as shown in Fig. 5.38.

Let us consider the two-port matrix $[F_n]$ relating the voltage e_n and current i_n for the n-th port;

$$\begin{pmatrix} e_n \\ i_n \end{pmatrix} = [F_n] \begin{pmatrix} e_{n+1} \\ i_{n+1} \end{pmatrix}. \tag{5.78}$$

In general, $[F_n]$ can be expressed in the form

$$[F_n] = \begin{pmatrix} (Z_i/Z_o)^{0.5} \cosh\phi & (Z_i Z_o)^{0.5} \sinh\phi \\ (Z_i Z_o)^{-0.5} \sinh\phi & (Z_o/Z_i)^{0.5} \cosh\phi \end{pmatrix}, \tag{5.79}$$

where Z_i, Z_o and ϕ are called image parameters. If we employ this expression for odd n, F_n for even n can be written as

$$[F_n] = \begin{pmatrix} (Z_o/Z_i)^{0.5} \cosh\phi & (Z_i Z_o)^{0.5} \sinh\phi \\ (Z_i Z_o)^{-0.5} \sinh\phi & (Z_i/Z_o)^{0.5} \cosh\phi \end{pmatrix}. \tag{5.80}$$

From the recursion relation of (5.78), the two-port matrix $[F]$ for the overall structure is given by

$$[F] = \prod_{n=1}^{N} [F_n] = \begin{cases} \begin{pmatrix} (Z_i/Z_o)^{0.5} \cosh(N\phi) & (Z_i Z_o)^{0.5} \sinh(N\phi) \\ (Z_i Z_o)^{-0.5} \sinh(N\phi) & (Z_o/Z_i)^{0.5} \cosh(N\phi) \end{pmatrix} \\ \qquad\qquad\qquad\qquad\qquad (N : \text{odd}) \\ \begin{pmatrix} \cosh(N\phi) & Z_i \sinh(N\phi) \\ Z_i \sinh(N\phi) & \cosh(N\phi) \end{pmatrix} \\ \qquad\qquad\qquad\qquad\qquad (N : \text{even}). \end{cases} \tag{5.81}$$

When this filter is connected to the peripheral circuits as shown in Fig. 5.38 and $R_{\text{out}} = R_{\text{in}}$ is assumed for simplicity, S_{21} is given by

$$
S_{21} = \begin{cases}
2\left[\left(\sqrt{Z_i/Z_o} + \sqrt{Z_o/Z_i}\right)\cosh(N\phi) \right. \\
\left. + \left(\sqrt{Z_iZ_o}/R_{\text{in}} + R_{\text{in}}/\sqrt{Z_oZ_i}\right)\sinh(N\phi)\right]^{-1} \quad (N : \text{odd}) \\[2ex]
2[2\cosh(N\phi) + (Z_i/R_{\text{in}} + R_{\text{in}}/Z_i)\sinh(N\phi)]^{-1} \quad (N : \text{even}).
\end{cases}
$$

$$(5.82)$$

If $\Re(N\phi)$ is sufficiently large, $|S_{21}|$ becomes very small because $|\sinh(N\phi)|$, $|\cosh(N\phi)| \gg 1$. On the other hand, if $\Re(N\phi)$ is sufficiently small and $|Z_i/Z_o| \cong 1$, $|S_{21}| \cong 1$ because $|\sinh(N\phi)|$, $|\cosh(N\phi)| \leq 1$. Namely, the former case corresponds to the rejection band whereas the latter corresponds to the passband.

Each component has the configuration shown in Fig. 5.39. The SAW ladder-type filters employ SAW resonators for either or both impedance elements. In the configuration, the image parameters are given by

$$Z_i/Z_o = 1 + Z_s/Z_p$$
$$Z_iZ_o = Z_sZ_p$$
$$\sinh^2\phi = Z_s/Z_p.$$

The passband where ϕ is purely imaginary appears at frequencies where $-1 \leq Z_s/Z_p \leq 0$. So if the capacitance is employed for Z_p or Z_s and the remaining is the resonator, the passband appears at frequencies between the resonance and antiresonance frequencies.

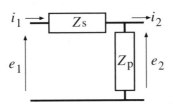

Fig. 5.39. Individual element for ladder-type filter

When the resonator is employed for both Z_p and Z_s and the antiresonance frequency ω_a^p of Z_p is chosen to coincide with the resonance frequency ω_r^s of Z_s, the achievable bandwidth is doubled. In addition, nulls are created at both sides of the passband. A typical frequency response of the ladder-type filter is shown in Fig. 5.40.

This figure is similar to Fig. 5.35, and the features described above are the same as those of the π-type filter described in Sect. 5.4.1. However, by the cascade-connection, we can improve the out-of-band rejection without

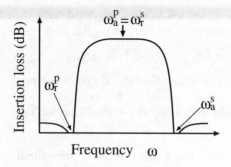

Fig. 5.40. Frequency response of ladder-type filter

decreasing the bandwidth provided that the resonant Qs of the resonators are sufficiently large. In fact, optimally designed ladder-type filters offer a relatively flat passband and sufficient out-of-band rejection.

When $0 < -Z_s/Z_p \ll 1$, (5.82) is simplified to

$$S_{21} \cong \frac{2}{2\cosh(N\phi) + \left(\sqrt{Z_p Z_s}/R_{in} + R_{in}/\sqrt{Z_p Z_s}\right)\sinh(N\phi)}$$
$$\cong \frac{2}{2 + N(Z_s/R_{in} + R_{in}/Z_p)}. \tag{5.83}$$

Thus $|S_{21}| \cong 1$ provided $\phi \cong 0$ and $Z_p Z_s = -R_{in}^2$.

Let us consider the case where both Z_p and Z_s are SAW resonators with the resonator Q of Q_r and capacitance ratio of γ and ω_r^s is equal to be ω_a^p. Then the width $2\delta\omega$ of the passband, identified by the relation where $-1 \le Z_p/Z_s \le 0$, is given by

$$\frac{2\delta\omega}{\omega_r^s} \cong \sqrt{1 + \frac{1}{\sqrt{\gamma(1+\gamma)(r^2+1)}}} - \sqrt{1 - \frac{1}{\sqrt{\gamma(1+\gamma)(r^2+1)}}}$$
$$\cong \frac{1}{\sqrt{\gamma(1+\gamma)(r^2+1)}}. \tag{5.84}$$

where $r = \sqrt{C_0^p/C_0^s}$. Then S_{21} at $\omega = \omega_r^s$ is given by

$$S_{21} \cong \frac{2}{2 + N(R_r^s/R_{in} - \omega^2 C_0^{p2} R_r^p R_{in})} \cong \frac{2}{2 + NrM^{-1}(\eta^{-1} + \eta)}, \tag{5.85}$$

where $\eta = R_{in}\omega\sqrt{C_0^s C_0^p}$ and $M = Q_r/\gamma$ is the figure of merit. Thus by designing $\eta = 1$, i.e., $R_{in}\omega\sqrt{C_0^s C_0^p} = 1$, $|S_{21}|$ takes a maximum value of $(1 + Nr/M)^{-1}$.

At the rejection band, ϕ is purely real. Then (5.82) becomes

$$
S_{21} \cong
\begin{cases}
\dfrac{4(\sqrt{r^2+1}-r)^N}{\sqrt{1+r^2}+1/\sqrt{1+r^2}+1/\eta+\eta} & (N : \text{odd}) \\[4mm]
\dfrac{4(\sqrt{r^2+1}-r)^N}{2+\sqrt{1+r^2}/\eta+\eta/\sqrt{1+r^2}} & (N : \text{even})
\end{cases}
$$

$$
= \frac{4(\sqrt{r^2+1}-r)^N}{2+\sqrt{1+r^2}+1/\sqrt{1+r^2}} \qquad (\text{when } \eta = 1). \tag{5.86}
$$

Thus, $|S_{21}|$ at the rejection band is simply determined by N and r. Figure 5.41 shows the calculated out-of-band rejection level as a function of r with N as a parameter. It is seen that the out-of-band rejection level is improved by choosing larger r and/or N. So r is usually chosen larger than 0.5.

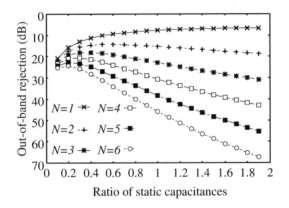

Fig. 5.41. r and N dependence of S_{21} in rejection band

Figure 5.42 shows the frequency response of S_{21} calculated by using (5.82) with r as a parameter. In the calculation, $\gamma = 10$, $\eta = 1$, $N = 5$, and $Q_r \to \infty$. It is seen that decreasing r enables us to increase the passband width and reduce the transition band width under the trade-off with deterioration of the out-of-band rejection level.

Figure 5.43 shows the Q_r dependence of the frequency response of S_{21}. In the calculation, $r = 0.4$, $\gamma = 10$, $\eta = 1$ and $N = 5$. With a decrease in Q_r, the insertion loss increases monotonically, and the large peaks at the passband edges diminish. The out-of-band characteristics are scarcely influenced by the finite Q_r. Note that r is also responsible for the insertion loss when Q_r is finite.

It is interesting to note that the coincidence between ω_r^s and ω_a^p is not necessary, and designing ω_r^s slightly larger than ω_a^p makes it possible to enlarge the bandwidth without badly affecting other characteristics [19]. Figure 5.44 shows the change in insertion loss and dip levels with the relative misfit

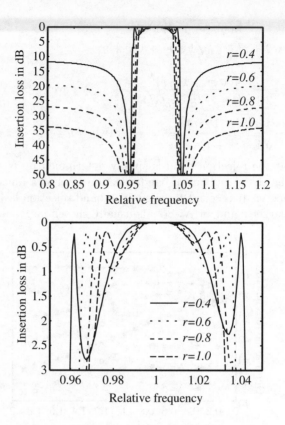

Fig. 5.42. r dependence of S_{21} for ladder-type filter

$\delta = (\omega_r^s - \omega_a^p)/\omega_c$ where $\omega_c = (\omega_r^s + \omega_a^p)/2$. When $\delta > 0$, $|S_{21}|$ becomes a maximum at $\omega_\pm = \omega_c(1 \pm \xi/2)$ instead of ω_r^s or ω_a^p where $\xi = \sqrt{\delta^2 + \delta\gamma^{-1}}$, and the increase in the minimum insertion loss with δ is very small. It is seen that by giving an appropriate δ, the effective bandwidth is increased and the transition bandwidth is effectively reduced. A negative δ simply reduces the passband width and increases the insertion loss.

Another factor we must pay attention to is the influence of parasitic circuit elements, such as stray capacitances, lead inductances, etc. Let us discuss this using the equivalent circuit for the one-port resonator shown in Fig. 5.1b. Because parallel-connected parasitic capacitance is equivalent to increasing the static capacitance C_0, it reduces the antiresonance frequency ω_a but does not influence the resonance frequency ω_r. Since this situation is equivalent to an increased capacitance ratio, the existence of parasitic capacitances simply deteriorates the achievable performance. On the other hand, a series-connected lead inductance decreases ω_r and scarcely affects ω_a, provided that Q_r is large. Since this situation is equivalent to a decreased capacitance ratio for

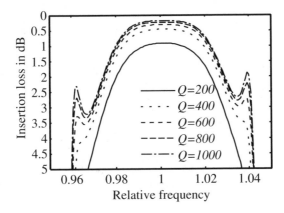

Fig. 5.43. Q_r dependence of S_{21} for ladder-type filter

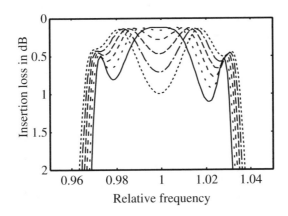

Fig. 5.44. Change in frequency response with relative frequency misfit δ. From upper traces, $\delta = 0, 0.025, 0.05, 0.075, 0.1, 0.125, 0.15$

this case, enhanced performance is achievable by taking the influence properly into account in the design [19]. Another interesting effect of lead inductances will be detailed in Sect. 8.2.2.

Figure 5.45 shows the frequency response of a ladder-type filter for GSM-Rx developed by Fujitsu Ltd. [20]. As a substrate, 42°YX-LiTaO$_3$ was employed [13]. The device was designed to minimize the insertion loss under trade-off with the out-of-band rejection. At the center frequency of 949 MHz, an insertion loss of 0.97 dB and -3 dB bandwidth of 51 MHz were achieved.

Figure 5.46 shows the frequency response of the ladder-type filter for a wireless LAN developed by Fujitsu Ltd. [21]. As a substrate, 42°YX-LiTaO$_3$ was employed [13]. This device was designed to maximize the absolute -5 dB bandwidth. At the center frequency of 2.45 GHz, an insertion

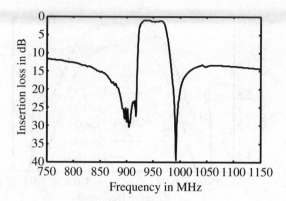

Fig. 5.45. Frequency response of ladder-type filter for GSM-Rx [20]

loss of 2.11 dB, out-of-band rejection level of 35 dB and absolute −5 dB bandwidth of 123 MHz were achieved.

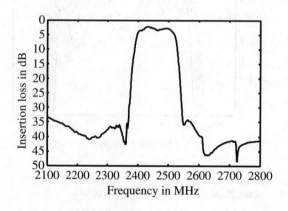

Fig. 5.46. Frequency response of ladder-type filter for wireless LAN [21]

Note that this device is mass-produced without giving any trimming in the fabrication process.

References

1. E.A. Ash: Surface Wave Grating Reflectors and Resonators, Digest of IEEE Microwave Symp. (1970) pp. 385–386.
2. E.A. Garber and A. Ballato (eds): Precision Frequency Control, Vol. 1, in Acoustic Resonators and Filters, Academic Press (1985).

3. D.T. Bell and R.C.M. Li: Surface-Acoustic-Wave Resonators, Proc. IEEE, **64** (1976) pp. 711–721.

4. E.J. Staples and R.C. Smythe: Surface Acoustic Wave Resonators on ST-Quartz, Proc. IEEE Ultrason. Symp. (1975) pp. 307–310.

5. W.H. Haydl, B. Dischler and P. Hiesinger: Multimode SAW Resonators – A Method to Study The Optimum Resonator Design, Proc. IEEE Ultrason. Symp. (1976) pp. 287–296.

6. Y. Yamamoto and S. Yoshimoto: SAW Transversely Guided Mode Spurious Elimination by Optimization of Conversion Efficiency Using W/W0 Electrode Structrue, Proc. IEEE Ultrason. Symp. (1998) pp. 229–234.

7. R.V. Schmidt and L.A. Coldren: Thin Film Acoustic Surface Waveguides on Anisotropic Media, IEEE Trans. Sonics and Ultrason., **SU-22**, 2 (1975) pp. 115.

8. K. Hashimoto, G. Endoh, M. Ohmaru, and M. Yamaguchi: Analysis of SAWs Obliquely Propagating under Metallic-Gratings with Finite Thickness, Jpn. J. Appl. Phys., **35** (1996) pp. 3006–3009.

9. Y. Ebata: Suppression of Bulk-Scattering Loss in SAW Resonator With Quasi-Constant Acoustic Reflection Periodicity, Proc. IEEE Ultrason. Symp. (1986) pp. 91–96.

10. G.K. Montress, T.E. Parker and D. Andres: Review of SAW Oscillator Performance, Proc. IEEE Ultrason. Symp. (1994) pp. 43–54.

11. T. Morita, Y. Watanabe, M. Tanaka and Y. Nakazawa: Wideband Low Loss Double Mode SAW Filters, Proc. IEEE Ultrason. Symp. (1992) pp. 95–104.

12. SAW device data sheet, Fujitsu Ltd., FAR-F5CE-947M50-D235.

13. O. Kawachi, G. Endoh, M. Ueda, O. Ikata, K. Hashimoto and M. Yamaguchi: Optimum Cut of LiTaO3 for High Performance Leaky Surface Acoustic Wave Filters, Proc. IEEE Ultrason. Symp. (1996) pp. 71–76.

14. M. Tanaka, T. Morita, K. Ono and Y. Nakazawa: Narrow Bandpass Filter Using Double-Mode SAW Resonators on Quartz, Proc. IEEE Freq. Contr. Symp. (1984) pp. 286–293.

15. J. Tsutsumi, O. Ikata and Y. Satoh: Transversely Coupled Resonator Filters with 0.1% Fractional Bandwidth in Quartz, Proc. IEEE Ultrason. Symp. (1996) pp. 65–69.

16. Y. Yamamoto and R. Kajihara: SAW Composite Longitudinal Mode Resonator (CLMR) Filters and Their Application to New Synthesized Resonator Filters, Proc. IEEE Ultrason. Symp. (1993) pp. 47–51.

17. S.N. Kondratiev, V.P. Plessky and M.A. Schwab: Compact Low Loss IF Balanced Bridge Filters, Proc. IEEE Ultrason. Symp. (1995) pp. 55–58.

18. S. Beaudin, S. Damphousse and T. Cameron: Shoulder Suppressing Technique for Dual Mode SAW Resonators, Proc. IEEE Ultrason. Symp. (1999) to be published.

19. O. Ikata, T. Miyashita, T. Matsuda, T. Nishikawa and Y. Satoh: Development of Low-Loss Band-Pass Filters Using SAW Resonators for Portable Telephone, Proc. IEEE Ultrason. Symp. (1992) pp. 112–115.

20. SAW device data sheet, Fujitsu Ltd., FAR-F5CJ-947M50-L211.

21. SAW device data sheet, Fujitsu Ltd., FAR-F6CE-2G4500-L2WA.

6. Selection of Substrate Material

This chapter describes a selection of substrate materials which determine the achievable performances of SAW devices. The desired properties for their use in SAW devices are discussed, along with the characterization of substrate materials by using the effective permittivity [1, 2]. Various characteristics of representative substrate materials are reviewed.

6.1 Substrate Material and Device Characteristics

6.1.1 Orientation

Selection of the substrate material is a determining factor for SAW device performance. Since piezoelectric materials commonly used as the substrate are intrinsically anisotropic, their properties are dependent upon not only the cut angle but also the SAW propagation direction.

In the SAW field, the coordinate system shown in Fig. 6.1 is commonly used. That is, x is taken parallel to the SAW propagation direction, z normal to the surface, and y parallel to the SAW wavefront. On the other hand, for each crystal structure, the crystal axes (X,Y,Z) are defined, and are identified by X-ray diffraction measurements.

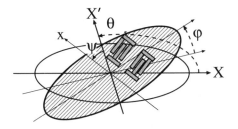

Fig. 6.1. Euler angles

To orient a material to the coordinate system (x, y, z), three successive rotation operations are required in general. Firstly, following the rotation by an angle θ about the Z axis, rotation by an angle ϕ about the X'-axis in Fig. 6.1 is applied. With these operations, the cut angle is specified. By rotation through an angle ψ about the Z'' ($= z$) axis, the propagation direction coincides with the x axis. The combination of these angles (θ, ϕ, ψ) is called the Euler angles, and are widely used to specify the cut angle and propagation direction.

By the way, the cut angles θ and ϕ are set by the mechanical stages, whose precision influences the device characteristics and price. In addition, when the propagation direction does not coincide with one of the crystal axes or their normal, beam steering occurs. For these reasons, single rotated cuts are desirable, and use of double rotated cuts is rare for SAW devices.

For a single rotated cut, the cut angle is often specified by the rotation angle from one of the crystal axes. For example, 128° rotated Y-cut X-propagation LiNbO$_3$, sometime written as 128°YX-LiNbO$_3$, indicates that the crystal is cut at the plane 128° rotated from the Y-plane about the X-axis; and X cut 112° off LiTaO$_3$, sometimes written as X-112°Y LiTaO$_3$, indicates that the propagation direction is rotated 112° from the Y-axis about the X-axis.

6.1.2 Influence of Substrate and Electrode Materials

SAW Velocity. The SAW velocity determines the resonance frequency. Usually, very fine lithography is required in the high-frequency range, and SAWs with fast velocity are desired to extend the upper limit of their applicable frequency range. On the other hand, an increase in device size becomes crucial in low-frequency applications, and SAWs with low velocity are desired to extend the lower limit.

For mass production, uniformity of the substrate is very important because variation of substrate properties directly influences device characteristics. The variation originates from that of the crystallographic properties such as lattice disorder and composition, and that of polarization if the materials are ferroelectric, such as LiNbO$_3$ and LiTaO$_3$. Since substrates employed for SAW devices are piezoelectric, the SAW velocity changes with the surface electrical boundary condition in the order of the electromechanical coupling factor. Thus, even if the variation in crystallographic properties is tiny, it may result in a large variation in the SAW velocity if the variation affects the polarization. Usually, SAWs with larger piezoelectricity exhibit larger variation in their velocity.

When one employs a layered structure where a piezoelectric thin film such as ZnO and AlN is deposited on the glass or sapphire substrate, the variation of the film thickness must be precisely controlled, adding to the material uniformity. So as to relax this requirement, the device structure should be designed so that changes in the SAW velocity with the wavelength, namely the frequency dispersion, are small.

Electromechanical Coupling Factor. The achievable insertion loss and the fractional -3 dB bandwidth are directly related to the electromechanical coupling factor. This is because the radiation conductance of IDTs is proportional to the electromechanical coupling factor as described in Sect. 3.3.1.

As will be shown in Sect. 6.2, the electromechanical coupling factor is dependent on the surface electrical boundary condition, i.e., the coupling factors K_{Sf}^2 and K_{Sm}^2 for the free and metallized surfaces, respectively, are given by [1]

$$K_{\mathrm{Sf}}^2 = -2\epsilon(\infty)\left[S_{\mathrm{f}}\frac{\partial\epsilon(S,\omega)}{\partial S}\bigg|_{S=S_{\mathrm{f}}}\right]^{-1}, \tag{6.1}$$

$$K_{\mathrm{Sm}}^2 = 2\epsilon(\infty)^{-1}\left[S_{\mathrm{m}}\frac{\partial\epsilon(S,\omega)^{-1}}{\partial S}\bigg|_{S=S_{\mathrm{m}}}\right]^{-1}, \tag{6.2}$$

where $\epsilon(S,\omega)$ is the effective permittivity [1, 2], and $S_{\mathrm{f}} = V_{\mathrm{f}}^{-1}$ and $S_{\mathrm{m}} = V_{\mathrm{m}}^{-1}$ are the slownesses of the SAW for the free and metallized surfaces, respectively. If these SAW velocities are sufficiently lower than the BAW velocities, K_{Sf}^2 and K_{Sm}^2 are almost equal to each other, and are estimated by

$$K^2 \cong (V_{\mathrm{f}}^2 - V_{\mathrm{m}}^2)/V_{\mathrm{f}}^2 \cong 2(V_{\mathrm{f}} - V_{\mathrm{m}})/V_{\mathrm{f}} \tag{6.3}$$

to good accuracy [3, 4]. Estimation of the electromechanical coupling factor by using (6.3) has been widely used, and this estimation is referred to as $2\Delta V/V$.

Temperature Stability. As an example, let us consider car radios. They must operate without adjustments under large variation of environment temperature from day time in hot summer to midnight in cold winter. Thus, even for consumer applications, the SAW characteristics are required to be invariant to temperature variation, for example, in a range from $-20\ ^\circ$C to $+60\ ^\circ$C.

The temperature dependence of the resonance frequency f_{r} is called the temperature coefficient of frequency (TCF), which is defined by

$$\mathrm{TCF} = f_{\mathrm{r}}^{-1}\frac{\partial f_{\mathrm{r}}}{\partial T}. \tag{6.4}$$

Since usually variation near room temperature is important, the gradient at $20\ ^\circ$C or $25\ ^\circ$C is employed. When $\mathrm{TCF} \cong 0$, f_{r} shows parabolic or cubic variation. Then,

$$\alpha^{(n)} = \frac{1}{f_{\mathrm{r}}n!}\frac{\partial^n f_{\mathrm{r}}}{\partial T^n} \tag{6.5}$$

are defined, and are called the n-th order temperature coefficients of frequency.

The temperature stability of SAW devices is often estimated by the temperature dependence of the time delay τ, which is called the temperature coefficient of delay (TCD). The TCD is evaluated by using equations equivalent to (6.4) and (6.5) where f_r is replaced by τ. Note that TCD $\cong$ $-$TCF if the frequency dispersion is small and the phase velocity is almost equal to the group velocity.

To calculate the TCD and TCF theoretically,

$$\text{TCF} = \text{TCV}_p - \alpha, \tag{6.6}$$
$$\text{TCD} = \alpha - \text{TCV}_g, \tag{6.7}$$

are used, where α is the thermal expansion coefficient for the SAW propagation direction, TCV_p is the temperature coefficient of the phase velocity, and TCV_g is that of the group velocity.

Usually materials with larger K^2 possess a worse TCD because a larger K^2 indicates that mechanical properties are more sensitive to perturbations.

Beam Steering. When substrate properties are not symmetric with respect to the propagation direction within the substrate surface, beam steering described in Sect. 1.1.2 occurs. Since the wavefront, namely, the equiphase plane, is not perpendicular to the power flow direction, the IDTs must be placed as shown in Fig. 6.2. This phenomenon does not directly affect device performance. However, if the power flow angle (PFA) is large, since a slight change in the propagation direction causes a considerable shift in the power flow angle, precise alignment must be given between the substrate orientation and device pattern. From this reason, substrates with large PFA are hard to use.

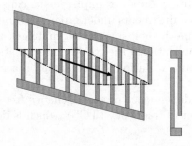

Fig. 6.2. IDT placement when beam steering exists

From the dependence of the phase velocity V_p on the Euler angle ψ, the PFA is estimated by using the relation

$$\text{PFA} = -V_p^{-1}\frac{\partial V_p}{\partial \psi}. \tag{6.8}$$

Propagation Loss. Materials with low propagation loss are desirable because it directly reflects the device insertion loss and/or Q. An intrinsic loss of the order of 10^{-3} dB/λ always exists due to thermal lattice vibration, surface roughness, leakage into air, etc. As described in Sect. 1.1.1, faster waves usually possess lower propagation loss than slower ones.

Leaky SAWs exhibit additional propagation loss due to coupling with the shear BAW. However, if the coupling is negligible, the propagation loss becomes very small. It is interesting to note that a propagation loss lower than the nonleaky SAW might be achievable by the use of leaky SAWs because of their larger velocities.

Note that in leaky SAWs, the coupling with the shear BAW is also dependent upon the surface boundary condition and other perturbations, such as temperature. Since the coupling is lost at the optimal condition and the propagation loss is always positive, it changes parabolically against any perturbations near the optimal condition [5, 6]. Thus, it is very important to reduce not only the achievable insertion loss but also variation through the fabrication process and/or temperature.

Permittivity ϵ. The permittivity of substrate materials is also important because it determines the impedance of the IDT which is usually designed to be about 50 Ω for impedance matching with peripheral circuits. Although the impedance can also be adjusted by the aperture of the IDT, a large aperture enlarges the device size and a small one may cause serious diffraction.

BAW Characteristics. As described in Sect. 4.3.2, suppression of spurious responses due to SSBWs [7] is crucial for transversal filters. If a spurious SSBW response exists near the SAW resonance frequency, it deteriorates the rejection band level and cannot be suppressed by IDT design. We may be able to use the multi-strip coupler (MSC) [8] which will be discussed in Sect. 7.1.1 under a trade-off with increased device size. So, the most practical solution is to use substrate materials with weak SSBW excitation. Note that, for resonators, although the SSBW radiation may reduce Q, the reduction is not significant for the fundamental resonance because the resonance frequency of the SSBW is usually somewhat different from that of the SAW. In addition, since resonance of the SSBW is weak, it scarcely affects the frequency response.

6.2 Evaluation of Acoustic Properties by Effective Permittivity

6.2.1 Effective Permittivity

In the previous sections, it was shown without giving any theoretical background that various parameters required for SAW device simulation are obtained by using the effective permittivity $\epsilon(S, \omega)$ [1, 2, 4]. Here we will discuss the evaluation of the parameters in detail. Its mathematical fundamentals are given in Appendix A.

Since the electrode thickness is usually much thinner than the SAW wavelength, the charge distribution $q(x_1)$ on the IDTs can be considered to be located on the substrate surface. Let us define the Fourier transforms $Q(\beta)$ and $\Phi(\beta)$ of $q(x_1)$ and the voltage at the substrate surface $\phi(x_1)$, respectively:

$$Q(\beta) = \frac{1}{2\pi} \int_{-\infty}^{+\infty} q(x_1) \exp(+j\beta x_1) dx_1, \tag{6.9}$$

$$\Phi(\beta) = \frac{1}{2\pi} \int_{-\infty}^{+\infty} \phi(x_1) \exp(+j\beta x_1) dx_1. \tag{6.10}$$

Then the effective permittivity $\epsilon(\beta, \omega)$ [2] is defined by

$$\epsilon(S, \omega) = \frac{Q(\beta)}{|\beta| \Phi(\beta)}, \tag{6.11}$$

where $S = \beta/\omega$ corresponds to the slowness along the surface.

If the substrate is not a layered structure, due to lack of frequency dispersion, $\epsilon(S, \omega)$ is dependent only on S.

The inverse transform of (6.10) gives

$$\phi(x_1) = \int_{-\infty}^{+\infty} \Phi(\beta) \exp(-j\beta x_1) d\beta. \tag{6.12}$$

By using (6.9) and (6.11), (6.12) can be rewritten in a convolution form of

$$\phi(x_1) = \int_{-\infty}^{+\infty} G_f(x_1 - x_1') q(x_1') dx_1', \tag{6.13}$$

where

$$G_f(x_1) = \frac{1}{2\pi} \int_{-\infty}^{+\infty} \frac{1}{|\beta| \epsilon(S, \omega)} \exp(-j\beta x_1) d\beta \tag{6.14}$$

is the Green function, that is to say, the spatial impulse response, which corresponds to $\phi(x_1)$ when a line source $q(x_1) = \delta(x_1)$ is placed on the free surface.

Since the product of voltage and current densities gives the power density, the radiated acoustic power P is given by

$$P = \frac{W}{2} \Re \left[\int_{-\infty}^{+\infty} \phi(x_1) \dot{q}(x_1)^* dx_1 \right] = \pi \omega W \Im \left[\int_{-\infty}^{+\infty} \frac{|Q(S\omega)|^2}{|S| \epsilon(S, \omega)} dS \right], \tag{6.15}$$

where W is the aperture.

On the other hand, for a discussion of SAWs on a metallized surface, we define the electric field $e(x_1)$ parallel to the surface by

$$e(x_1) = -\frac{\partial \phi(x_1)}{\partial x_1}, \tag{6.16}$$

and the total charge $q(x_1)$ by

$$h(x_1) = \int_{-\infty}^{x_1} q(x_1)dx_1. \tag{6.17}$$

Then the following relation holds between their Fourier transforms $E(\beta)$ and $H(\beta)$:

$$\epsilon(S, \omega) = \frac{|\beta|H(\beta)}{E(\beta)}. \tag{6.18}$$

Following to the procedure mentioned above, we may obtain its convolution form of

$$h(x_1) = \int_{-\infty}^{+\infty} G_m(x_1 - x_1')e(x_1')dx_1', \tag{6.19}$$

where

$$G_m(x_1) = \frac{1}{2\pi} \int_{-\infty}^{+\infty} \frac{\epsilon(S, \omega)}{|\beta|} \exp(-j\beta x_1)d\beta \tag{6.20}$$

is the Green function. It corresponds to $h(x_1)$ when an infinitesimally narrow slot $e(x_1) = \delta(x_1)$ is placed on the metallized surface, and the radiated acoustic power P is given by

$$P = \frac{W}{2}\Re\left[\int_{-\infty}^{+\infty} \phi(x_1)\dot{q}(x_1)^* dx_1\right]$$

$$= \pi\omega W\Im\left[\int_{-\infty}^{+\infty} \frac{\epsilon(S, \omega)^*|E(S\omega)|^2}{|S|}dS\right]. \tag{6.21}$$

Figure 6.3 shows the effective permittivity for $128°\text{YX-LiNbO}_3$. In the figure, the horizontal axis is $V = S^{-1}$. In $\epsilon(S)^{-1}$, there exists a pole at

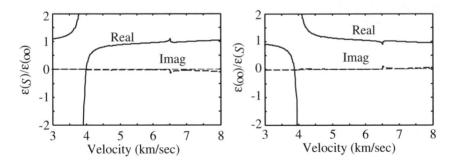

Fig. 6.3. Effective permittivity for $128°\text{YX-LiNbO}_3$

$V = 3992$ m/sec. Since the pole indicates that a nonzero $\Phi(\beta)$ can exist even though no wave source exists, i.e., $Q(\beta) = 0$, this indicates the existence of a SAW on the free surface, which may come from infinitely far away. On the other hand, there exists a pole at $V = 3886$ m/sec in $\epsilon(S)$. This indicates that a nonzero $Q(\beta)$ can exist even though no wave source exists, i.e., $\Phi(\beta) = 0$, and implies the existence of a SAW on the metallized surface.

Applying the residue theorem to (6.14) and (6.20), the Green functions G_{Sf} and G_{Sm} associated with the SAW radiation are given by [1, 2, 4]

$$G_{Sf}(x_1) = j\frac{K_{Sf}^2}{2\epsilon(\infty)}\exp(-j\omega S_{Sf}|x_1|), \tag{6.22}$$

$$G_{Sm}(x_1) = -j\frac{K_{Sm}^2\epsilon(\infty)}{2}\exp(-j\omega S_{Sm}|x_1|), \tag{6.23}$$

where $S_{Sf} = V_{Sf}^{-1}$, $S_{Sm} = V_{Sm}^{-1}$, and K_{Sf}^2 and K_{Sm}^2 are the electromechanical coupling factors given by (6.1) and (6.2), respectively. Note that

$$\epsilon(\infty, \omega) = \epsilon_0 + \sqrt{\epsilon_{11}^T\epsilon_{33}^T - \epsilon_{13}^{T2}}, \tag{6.24}$$

corresponds to the permittivity required for the calculation of the static capacitance.

For SAW excitation by an IDT, application of the residue theorem to (6.15) or (6.21) gives the radiation conductance $G_S(\omega)$ as

$$G_S(\omega) = 2\pi^2\omega WV^{-2}K_{Sf}^2\epsilon(\infty)^{-1}|Q(\omega S_{Sf})|^2, \tag{6.25}$$

or

$$G_S(\omega) = 2\pi^2\omega WV^{-2}K_{Sm}^2\epsilon(\infty)|E(\omega S_{Sm})|^2, \tag{6.26}$$

where V is the voltage applied to the IDT.

By the way, in Fig. 6.3, the imaginary part appears in $\epsilon(S)^{-1}$ and $\epsilon(S)$ for $S^{-1} > 4070$ m/sec. As is clear from (6.15) and (6.21), the imaginary part corresponds to the power radiation into the bulk from the IDT, namely excitation of the BAW.

From (6.15) and (6.21), the radiation conductance $G_B(\omega)$ for the BAW is given by [1]

$$G_B(\omega) = 2\pi\omega WV^{-2}\int_{-S_B}^{+S_B}\Im[\epsilon(S)^{-1}]\frac{|Q(S\omega)|^2}{|S|}dS, \tag{6.27}$$

or

$$G_B(\omega) = 2\pi\omega WV^{-2}\int_{-S_B}^{+S_B}\Im[\epsilon(S)^*]\frac{|E(S\omega)|^2}{|S|}dS. \tag{6.28}$$

Then substitution of (6.25) and (6.26) gives

$$G_B(\omega) = \frac{1}{\pi}\int_{-S_B}^{+S_B}\Im\left[\frac{\epsilon(\infty)}{\epsilon(S)}\right]\frac{G_S(\omega S/S_f)}{|S|K_{Sf}^2}dS, \tag{6.29}$$

and

$$G_B(\omega) = \frac{1}{\pi}\int_{-S_B}^{+S_B}\Im\left[\frac{\epsilon(S)^*}{\epsilon(\infty)}\right]\frac{G_S(\omega S/S_m)}{|S|K_{Sm}^2}dS. \tag{6.30}$$

Hence, provided the effective permittivity is given, we can estimate the radiation conductance for the BAW quantitatively by the convolution-like operation between the radiation conductance $G_S(\omega)$ for the SAW and $\Im[\epsilon(S)]$ (see Fig. 6.4).

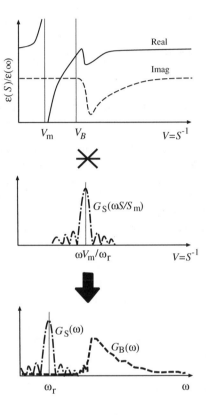

Fig. 6.4. Relation between effective permittivity and radiation conductance for a BAW

Then Fig. 6.3 suggests that, in 128°YX-LiNbO$_3$, not so strong radiation of the L-type BAW occurs at a frequency $1.7(\cong 752/3886)$ times higher than the resonance frequency for the SAW. On the other hand, radiation of the slow-shear BAW is negligible.

Figure 6.5 shows the effective permittivity for AT-cut quartz. A pole due to SAW excitation exists at 3151 m/sec, and its width is quite small due to the very weak piezoelectricity. Strong peaks due to BAW radiation are seen at 5101 m/sec and 5746 m/sec. At $V \cong 5746$ m/sec, the imaginary part changes rapidly, and is asymmetric with respect to $S = V^{-1}$. Since this represents a cut-off nature, we can identify this as the SSBW radiation. On the other hand, the imaginary part at $V \cong 5101$ m/sec is symmetric with respect to V.

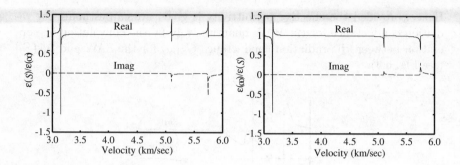

Fig. 6.5. Effective permittivity for AT-cut quartz

This indicates that the peak is due to the complex pole, and represents the SAW with BAW leakage, i.e, a leaky SAW. Although its propagation loss is small, the electromechanical coupling is also very small. Spurious responses due to the leaky SAW and the SSBW are often observed for SAW devices employing AT-cut or ST-cut quartz.

Next, let us estimate the values which characterize the SSBW properties [9]. Assume that $\epsilon(S)$ at $S \cong S_B$ can be approximately represented in the form of

$$|S|\epsilon(S) \cong S_B \times c \frac{\sqrt{2(S/S_B - 1)} + a}{\sqrt{2(S/S_B - 1)} + b} \tag{6.31}$$

where a and b are expansion coefficients, and are numerically evaluated from $\epsilon(S)$ and $\epsilon(S)^{-1}$, respectively, by expanding them as a polynomial in $\sqrt{2(S/S_B - 1)}$ at $S \cong S_B$.

By using this, one may obtain the Green function $G_{Bf}(x_1)$ for the SSBW on the free surface as

$$G_{Bf}(x_1) = -\frac{K_{Bf}^2}{2\pi\epsilon(\infty)} H(S_B \omega x_1) U(x_1/x_{cf}), \tag{6.32}$$

where

$$K_{Bf}^2 = \epsilon(\infty)(a - b)/c \tag{6.33}$$

corresponds to the electromechanical coupling factor for the BAWs, and

$$H(\theta) = \sqrt{2/\pi\theta} \exp(-j\theta - j\pi/4) \cong H_0^{(2)}(\theta), \tag{6.34}$$

where $H_0^{(2)}(\theta)$ is the 0th-order Hankel function of the second kind, and

$$U(r) = \frac{2}{\sqrt{\pi}} \int_0^1 \frac{t^2 \exp(-t^2)}{t^2 - jr} dt, \tag{6.35}$$

whose behavior has already been discussed in Sect. 4.3.2. In addition,

$$x_{cf} = a^{-2}/2S_B \omega \tag{6.36}$$

is the critical propagation length representing how the radiated SSBW amplitudes (on the substrate surface) attenuate with propagation. When $x_1 \ll x_{cf}$, the attenuation is simply proportional to $x_1^{-0.5}$ like the DBAWs. For $x_1 \gg x_{cf}$, because of the effect of surface boundary conditions, the amplitude attenuation of the SSBW on the substrate surface becomes more significant, typically proportional to $x_1^{-1.5}$.

The Green function $G_{Bm}(x_1)$ for the metallized surface is given by

$$G_{Bm}(x_1) = K_{Bm}^2 \epsilon(\infty) H(S_B \omega x_1) U(x_1/x_{cm}), \tag{6.37}$$

where

$$K_{Bm}^2 = \epsilon(\infty)^{-1} c(b-a), \tag{6.38}$$

$$x_{cm} = b^{-2}/2S_B\omega. \tag{6.39}$$

6.2.2 Approximate Expressions

At the SAW velocities $V_{Sf} = S_{Sf}^{-1}$ and $V_{Sm} = S_{Sm}^{-1}$ on the free and metallized surfaces, the relations $\epsilon(S_{Sf}, \omega) = 0$ and $\epsilon(S_{Sm}, \omega)^{-1} = 0$ hold. Then, it is expected that $\epsilon(S, \omega)$ can be approximated by [4]

$$\epsilon(S, \omega) \cong \epsilon(\infty) \frac{S^2 - S_{Sf}^2}{S^2 - S_{Sm}^2}. \tag{6.40}$$

Under this approximation, (6.1) and (6.2) give

$$K_{Sf}^2 \cong -\frac{S_{Sf}^2 - S_{Sm}^2}{S_{Sf}^2} = \frac{V_{Sf}^2 - V_{Sm}^2}{V_{Sm}^2}, \tag{6.41}$$

$$K_{Sm}^2 \cong \frac{S_{Sm}^2 - S_{Sf}^2}{S_{Sm}^2} = \frac{V_{Sf}^2 - V_{Sm}^2}{V_{Sf}^2}. \tag{6.42}$$

Thus both K_{Sf}^2 and K_{Sm}^2 are mostly equal to

$$K_V^2 = \frac{V_{Sf}^2 - V_{Sm}^2}{V_{Sf}^2}. \tag{6.43}$$

Since the estimation of K_V^2 requires only two SAW velocities which can be determined both experimentally and theoretically, it is widely adopted for the evaluation of the electromechanical coupling factor.

Next let us consider a leaky SAW. Since $\epsilon(S, \omega)$ and $\epsilon^{-1}(S, \omega)$ are continuous and smooth at $S \cong S_{Sf}$ and $S \cong S_{Sm}$, respectively, we can estimate approximate values of $\Re(S_{Sf}) \equiv \hat{S}_{Sf}$ and $\Re(S_{Sm}) \equiv \hat{S}_{Sm}$ by $\Re[\epsilon(\hat{S}_{Sf}, \omega)] \cong 0$ and $\Re[\epsilon(\hat{S}_{Sm}, \omega)^{-1}] \cong 0$, respectively. By using the determined $\hat{S}_{Sf}$ and $\hat{S}_{Sm}$, we can estimate K_{Sf}^2 and K_{Sm}^2 by using (6.1) and (6.2), respectively, with high accuracy provided that their propagation loss is not too large.

In addition, by using the approximate expression given by (6.40), the propagation losses α_f and α_m of leaky SAWs for free and metallized surfaces, respectively, are estimated by

$$\alpha_f = 27.287 K_{Sf}^2 \Im[\epsilon(\infty)/\epsilon(\hat{S}_f)^*]^{-1} \quad (dB/\lambda), \qquad (6.44)$$

$$\alpha_m = 27.287 K_{Sm}^2 \Im[\epsilon(\hat{S}_m)/\epsilon(\infty)]^{-1} \quad (dB/\lambda), \qquad (6.45)$$

to sufficient accuracy.

6.3 Single Crystals

6.3.1 Quartz

The hydrothermal growth of an α-quartz crystal has already been well established, and high-quality substrates are commercially available at reasonable price. The variation of material properties with lots and wafers is quite small, and the material constants with sufficient accuracy have been reported [10, 11], which enables us to perform precise device simulations.

Note that since α-quartz is paraelectric, the dielectric constant and piezoelectricity are small. Thus its application is limited to the use for narrow-bandwidth devices, and the choice of substrate orientation with high temperature stability is crucial.

Figure 6.6 shows the velocity V, electromechanical coupling factor K^2 and temperature coefficient of velocity TCV for a nonleaky SAW on a metallized surface of rotated Y-cut quartz. Since the piezoelectricity of quartz is small, the properties are scarcely affected by the surface electrical boundary condition.

In the substrate, the TCV is always positive for all rotation angles. Note that most materials have negative TCV. Since the thermal expansion coefficient for the X direction is 13.71 ppm/°C, it is seen that a zero TCF can be realizable by using 42° Y-cut quartz. This is called ST-cut quartz, and is frequently used to realize highly stable devices [12]. Note that the optimal angle is dependent upon the electrode material and thickness, and it sometimes reaches to AT-cut (36°YX-) quartz, which is widely used for highly stable BAW resonators. The effective permittivity for the substrate has already been shown in Fig. 6.5.

Figure 6.6 also shows the velocity V, electromechanical coupling factor K^2, propagation loss α and temperature coefficient of velocity TCV for a leaky SAW on rotated Y-cut quartz. At 105.7°, zero TCF is achieved, and this substrate is called LST-cut quartz [13]. The effective permittivity for the substrate is shown in Fig. 6.7. At $V = 3992$ m/sec, there is a complex pole due to the leaky SAW, which associates a very narrow peak of the imaginary part corresponding to BAW radiation.

When the propagation direction is rotated by 90° from the X-axis on AT-cut quartz, the L and SV components decouple with the piezoelectricity. Due to the small piezoelectricity, SH-type SAWs do not exist, and only an SH-type SSBW is electrically active. This wave is very fast ($\cong 5000$ m/sec) and possesses zero TCF [7]. Figure 6.8 shows the effective permittivity for the

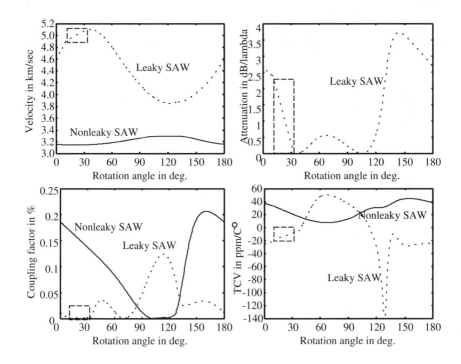

Fig. 6.6. Properties of nonleaky and leaky SAWs on rotated Y-cut quartz. The dashed box indicates that SAW properties were not determined accurately due to very small piezoelectricity

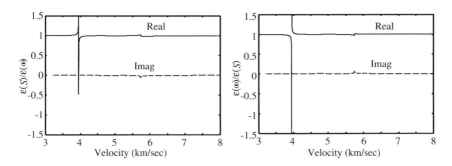

Fig. 6.7. Effective permittivity for LST-cut quartz

substrate. It is seen that only the SSBW is excited. Since the piezoelectricity of quartz is small, SSBW excitation characteristics are mostly independent of the surface electrical boundary condition.

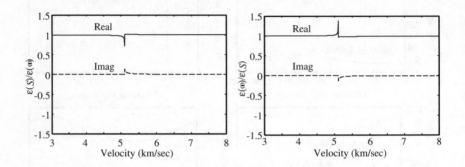

Fig. 6.8. Effective permittivity for AT-cut 90°-off quartz

It is known that the SSBW is guided by applying a grating on the propagation surface. This type of wave is called a surface transverse wave (STW) [14, 15]. The STW will be discussed in Sect. 8.2.2.

6.3.2 LiNbO$_3$

LiNbO$_3$ is ferroelectric and grown by the Czochralski method. In addition to a very large dielectric constant and piezoelectricity, LiNbO$_3$ possesses a large pyroelectricity and electro-optic effect. It has been applied not only to SAW devices but also to optical devices and microwave sensors, etc.

This material is also commercially available, and material constants with reasonable accuracy have been reported [16–19]. Note that there are considerable differences among the reported material constants, and the latest ones exhibit larger piezoelectricity than former ones. This may be due to the improvement of the polarization associated with that of the crystal uniformity. So it seems better to use the latest material constants for numerical analysis.

Figure 6.9 shows the velocity V, electromechanical coupling factor K^2 and temperature coefficient of velocity TCV for a nonleaky SAW on rotated Y-cut LiNbO$_3$. Due to the large piezoelectricity, their properties are very sensitive to the surface electrical boundary condition. So those for both the free and metallized surfaces are shown in the figure. In this case, TCV is always negative, and TCF is worse than $-$TCV by the thermal expansion coefficient 15.4 ppm/°C.

Since temperature stability is not so sensitive to the rotation angle θ for this case, it seems most useful to choose $\theta \cong 130°$ due to its largest electromechanical coupling factor, up to 5.5%.

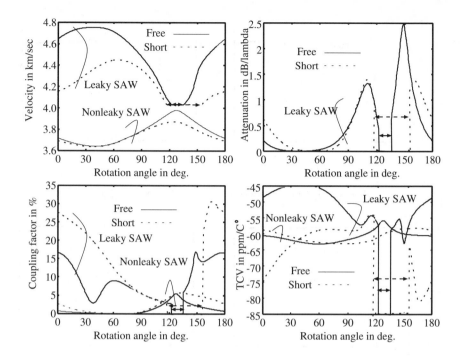

Fig. 6.9. Properties of leaky and nonleaky SAWs on rotated Y-cut LiNbO₃. A leaky SAW does not exist within the region indicated by arrows

Note that the radiation characteristics of the slow-shear SSBW change significantly with rotation angle in the region from 126° to 130°. Figures 6.10–6.12 show the change in the effective permittivity with rotation angle. As shown in Fig. 6.9, a leaky SAW does not exist at these angles. However, the SSBW which is slightly faster than the nonleaky SAW is strongly excited. At 128°, the SSBW radiation is suppressed anomalously since the slow-shear component decouples with the piezoelectricity. For this reason, 128°YX-LiNbO₃ [20] has been widely used. The effective permittivity for the substrate has already been shown in Fig. 6.3.

Figure 6.9 also shows the velocity V, electromechanical coupling factor K^2, propagation loss α and temperature coefficient of velocity TCV for the leaky SAW on rotated Y-cut LiNbO₃. Although a zero TCF cannot be realized, the electromechanical coupling factor becomes large at smaller rotation angels. At 41° and 64°, the propagation loss becomes almost zero, respectively, for the free and metallized surfaces [21]. These substrates for highly piezoelectric leaky SAWs have been applied to SAW devices for mobile communications, and are commercially available. Leaky SAWs will be discussed again in detail in Chap. 8.

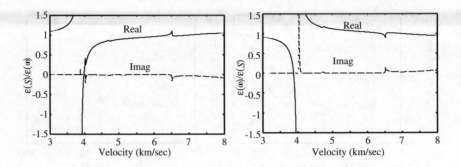

Fig. 6.10. Effective permittivity for 125°YX-LiNbO₃

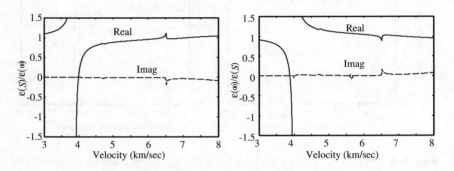

Fig. 6.11. Effective permittivity for 128°YX-LiNbO₃

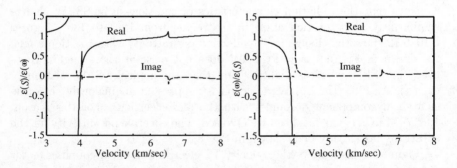

Fig. 6.12. Effective permittivity for 130°YX-LiNbO₃

When the rotation angle is about $0°$, a very large electromechanical coupling factor up to 25% is obtained although the propagation loss is large. The leaky SAW can be transformed to a nonleaky SAW by depositing thick Au gratings and reducing the SAW velocity, and this technique was applied for the development of resonators with low capacitance ratio [22].

6.3.3 LiTaO$_3$

LiTaO$_3$ is also ferroelectric and grown by the Czochralski method. Added to the large dielectric constant and piezoelectricity, LiTaO$_3$ possesses the pyroelectricity and electro-optic effect. Although these values are smaller than those of LiNbO$_3$, their temperature stability is better. It has been widely applied for TV-IF filters [23, 24], etc., and material constants with reasonable accuracy has been reported [16, 17, 19].

Figure 6.13 also shows the velocity V, electromechanical coupling factor K^2, propagation loss α and temperature coefficient of velocity TCV for leaky and nonleaky SAWs on rotated Y-cut LiTaO$_3$. Properties for both the free and metallized surfaces are shown. In this case, the thermal expansion coefficient for the X direction is 16.1 ppm/$°$C.

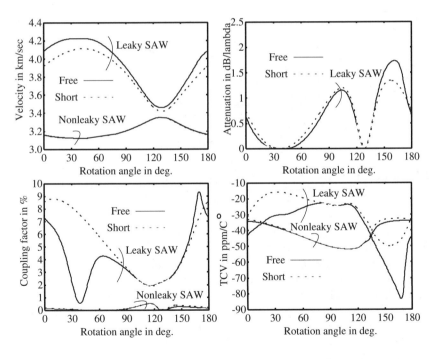

Fig. 6.13. Leaky and nonleaky SAWs on rotated Y-cut LiTaO$_3$

The rotation angle dependence for LiTaO$_3$ is qualitatively similar to that of LiNbO$_3$ except for two significant differences: (1) the electromechanical coupling factor for the nonleaky SAW is small, and (2) a leaky SAW exists for all rotation angles. Although no interesting cut exists for a nonleaky SAW on rotated Y-cut LiTaO$_3$, a leaky SAW at 36° possesses a large electromechanical coupling factor with negligible propagation loss [25]. This highly piezoelectric leaky SAW cut has been widely used in mobile communications, and is commercially available. Its behavior will again be discussed in Chap. 8.

Figure 6.14 shows the phase velocity of SAWs on X-cut LiTaO$_3$ when the propagation direction is rotated from the Y axis. Added to the nonleaky Rayleigh-type SAW, there exists a leaky SAW and a nonleaky SH-type SAW for a limited range of rotation angle [26]. When the propagation direction is shifted from the Y direction, there exists a nonzero power flow angle which is given by the gradient of the phase velocity with respect to the rotation angle (see (6.8)). Properties on the metallized surface are shown in the figure. Since the leaky SAW and nonleaky SH-type SAW are guided predominantly by the surface electrical boundary condition, they exist within a very narrow region of the rotation angle for the free surface.

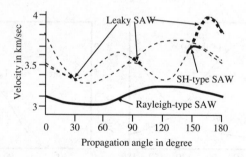

Fig. 6.14. Phase velocities of nonleaky Rayleigh-type SAW, leaky SAW and non-leaky SH-type SAW on X-cut LiTaO$_3$

Figures 6.15 and 6.16 show, respectively, the electromechanical factor and TCF for the Rayleigh-type SAW on X-cut LiTaO$_3$. Although no interesting orientation exists for the SH-type SAW, a relatively small TCF and large electromechanical factor are simultaneously achieved when the rotation angle is about 110°. In addition, due to the lack of beam steering and sufficient suppression of the spurious BAW response, X-cut 112°Y-off (X-112°Y) LiTaO$_3$ is widely used as a substrate for Rayleigh-type SAWs.

Figure 6.17 shows the effective permittivity for the substrate. There is a pole at 3200 m/sec corresponding to the SAW excitation, and its electromechanical coupling factor K^2 is about 0.72%. As is clear from the figure, the

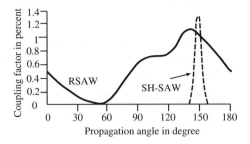

Fig. 6.15. Propagation direction dependence of electromechanical coupling factor for X-cut LiTaO₃

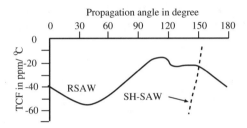

Fig. 6.16. Propagation direction dependence of temperature coefficient of frequency for X-cut LiTaO₃

spurious response due to the slow-shear SSBW is very small. Added to this, the TCF is relatively small ($\cong 18$ ppm/°C).

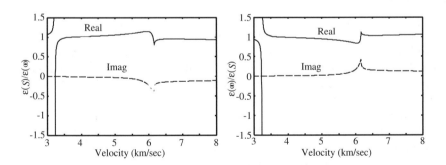

Fig. 6.17. Effective permittivity for X-112°Y LiTaO₃

6.3.4 Li$_2$B$_4$O$_7$

Although Li$_2$B$_4$O$_7$ [27] is paraelectric, it possess a relatively large piezoelectricity and zero TCF, and has been used for resonators [28]. The material
belongs to the 4mm crystal class, and materials in this class often exhibit
anomalously large anisotropy. For example, the thermal expansion coefficient
for the Z direction is negative at room temperature. Material constants with
reasonable accuracy have been reported [27, 29].

Figure 6.18 shows the effective permittivity for 45°YZ-Li$_2$B$_4$O$_7$, which
offers zero TCF at room temperature. There exists a pole at 3391 m/sec due
to the nonleaky SAW, whose electromechanical coupling factor is about 1.0%.
Strong SSBW excitation occurs at above 3777 m/sec, which deteriorates the
rejection band level near the passband. From this reason, this substrate has
not been applied for transversal filters but for resonators.

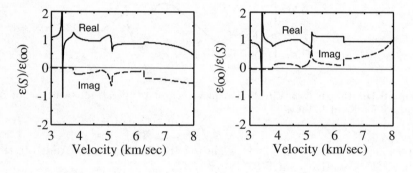

Fig. 6.18. Effective permittivity for 45°YZ-Li$_2$B$_4$O$_7$

Figure 6.19 shows the effective permittivity for (0, 47.3°, 90°) Li$_2$B$_4$O$_7$.
Added to a pole at 3186 m/sec due to the nonleaky SAW, there exists another
pole at 6635 m/sec due to the leaky SAW. In this case, the SH component is
isolated from the other components, and both SAWs are composed of L and
SV components. In the rotation angle, the effective Poisson ratio is small, and
the coupling between the components becomes small. The nonleaky SAW is
SV-type whereas the leaky SAW is L-type. For this reason, the leaky SAW
is called a longitudinal-type leaky SAW [30].

Figure 6.20 shows the Al thickness h dependence of the phase velocity,
electromechanical coupling factor and propagation loss for a longitudinal-
type leaky SAW on (0, 47.3°, 90°) Li$_2$B$_4$O$_7$. With an increase in h, the phase
velocity decreases monotonically and the propagation loss becomes very small
at $h/\lambda \cong 0.03$. In this configuration, low loss devices can be realized in the
high-frequency range [31].

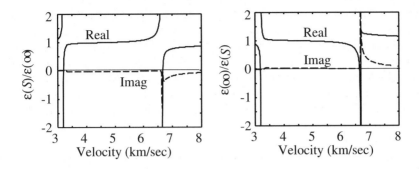

Fig. 6.19. Effective permittivity for $(0, 47.3°, 90°)$ $Li_2B_4O_7$

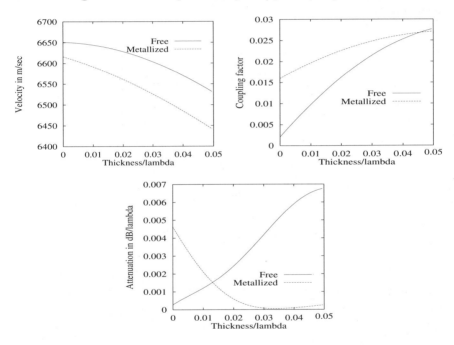

Fig. 6.20. Longitudinal-type leaky SAW on $(0, 47.3°, 90°)$ $Li_2B_4O_7$

6.3.5 Langasite

Recently langasite $(La_3Ga_5SiO_{14})$ [32] has been paid much attention because of its temperature stability comparable to ST-cut quartz and relatively large piezoelectricity. Like quartz, langasite belongs to the crystal class 32, and is paraelectric. Presently, large size crystals with high quality are successfully grown by the Czochralski method [33, 34]. Although material constants were reported by several groups [34, 35, 36], there still exists differences among them.

Many groups have tried to find optimal substrate orientations. Unfortunately zero TCF is not obtained on simple cut angles and/or propagation directions. As far as the author knows, a nonleaky SAW on $(0°, 140°, 24°)$ (and its neighborhood) seems most promising [34, 37]. It exhibits zero TCF around room temperature and an electromechanical coupling factor of 0.36%. It should be noted that the cut also exhibits the natural unidirectionality [38] described in Sect. 3.1.2.

Figure 6.21 shows the effective permittivity for the cut. There is a pole at 2742 m/sec due to the nonleaky SAW. Although the velocity of the slow-shear SSBW is relatively close (2857 m/sec) to the SAW velocity, its amplitude is relatively weak.

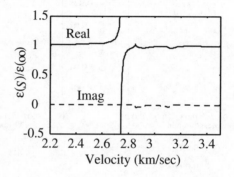

Fig. 6.21. Effective permittivity for $(0°, 140°, 24°)$ $La_3Ga_5SiO_{14}$

Figure 6.22 shows the frequency response of the IF SAW filter for IS-95 fabricated on the substrate. The device was developed by Mitsubishi Material Co. Ltd. [34]. Owing to the large dielectric constant $\epsilon(\infty) = 29.1\epsilon_0$ and small SAW velocity, the tip can be mounted into a ceramic package of $15.4 \times 6.5 \times 2.0$ mm^3. In addition, due to the large K^2, the input impedance of the device was six times smaller than that of the device on ST-cut quartz.

6.4 Thin Films

SAW devices are also realized by using piezoelectric thin films such as zinc oxide (ZnO) and aluminum nitride (AlN). ZnO has already been used for the development of TV-IF filters [39, 40] and front-end filters in the 1.5 GHz range for mobile communications [41].

In layered structures, SAW properties significantly depend on allocation of the IDTs and counter-electrode. It is interesting to note that the achievable

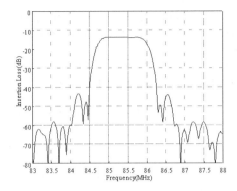

Fig. 6.22. IF SAW filter for IS-95 on $(0°, 140°, 24°)$ $La_3Ga_5SiO_{14}$ [34]

electromechanical coupling factor in the layered structure is sometimes larger than that of individual materials under proper combination of materials and film thickness. This is because the structure can be designed so that the SAW field concentrates on the piezoelectric film, and the IDTs can be placed appropriately for efficient SAW excitation.

Let us consider the structures shown in Fig. 6.23. In configurations (c) and (d) shown in Fig. 6.23, the crossed electric field is dominant due to the existence of the counter-electrode. On the other hand, the in-line field is dominant in configurations (a) and (b). Then the electromechanical coupling factor is determined according to which direction of the electric field the piezoelectricity of the substrate is most sensitive.

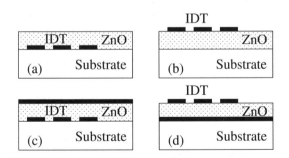

Fig. 6.23. Counter-electrode and IDT

A combination of substrate materials with opposite TCF enables us to realize superior temperature stability. On the other hand, a combination with high-speed materials, such as sapphire and diamond, offers very fast SAW velocities, and much effort has been paid for the application to high-frequency SAW devices [41, 42].

Although deposition of high-quality piezoelectric thin films is crucial to realizing this device structure, related technologies are too wide and deep to be included in this book.

In layered structures, since piezoelectric thin films have comparable thickness with the SAW wavelength, the electromechanical coupling factor and phase velocity are frequency dispersive. The frequency dispersion makes the group delay deviate and reduces the frequency bandwidth. Note that since the SAW properties are determined by the ratio between the physical dimension and wavelength, the existence of frequency dispersion results in a variation of device characteristics by a slight variation in the film thickness. In order to guarantee sufficient production yield, structures with large frequency dispersion are hard to use for mass production.

Figure 6.24a shows the phase velocity of a Rayleigh-type SAW on a ZnO/Pyrex-glass structure. The horizontal axis is the ratio between the ZnO thickness h and the SAW wavelength λ. In the analysis, the C-axis of ZnO is assumed to be aligned normal to the substrate surface. Due to the relatively small piezoelectricity of ZnO, the phase velocity is not so sensitive to the electrical condition.

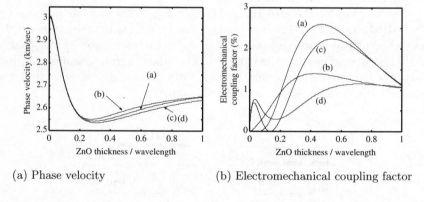

(a) Phase velocity   (b) Electromechanical coupling factor

Fig. 6.24. SAWs on ZnO/Pyrex-glass structure

Figure 6.24b shows the electromechanical coupling factor for the lowest mode as a function of h/λ. In configurations (a) and (b) without the counterelectrode, the coupling factor takes a maximum at $h/\lambda \cong 0.4 - 0.6$. This is because the SAW field tends to be involved in the piezoelectric film with an increase in h/λ, and concentration of the SAW field becomes strongest at $h/\lambda \cong 0.4 - 0.6$. After taking the maximum, the coupling factor decreases and converges to that for ZnO.

In configurations (c) and (d) with the counter-electrode, there are two peaks at $h/\lambda \cong 0.05$ and at $h/\lambda \cong 0.5 - 0.7$. The existence of the second peak can be explained in the same way as configurations (a) and (b).

To discuss the behavior of the first peak, let us consider propagation of the lowest-order symmetric (S_0) Lamb wave [43] in the plate shown in Fig. 6.25. The plate expands and deflates by the crossed electric field, and then the vibration propagates as the Lamb wave. When the plate is much thinner than the wavelength, the Lamb wave can be excited efficiently by the crossed field applied between the IDT and counter-electrode because its field is uniform throughout the thickness (see Fig. 6.25a). With an increase in the film thickness, the piezoelectricity reduces due to the field variation within the thickness. On the other hand, when the counter-electrode does not exist as shown in Fig. 6.25b, the Lamb wave cannot be excited efficiently due to the leakage of the electric field into the nonpiezoelectric region. When the film thickness is small, the existence of the substrate usually reduces the piezoelectricity due to the penetration of the SAW field into the substrate.

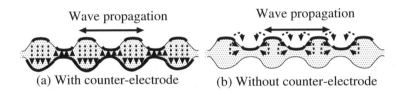

Wave propagation   Wave propagation

(a) With counter-electrode   (b) Without counter-electrode

Fig. 6.25. Excitation of lowest-order symmetric (S_0) Lamb wave

In the structure, since the SAW velocity in the film is lower than that in the substrate, the SAW velocity in the composite structure decreases with an increase in the film thickness, and higher modes appear over the cut-off film thicknesses. Figure 6.26 shows the frequency dispersion of higher modes in the ZnO/Pyrex-glass structure with configuration (a). For the lowest or fundamental mode, the SAW velocity in the composite structure coincides with that in the substrate at zero film thickness and converges into that in the ZnO film. On the other hand, the phase velocity of the higher modes converges into the slow-shear wave velocity in the film.

Figure 6.27 shows the dispersion relation for the Rayleigh-type SAW on the ZnO/diamond structure as a function of h/λ. Structure (b) was assumed. In this configuration, the second mode, called the Sezawa mode [43], exhibits an electromechanical coupling factor of 3.5% at $h/\lambda \cong 0.6$, where its phase velocity attains 5000 m/sec, and the velocity dispersion is small.

In the structure, at smaller h/λ, velocities higher than 10000 m/sec are achievable, but the piezoelectricity is small. On the other hand, a relatively large electromechanical coupling factor can be realized by using the configurations shown in Figs. 6.23c and d.

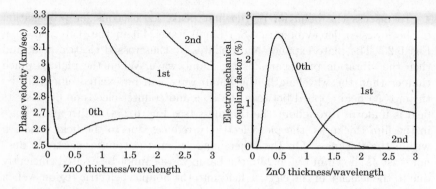

Fig. 6.26. Electromechanical coupling factor and phase velocity for ZnO/Pyrex-glass structure

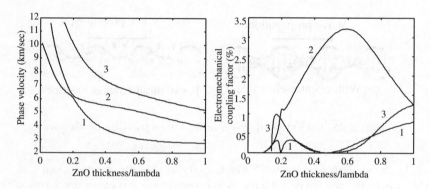

Fig. 6.27. Electromechanical coupling factor and phase velocity for ZnO/diamond structure

References

1. R.F. Milsom, N.H.C. Reilly and M. Redwood: The Interdigital Transducer, in Surface Wave Filters, H.Matthews (ed), Wiley (1977) pp. 55–108.
2. R.F. Milsom, N.H.C. Reilly and M. Redwood: Analysis of Generation and Detection of Surface and Bulk Acoustic Waves by Interdigital Transducers, IEEE Trans. Sonics and Ultrason., **SU-24** (1977) pp. 147–166.
3. J.J. Campbell and W.R. Jones: A Method for Optimal Crystal Cuts and Propagation Directions for Excitation of Piezoelectric Surface Waves, IEEE Trans. Sonics and Ultrason., **SU-15** (1968) pp. 209–217.
4. K.A. Ingebrigtsen: Surface Waves in Piezoelectrics, J. Appl. Phys., **40** (1969) pp. 2681–2686.
5. C.S. Lam, D.E. Halt and K. Hashimoto: The Temperature Dependency of Power Leakage in LST-cut Quartz Surface Acoustic Wave Filters , Proc. IEEE Ultrason. Symp. (1989) pp. 275–279.
6. O. Kawachi, G. Endoh, M. Ueda, O. Ikata, K. Hashimoto and M. Yamaguchi: Optimum Cut of LiTaO$_3$ for High Performance Leaky Surface Acoustic Wave Filters, Proc. IEEE Ultrason. Symp. (1996) pp. 71–76.

7. M. Lewis: Surface skimming bulk waves, SSBW, Proc. IEEE Ultrason. Symp. (1977) pp. 744–752.
8. F.G. Marshall and E.G.S. Paige: Novel Acoustic-Surface-Wave Directional Coupler with Diverse Applications, Electron. Lett., **7** (1971) pp. 460–464.
9. K. Hashimoto and M. Yamaguchi: Delta Function Model Analysis of SSBW Spurious Response in SAW Devices, Proc. IEEE Freq. Contr. Symp. (1993) pp. 639–644.
10. R. Bechmann: Elastic and Piezoelectric Constants of Alpha Quartz, Phys. Rev., **110** (1958) pp. 1060–1061.
11. R. Bechmann, A.D. Ballato and T.J. Lukaszek: Higher-Order Temperature Coefficients of the Elastic Stiffnesses and Compliances of Alpha-Quartz, Proc. IRE, **50** (1962) pp. 1812–1822.
12. M.B. Schulz, M.J. Matsinger and M.G. Holland: Temperature Dependence of Surface Acoustic Wave Velocity on α-Quartz, J. Appl. Phys., **41** (1970) pp. 2755–2765.
13. Y. Shimizu, M. Tanaka and T. Watanabe: A New Cut of Quartz with Extremely Small Temperature Coefficient by Leaky Surface Wave, Proc. IEEE Ultrason. Symp. (1985) pp. 233–236.
14. B.A. Auld, J.J. Gagnepain and M. Tan: Horizontal Shear Surface Waves on Corrugated Surface, Electron. Lett., **12** (1976) pp. 650–651.
15. Y.V. Gulyaev and V.P. Plessky: Slow Surface Acoustic Waves in Solids, Sov. Tech. Phys. Lett. **3** (1977) pp. 220–223.
16. A.W. Warner, M. Onoe and G.A. Coquin: Determination of Elastic and Piezoelectric Constants for Crystals in Class (3m), J. Acoust. Soc. Am., **42** (1967) pp. 1223–1231.
17. R.T. Smith and F.J. Welsh: Temperature Dependence of the Elastic, Piezoelectric and Dielectric Constants of Lithium Tantalate and Lithium Niobate, J. Appl. Phys., **42** (1971) pp. 2219–2230.
18. Y. Nakagawa, K. Yamanouchi and K. Shibayama: Third-Order Elastic Constants of Lithium Niobate, J. Appl. Phys., **44** (1973) pp. 3969–3974.
19. G. Kovacs, M. Anhorn, H.E. Engan, G. Visintini, and C.C.W. Ruppel: Improved Material Constants for LiNbO$_3$ and LiTaO$_3$, Proc. IEEE Ultrason. Symp. (1990) pp. 435–438.
20. K. Shibayama, K. Yamanouchi, H. Sato and T. Meguro: Optimal Cut for Rotated Y-cut LiNbO$_3$ Crystal Used as the Substrate of Acoustic-Surface-Wave Filters, Proc. IEEE, **64** (1976) pp. 595–597.
21. K. Yamanouchi and K. Shibayama: Propagation and Amplification of Rayleigh Waves and Piezoelectric Leaky Surface Waves in LiNbO$_3$, J. Appl. Phys., **43** (1970) pp. 856–862.
22. H. Shimizu, Y. Suzuki and T. Kanda: Love-Type-SAW Resonator of Small Size with Low Capacitance Ratio and Its Application to VCO, Proc. IEEE Ultrason. Symp. (1990) pp. 103–108.
23. S. Takahashi, H. Hirano, T. Kodama, F. Miyashiro, B. Suzuki, A. Onoe, T. Adachi and K. Fujinuma: SAW IF Filter on LiTaO$_3$ for Color TV Receivers, IEEE Trans. Consumer Electron., **CE-24**, 3 (1978) pp. 337–346.
24. Y. Ebata, K. Sato and S. Morishita: A LiTaO$_3$ SAW Resonator and its Application to Video Cassette Recorder, Proc. IEEE Ultrason. Symp. (1981) pp. 89–93.
25. K. Nakamura, M. Kazumi and H. Shimizu: SH-type and Rayleigh-Type Surface Waves on Rotated Y-cut LitaO$_3$, Proc. IEEE Ultrason. Symp. (1985) pp. 510–518.
26. K. Hashimoto and M. Yamaguchi: Non-Leaky, Piezoelectric, Quasi-Shear-Horizontal Type SAW on X-cut LiTaO$_3$, Proc. IEEE Ultrason. Symp. (1988) pp. 97–101.

27. N.M. Shorrocks, R.M. Whatmore and F.W. Ainger: Lithium Tetraborate – A New Temperature Compensated Piezoelectric Substrates Material for Surface Acoustic Wave Devices, Proc. IEEE Ultrason. Symp. (1981) pp. 337–340.

28. Y. Ebata and H. Satoh: Current Applications and Future Trends for SAW in Asia, Proc. IEEE Ultrason. Symp. (1986) pp. 195–202.

29. M. Adachi, T. Shiosaki, H. Kobayashi, O. Ohnishi and A. Kawabata: Temperature Compensated Piezoelectric Lithium Tetraborate Crystal for High Frequency Surface Acoustic and Bulk Wave Device Applications, Proc. IEEE Ultrason. Symp. (1985) pp. 228–232.

30. T. Sato and H. Abe: Propagation Properties of Longitudinal Leaky Surface Waves on Lithium Tetraborate, Proc. IEEE Ultrason. Symp. (1994) pp. 287–292.

31. T. Sato and H. Abe: Longitudinal Leaky Surface Waves for High Frequency SAW Device Application, Proc. IEEE Ultrason.Symp. (1995) pp. 305–315.

32. B.V. Mill, A.V. Butashin, G.G. Khodzhabagyan, E.L. Belokoneva and N.V.Belov: Modified Gallates with the Structure $La_3Ga_5SiO_{14}$, Sov. Phys. Dokl, **27** (1982) pp. 434.

33. B. Chai, J.J. Lefaucheur, Y.Y. Ji and H. Qiu: Growth and Evaluation of Large Size LGS ($La_3Ga_5SiO_{14}$), LGN ($La_3Ga_{5.5}Nb_{0.5}O_{14}$) and LGT ($La_3Ga_{5.5}Ta_{0.5}O_{14}$) Single Crystals, Proc. IEEE Freq. Contr. Symp. (1998) pp. 748–758.

34. A. Bungo, C. Jian, K. Yamaguchi, Y. Sawada, S. Uda and Yu.P. Pisarevsky: Analysis of Surface Acoustic Wave Properties of Rotated Y-cut Langasite Substrate, Jpn. J. Appl. Phys. **38** (1999) pp. 3239–3243.

35. A.A. Kaminskii, I.M. Silvestrova, S.A. Sarkisov and G.A. Denisenko: Investigation of Trigonal ($La_{1-x}Nd_x$)Ga_5SiO_{14} (II Spectral Laser and Electromechanical Properties), Phys. Stat. Sol. (a), **80** (1983) pp. 607–620.

36. S. Sakharov, P. Senushenkov, A. Medvedev and Yu. Pisarevsky: New Data on Temperature Stability and Acoustic Losses of Langasite Crystals, Proc. IEEE Freq. Contr. (1995) pp. 647–652.

37. N.F. Naumenko and L.P. Solie: Optimal Cut of Langasite for High Performance SAW Devices, Proc. IEEE Ultrason. Symp. (1999) to be published.

38. A. Bungo, C. Jian, K. Yamaguchi, Y. Sawada, R. Kimura and S. Uda: Experimental and Theoretical Analysis of SAW Properties of the Langasite Substrate with Euler Angle $(0°,140°,\phi)$, Proc. IEEE Ultrason. Symp. (1999) to be published.

39. S. Fujishima, H. Ishiyama, A. Inoue, H. Ieki: Surface Acoustic Wave VIF Filters for TV Using ZnO Sputtered film, Proc. Freq. Contr. Symp. (1976) pp. 119–122.

40. S. Fujishima: Piezoelectric Devices for Frequency Control and Selection in Japan, Proc. IEEE Ultrason. Symp. (1990) pp. 87–94.

41. J. Koike, H. Tanaka and H. Ieki: Quasi-Microwave Band Longitudinally Coupled Surface Acoustic Wave Resonator Filters Using ZnO/Sapphire Substrate, Jpn. J. Appl. Phys., **34**, Part 1, 5B (1995) pp. 2678–2682.

42. H. Nakahata, K. Higaki, S. Fujii, A. Hachigo, H. Kitabayashi, K. Tanabe, Y. Seki and S. Shikata: SAW Devices on Diamond, Proc. IEEE Ultrason. Symp. (1995) pp. 361–370.

43. B.A. Auld: Acoustic Waves and Fields in Solids, Vol. II, Chap. 5, Wiley, New York (1973) pp. 135–161.

7. Coupling-of-Modes Theory

This chapter fully describes the coupling-of-modes (COM) theory, which is now widely used in SAW device simulation. Starting from its fundamentals, the construction of the COM-based simulator and the adjustment of parameters required for the simulation are described in detail.

7.1 Fundamentals

7.1.1 Collinear Coupling

Normal modes or eigenmodes are waves propagating without changing their amplitude, and they do not exchange their energy among them. Hence, the amplitude $u_n(x)$ of the n-th eigenmode propagating toward the $+x$ direction satisfies the following differential equation [1],

$$\frac{\partial u_n(x)}{\partial x} = -j\beta u_n(x), \tag{7.1}$$

where β is the wavenumber of the mode. In the following discussion, the mode amplitude is implicitly assumed to be normalized so that modes with $|u_n|^2 = 1$ carry unit power.

Let us consider the case where two optical fibers are placed in proximity as shown in Fig. 7.1b. Then coupling between two modes u_1 and u_2 may arise. If the coupling is small, its contribution may be expressed by adding linear perturbation terms in (7.1);

$$\frac{\partial u_1(x)}{\partial x} = -j\beta u_1(x) - j\kappa u_2(x), \tag{7.2}$$

$$\frac{\partial u_2(x)}{\partial x} = -j\beta u_2(x) - j\kappa' u_1(x). \tag{7.3}$$

These types of simultaneous differential equations with linear mutual coupling terms are called coupling-of-modes (COM) equations. In these equations, κ and κ' are parameters responsible for the mutual coupling, and are called mutual coupling coefficients. As an extension, mutual coupling among multiple modes can be constructed in the same manner.

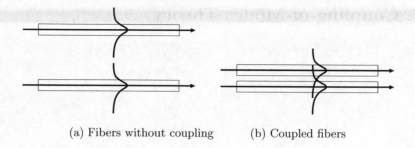

(a) Fibers without coupling   (b) Coupled fibers

Fig. 7.1. Two fibers

Let us assume the system is lossless, namely unitary. Since the mode amplitudes are normalized, the unitary condition for the section shown in Fig. 7.2 is given by

$$\frac{|u_1(x + \Delta x)|^2 + |u_2(x + \Delta x)|^2 - |u_1(x)|^2 - |u_2(x)|^2}{\Delta x} = 0.$$

Since this condition must be fulfilled for arbitrary x and Δx, setting $\Delta x \to 0$

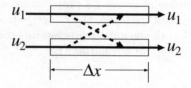

Fig. 7.2. Input and output for an arbitrary section

gives

$$\frac{\partial(|u_1(x)|^2 + |u_2(x)|^2)}{\partial x} = 0. \tag{7.4}$$

Substitution of (7.2) and (7.3) into (7.4) gives

$$0 = u_1 \frac{\partial u_1(x)^*}{\partial x} + u_1^* \frac{\partial u_1(x)}{\partial x} + u_2 \frac{\partial u_2(x)^*}{\partial x} + u_2^* \frac{\partial u_2(x)}{\partial x}$$

$$= 2\Im(\beta)(|u_1(x)|^2 + |u_2(x)|^2) + 2\Im\{(\kappa - \kappa'^*)u_1(x)^* u_2(x)\}. \tag{7.5}$$

So that the condition of (7.5) holds for arbitrary u_i, the relations $\Im(\beta) = 0$ and $\kappa' = \kappa^*$ must be satisfied. Then substitution of these relations into (7.2) and (7.3) gives the final COM equation, i.e.,

$$\frac{\partial u_1(x)}{\partial x} = -j\beta u_1(x) - j\kappa u_2(x) \tag{7.6}$$

$$\frac{\partial u_2(x)}{\partial x} = -j\kappa^* u_1(x) - j\beta u_2(x). \tag{7.7}$$

If the two fibers are equivalent to each other, the subscripts 1 and 2 are exchangeable. Hence, from (7.6) and (7.7), we get $\kappa^* = \kappa$.

Let us solve the simultaneous linear differential equations (7.6) and (7.7). When we define a differential operator $\partial/\partial x \equiv \Lambda$, it is written in matrix form

$$\begin{pmatrix} \Lambda + j\beta & j\kappa \\ j\kappa^* & \Lambda + j\beta \end{pmatrix} \begin{pmatrix} u_1(x) \\ u_2(x) \end{pmatrix} = \begin{pmatrix} 0 \\ 0 \end{pmatrix}. \tag{7.8}$$

For existence of nontrivial solutions, the determinant of the matrix in the left-hand side must be zero. Then we get

$$(\Lambda + j\beta)^2 + |\kappa|^2 = 0,$$

and its solutions are given by

$$\Lambda = -j\beta \mp j|\kappa| \equiv -j\theta_\pm. \tag{7.9}$$

The solutions of (7.8) can be written as a linear combination of solutions of the following linear differential equation;

$$\Lambda u_i = \frac{\partial u_i}{\partial x} = -j\theta_\pm u_i.$$

Thus, we get

$$u_1(x) = c_1 \exp(-j\theta_+ x) + c_2 \exp(-j\theta_- x), \tag{7.10}$$

$$u_2(x) = r_+ c_1 \exp(-j\theta_+ x) + r_- c_2 \exp(-j\theta_- x), \tag{7.11}$$

where $c_\pm$ is a constant determined by the boundary condition, and $r_\pm = (\theta_\pm - \beta)/\kappa = \pm|\kappa|/\kappa$. As is clear from (7.10) and (7.11), $\theta_\pm$ defined in (7.9) corresponds to the wavenumbers of two modes when coupling occurs.

If the two fibers are equivalent to each other, $r_\pm = \pm 1$ because $\kappa^* = \kappa$. Then, it is clear that $u_1(x) = u_2(x)$ for θ_+ whereas $u_1(x) = -u_2(x)$ for θ_-. Namely, the former and latter correspond to symmetric and antisymmetric modes, respectively, as shown in Fig. 7.3.

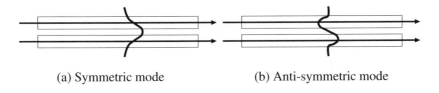

(a) Symmetric mode   (b) Anti-symmetric mode

Fig. 7.3. Symmetric and antisymmetric modes

Let us apply this technique to practical problems. When there is an input A_{in} only to fiber 1 as shown in Fig. 7.4, the boundary conditions are given by $u_1(0) = A_{\text{in}}$ and $u_2(0) = 0$. Then (7.10) and (7.11) give $c_1 = c_2 = A_{\text{in}}/2$. Substituting this relation into (7.10) and (7.11) gives

$$u_1(x) = A_{\text{in}} \cos(\kappa x) \exp(-j\beta x), \qquad\qquad (7.12)$$
$$u_2(x) = -jA_{\text{in}} \sin(\kappa x) \exp(-j\beta x). \qquad\qquad (7.13)$$

Fig. 7.4. Mode incidence to fiber 1

Figure 7.5 shows $|u_i(x)|$ as a function of x. All power of u_1 is transferred to u_2, and is then returned to u_1 after propagation length of each $\pi/2|\kappa| \equiv x_c$. This x_c is called the coupling length.

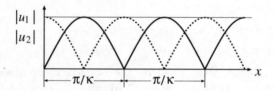

Fig. 7.5. Change in mode amplitudes within coupled fibers

This type of the energy transfer between two modes is also employed in SAW devices. Figure 7.6 shows an array of periodic metallic strips called a multi-strip coupler (MSC) [2].

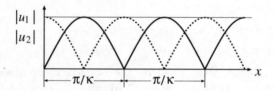

Fig. 7.6. Multi-strip coupler

Let us consider the case where the SAW is incident upon the upper track from the left as shown in Fig. 7.6. The incident wave can be expressed as a sum of two modes with equal amplitude where one of them is symmetric and the other is antisymmetric with respect to the center. For the symmetric mode, since its surface electrical potential is uniform throughout its wavefront, the MSC is equivalent to an open-circuited grating. On the other hand, for the antisymmetric mode, since the surface electrical potential is canceled out within each metal strip, the MSC is equivalent to a short-circuited grating. Hence, due to piezoelectricity, these two modes have different phase velocities, and when the phase difference reaches 180° after propagation, the entire SAW power is transferred to the lower track. Note that their phase velocities can be derived by using the technique described in Sects. 2.4.2 and 2.4.3 [3].

Figure 7.7 shows the configuration of a transversal filter where the MSC is inserted between the input and output IDTs [4]. By the insertion of the MSC, the BAW spurious response is significantly reduced because the SAW changes its propagation path to the lower track whereas the BAWs pass straight through the MSC. It is interesting to note that, even when the SAW is excited by the apodized IDT and its amplitude on the wavefront is not uniform, it becomes uniform when the SAW is regenerated by the MSC. Hence, the product of the transfer functions (see Sect. 4.1.1) still holds when the apodization is applied to both the input and output IDTs [4].

Fig. 7.7. SAW filter employing multi-strip coupler

In the resonator filter structure shown in Fig. 7.8 [5], the MSC is designed to cause a phase difference of 90° between the modes. Thus, SAWs with the same amplitude are incident at the reflectors, and after reflection, the SAWs cause a phase difference of 90° again at the MSC; then the SAW exits only from the lower side of the MSC. Thus, the SAWs go forward and back between the upper and lower tracks, and resonate in the structure [5].

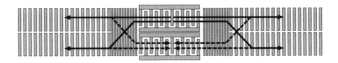

Fig. 7.8. Resonator filter employing multi-strip coupler

7.1.2 Periodic Structures

Let us consider the coupling of two modes propagating in the periodic structure shown in Fig. 7.9 with periodicity p, where one of them propagates toward the $+x$ direction whereas the other propagates toward the $-x$ direction.

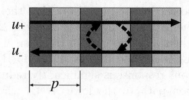

Fig. 7.9. Periodic structure

As described in Sect. 2.2.1, the Floquet theorem suggests that fields within the periodic structure can be expressed in the form of

$$u \propto \sum_{-\infty}^{+\infty} U_m \exp\{-j(\beta + 2\pi m/p)x\}.$$

The M-th order Bragg reflection occurs at $|\beta - 2\pi M/p| \cong |\beta|$. This means that the most significant coupling is what between the component with wavenumber β and that with wavenumber $\beta - 2\pi M/p$, and contributions by the others are relatively small.

Thus, by taking only these two components into account, we obtain the COM equation of

$$\frac{\partial u_+(x)}{\partial x} = -j\beta u_+(x) - j\kappa_{12}u_-(x)\exp(-2\pi jMx/p), \qquad (7.14)$$

$$\frac{\partial u_-(x)}{\partial x} = +j\beta u_-(x) + j\kappa'_{12}u_+(x)\exp(+2\pi jMx/p), \qquad (7.15)$$

where $u_\pm$ is the mode amplitude propagating toward the $\pm x$ direction, and κ_{12} and κ'_{12} are the mutual coupling coefficients.

By using the expression

$$u_\pm(x) = U_\pm(x)\exp(\mp\pi jMx/p), \qquad (7.16)$$

(7.14) and (7.15) can be rewritten as

$$\frac{\partial U_+(x)}{\partial x} = -j\theta_u U_+(x) - j\kappa_{12}U_-(x), \qquad (7.17)$$

$$\frac{\partial U_-(x)}{\partial x} = +j\kappa'_{12}U_+(x) + j\theta_u U_-(x), \qquad (7.18)$$

where $\theta_u = \beta - \pi M/p$ is the deviation of the wavenumber β from the Bragg condition, and is called the detuning factor.

Next let us derive the unitary condition by using the procedure described in Sect. 7.1.1. Since

$$0 = \frac{\partial(|U_+(x)|^2 - |U_-(x)|^2)}{\partial x}$$

$$= U_+ \frac{\partial U_+(x)^*}{\partial x} + U_+^* \frac{\partial U_+(x)}{\partial x} - U_- \frac{\partial U_-(x)^*}{\partial x} - U_-^* \frac{\partial U_-(x)}{\partial x}$$

$$= 2\Im(\theta_u)\{|U_+(x)|^2 + |U_-(x)|^2\} + 2\Im\{(\kappa_{12} - \kappa_{12}'^*)U_+(x)^*U_-(x)\},$$

we obtain the unitary condition of $\Im(\theta_u) = 0$ and $\kappa_{12}' = \kappa_{12}^*$. Thus, the final COM equation is given by

$$\frac{\partial U_+(x)}{\partial x} = -j\theta_u U_+(x) - j\kappa_{12}U_-(x), \tag{7.19}$$

$$\frac{\partial U_-(x)}{\partial x} = j\kappa_{12}^* U_+(x) + j\theta_u U_-(x). \tag{7.20}$$

If the structure is symmetric, since (7.19) and (7.20) must be invariant against exchanging $\pm x \leftrightarrow \mp x$ and $U_\pm \leftrightarrow U_\mp$, we get $\kappa_{12}^* = \kappa_{12}$.

Next, let us solve (7.19) and (7.20). Setting $\partial/\partial x \to \Lambda$ gives

$$\begin{pmatrix} \Lambda + j\theta_u & j\kappa_{12} \\ -j\kappa_{12}^* & \Lambda - j\theta_u \end{pmatrix} \begin{pmatrix} U_1(x) \\ U_2(x) \end{pmatrix} = \begin{pmatrix} 0 \\ 0 \end{pmatrix}. \tag{7.21}$$

Then nontrivial solutions are obtained when the determinant of the left-hand side matrix is zero, i.e.,

$$\Lambda^2 + \theta_u^2 - |\kappa_{12}|^2 = 0,$$

namely, the condition is given by

$$\Lambda = \mp j\sqrt{\theta_u^2 - |\kappa_{12}|^2} \equiv \mp j\theta_p.$$

By using this, the solution can be expressed in the form

$$U_+(x) = c_+ \exp(-j\theta_p x) + \Gamma_- c_- \exp(+j\theta_p x), \tag{7.22}$$

$$U_-(x) = \Gamma_+ c_+ \exp(-j\theta_p x) + c_- \exp(+j\theta_p x), \tag{7.23}$$

where $c_\pm$ are unknowns determined by the boundary conditions, and

$$\Gamma_+ = \frac{\theta_p - \theta_u}{\kappa_{12}}, \tag{7.24}$$

$$\Gamma_- = \frac{\theta_p - \theta_u}{\kappa_{12}^*}. \tag{7.25}$$

Equations (7.22) and (7.23) suggest that $\theta_p + \pi M/p$ corresponds to the wavenumber of the perturbed mode near the M-th resonance. Figure 7.10 shows θ_p as a function of θ_u. The reflection coefficient takes a maximum at $\theta_u = 0$, and its amplitude and stopband width are determined by $|\kappa_{12}|$.

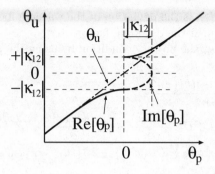

Fig. 7.10. Dispersion characteristics of grating mode

It is clear from (7.22) and (7.23) that $\Gamma_\pm$ corresponds to the reflection coefficient for SAW incidence at a semi-infinite grating from the $\mp x$ direction. Figure 7.11 shows $\Gamma_\pm$ as a function of frequency. In the figure, $\phi = \angle \kappa_{12}$. In

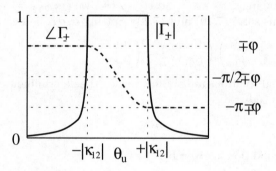

Fig. 7.11. Reflection coefficient $\Gamma_\pm$ of semi-infinite grating

the stopband, $|\Gamma_\pm| = 1$, and $\angle\Gamma_\pm$ changes gradually as

$$\angle\Gamma_\pm = -\sin^{-1}(\theta_u/|\kappa_{12}|) - \pi/2 \mp \phi. \tag{7.26}$$

On the other hand, out of the stopband, $\angle\Gamma_\pm$ is constant and $|\Gamma_\pm|$ changes gradually as

$$|\Gamma_\pm| = |\theta_u|/|\kappa_{12}| - \sqrt{(\theta_u/|\kappa_{12}|)^2 - 1}. \tag{7.27}$$

Let us consider the case where the SAW is incident at a grating with finite length as shown in Fig. 7.12. Since the boundary condition is given by $u_+(0) = A_{\rm in}$ and $u_-(L) = 0$, we get

$$c_+ = A_{\rm in}\frac{1}{1 - \Gamma_+\Gamma_- \exp(-2j\theta_p L)},$$

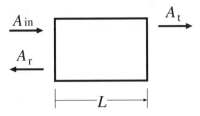

Fig. 7.12. SAW incidence at grating of finite length

$$c_- = -A_{\text{in}} \frac{\Gamma_+ \exp(-2j\theta_p L)}{1 - \Gamma_+\Gamma_- \exp(-2j\theta_p L)},$$

and then

$$U_+(x) = A_{\text{in}} \frac{\exp(-j\theta_p x) - \Gamma_+\Gamma_- \exp\{j\theta_p(x - 2L)\}}{1 - \Gamma_+\Gamma_- \exp(-2j\theta_p L)}, \tag{7.28}$$

$$U_-(x) = \Gamma_+ A_{\text{in}} \frac{\exp(-j\theta_p x) - \exp\{j\theta_p(x - 2L)\}}{1 - \Gamma_+\Gamma_- \exp(-2j\theta_p L)}. \tag{7.29}$$

Hence, the reflection coefficient Γ and transmission coefficient T are given by

$$T = \frac{A_t}{A_{\text{in}}} = \frac{(1 - \Gamma_+\Gamma_-) \exp(-j\theta_p L)}{1 - \Gamma_+\Gamma_- \exp(-2j\theta_p L)}. \tag{7.30}$$

$$\Gamma = \frac{A_r}{A_{\text{in}}} = \Gamma_+ \frac{1 - \exp(-2j\theta_p L)}{1 - \Gamma_+\Gamma_- \exp(-2j\theta_p L)}. \tag{7.31}$$

Figure 7.13 shows Γ given by (7.31). Comparison with Fig. 7.11 suggests

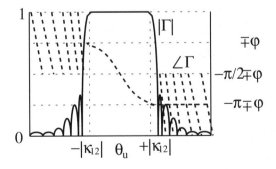

Fig. 7.13. Reflection coefficient Γ for grating with finite length

that the difference between Γ and $\Gamma_\pm$ is negligible in the stopband. This is because the SAW field is evanescent in the grating, and is scarcely influenced by the right-side end. On the other hand, considerable ripple occurs out of the stopband. As discussed in Sect. 2.3.2, this is due to interference between SAWs reflected at the grating edges, and the locations of corresponding nulls are given by

$$\theta_u = \pm\sqrt{|\kappa_{12}|^2 + (n\pi/L)^2}, \tag{7.32}$$

and are determined by $|\kappa_{12}|$. On the other hand, the maximum of the reflection coefficient is given by

$$\max(|\Gamma|) = \tanh(|\kappa_{12}|L), \tag{7.33}$$

and is again determined by $|\kappa_{12}|$.

7.1.3 Excitation

Let us consider wave excitation without reflection. In this case, the COM equation is given by

$$\frac{\partial u(x)}{\partial x} = -j\beta u(x) + f(x), \tag{7.34}$$

where β is the wavenumber of the mode, and $f(x)$ is the source distribution. Setting $u(x) = U(x)\exp(-j\beta x)$, (7.34) can be rewritten as

$$\frac{\partial U(x)}{\partial x} = f(x)\exp(j\beta x). \tag{7.35}$$

Thus the general solution of this differential equation is given by

$$u(x) = \exp(-j\beta x)\left[c_0 + \int_{-\infty}^{x} f(x')\exp(j\beta x')dx'\right], \tag{7.36}$$

where c_0 is a constant. If $f(x)$ is bounded within a certain region and the observation point x is chosen outside of this region, the integral interval in (7.36) can be replaced by $[-\infty, +\infty]$, and then the integral corresponds to the Fourier integral of $f(x)$. This suggests that only the Fourier component with wavenumber β is responsible for the radiation of the corresponding wave.

7.2 COM Theory for SAW Devices

7.2.1 Derivation

Let us consider a portion of an IDT as shown in Fig. 7.14, where V is the voltage applied to the IDT.

Following the general derivation procedure for the COM equations described previously, we may obtain the COM equation for the IDT in the form of

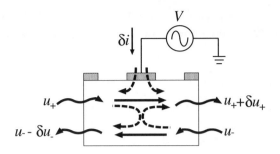

Fig. 7.14. Unit section of IDT

$$\frac{\partial u_+(x)}{\partial x} = -j\beta u_+(x) - j\kappa_{12}u_-(x)\exp(-2\pi jMx/p)$$
$$+j\zeta V\exp(-\pi jMx/p), \tag{7.37}$$

$$\frac{\partial u_-(x)}{\partial x} = +j\kappa'_{12}u_+(x)\exp(+2\pi jMx/p) + j\beta u_-(x)$$
$$-j\zeta'V\exp(+\pi jMx/p), \tag{7.38}$$

where ζ is a coefficient responsible for the excitation efficiency, and is called the transduction coefficient.

By setting $u_\pm(x) = U_\pm(x)\exp(\mp\pi jMx/p)$, we obtain

$$\frac{\partial U_+(x)}{\partial x} = -j\theta_\mathrm{u}U_+(x) - j\kappa_{12}U_-(x) + j\zeta V, \tag{7.39}$$

$$\frac{\partial U_-(x)}{\partial x} = +j\kappa'_{12}U_+(x) + j\theta_\mathrm{u}U_-(x) - j\zeta'V, \tag{7.40}$$

where $\beta - \pi M/p = \theta_\mathrm{u}$.

Let us derive the current $I(x)$ on the bus-bar. Figure 7.15 suggests that its increment within a period p_I of the IDT is given by

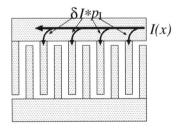

Fig. 7.15. Current on bus-bar

$$I(x + p_\mathrm{I}) - I(x) = p_\mathrm{I}\{-j\eta U_+(x) - j\eta'U_-(x) + j\omega CV\}, \tag{7.41}$$

where C is the static capacitance per unit length, and η and η' are coefficients responsible for the receiving efficiency. By taking the limit with respect to p_I, we obtain

$$\frac{\partial I(x)}{\partial x} = -j\eta U_+(x) - j\eta' U_-(x) + j\omega CV. \tag{7.42}$$

Let us derive the unitary condition from (7.39)–(7.42). Although $u_\pm$ is an effective value, we express V and $I(x)$ at their peak amplitudes. Then from the power conservation relation of

$$\frac{\partial(|u_+(x)|^2 - |u_-(x)|^2)}{\partial x} = \frac{1}{2}\frac{\partial \Re[VI(x)^*]}{\partial x}, \tag{7.43}$$

we obtain the relations $\Im(\beta) = 0$, $\kappa'_{12} = \kappa^*_{12}$, $\eta = 4\zeta$ and $\eta' = 4\zeta'$.

Added to this, in the case where the IDT is driven by a single-phase electrical source, the relation $\zeta' = \zeta^*$ holds.

Thus, the final COM equation is given by

$$\frac{\partial U_+(x)}{\partial x} = -j\theta_\mathrm{u}U_+(x) - j\kappa_{12}U_-(x) + j\zeta V, \tag{7.44}$$

$$\frac{\partial U_-(x)}{\partial x} = +j\kappa^*_{12}U_+(x) + j\theta_\mathrm{u}U_-(x) - j\zeta^*V, \tag{7.45}$$

$$\frac{\partial I(x)}{\partial x} = -4j\zeta^*U_+(x) - 4j\zeta U_-(x) + j\omega CV. \tag{7.46}$$

Note that the propagation loss can be included by giving a complex value to θ_u although the unitary condition was employed for the derivation of (7.44)–(7.46).

Solving (7.44)–(7.46) gives the general solutions of

$$U_+(x) = c_+ \exp(-j\theta_\mathrm{p}x) + \Gamma_-c_- \exp(+j\theta_\mathrm{p}x) + \xi_+V, \tag{7.47}$$

$$U_-(x) = \Gamma_+c_+ \exp(-j\theta_\mathrm{p}x) + c_- \exp(+j\theta_\mathrm{p}x) + \xi_-V, \tag{7.48}$$

$$I(x) = \int (-4j\zeta^*U_+(x) - 4j\zeta U_-(x) + j\omega CV)dx, \tag{7.49}$$

where θ_p is the wavenumber of the perturbed mode coupled with the reflected wave and is given by

$$\theta_\mathrm{p} = \begin{cases} -\sqrt{\theta_\mathrm{u}^2 - |\kappa_{12}|^2} & (\theta_\mathrm{u} < -|\kappa_{12}|) \\ -j\sqrt{|\kappa_{12}|^2 - \theta_\mathrm{u}^2} & (|\theta_\mathrm{u}| < |\kappa_{12}|) \\ \sqrt{\theta_\mathrm{u}^2 - |\kappa_{12}|^2} & (\theta_\mathrm{u} > |\kappa_{12}|), \end{cases} \tag{7.50}$$

and

$$\Gamma_+ = \frac{\theta_\mathrm{p} - \theta_\mathrm{u}}{\kappa_{12}} \tag{7.51}$$

is the reflection coefficient for the $+x$ propagating SAW whereas

$$\Gamma_- = \frac{\theta_\mathrm{p} - \theta_\mathrm{u}}{\kappa^*_{12}} \tag{7.52}$$

is that for the $-x$ propagating SAW. In addition,

$$\xi_+ = \frac{\zeta\theta_{\mathrm{u}} - \zeta^*\kappa_{12}}{\theta_{\mathrm{p}}^2} \tag{7.53}$$

is the excitation efficiency for the $+x$ propagating SAW whereas

$$\xi_- = \frac{\zeta^*\theta_{\mathrm{u}} - \zeta\kappa_{12}^*}{\theta_{\mathrm{p}}^2} \tag{7.54}$$

is that for the $-x$ propagating SAW.

Hereafter, as the mode amplitude, we use the normalized surface electrical potential associating with the SAW.

If a unit period of the IDT is symmetric as shown in Fig. 7.16, and also the natural unidirectionality [6] described in Sect. 3.1.2 does not exist, then κ_{12} and ζ are purely real because of the symmetry in the normalized surface electrical potential.

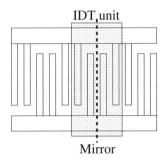

Fig. 7.16. IDT with mirror symmetry for each period

In this case,

$$\Gamma_0 = \Gamma_+ = \Gamma_- = \frac{\theta_{\mathrm{p}} - \theta_{\mathrm{u}}}{\kappa_{12}}, \tag{7.55}$$

and

$$\xi_0 = \xi_+ = \xi_- = \frac{\zeta}{\theta_{\mathrm{u}} + \kappa_{12}}. \tag{7.56}$$

7.2.2 COM Equations in Other Forms

In the COM equation derived previously, $U_\pm(x)$ is defined as the effective value whereas V and $I(x)$ are defined as the peak amplitudes. If we redefine the mode amplitude $u_\pm(x)$ by its peak amplitude, the COM equation becomes

$$\frac{\partial \hat{U}_+(x)}{\partial x} = -j\theta_u \hat{U}_+(x) - j\kappa_{12}\hat{U}_-(x) + j\hat{\zeta}V, \tag{7.57}$$

$$\frac{\partial \hat{U}_-(x)}{\partial x} = +j\kappa_{12}^* \hat{U}_+(x) + j\theta_u \hat{U}_-(x) - j\hat{\zeta}^* V, \tag{7.58}$$

$$\frac{\partial I(x)}{\partial x} = -2j\hat{\zeta}^* \hat{U}_+(x) - 2j\hat{\zeta}\hat{U}_-(x) + j\omega CV, \tag{7.59}$$

by setting $\hat{U}_+(x) = \sqrt{2}U_+(x)$ and $\hat{\zeta} = \sqrt{2}\zeta$.

In Fig. 7.15, the direction of the current flow is opposite to that of the co-ordinate system. If we define $\hat{I}(x) = -I(x)$ so as to remove this contradiction, the resulting COM equation is given by

$$\frac{\partial \hat{U}_+(x)}{\partial x} = -j\theta_u \hat{U}_+(x) - j\kappa_{12}\hat{U}_-(x) + j\hat{\zeta}V, \tag{7.60}$$

$$\frac{\partial \hat{U}_-(x)}{\partial x} = +j\kappa_{12}^* \hat{U}_+(x) + j\theta_u \hat{U}_-(x) - j\hat{\zeta}^* V, \tag{7.61}$$

$$\frac{\partial \hat{I}(x)}{\partial x} = +2j\hat{\zeta}^* \hat{U}_+(x) + 2j\hat{\zeta}\hat{U}_-(x) - j\omega CV. \tag{7.62}$$

This form has been widely used in the USA [7].

7.2.3 Inclusion of Electrode Resistivity

The effects of the finger resistivity can be introduced into the COM equation by the series-connection of resistors among the bus-bars and strips (see Fig. 7.17) [8].

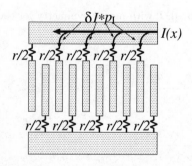

Fig. 7.17. Equivalent circuit of IDT including electrical resistivity

Let V denote the voltage between the bus-bars, V' the voltage applied to the fingers, and $R = rp_{\rm I}$. Since $V - V' = r\delta I$, we obtain

$$\frac{\partial U_+(x)}{\partial x} = -j\theta_u U_+(x) - j\kappa_{12}U_-(x) + j\zeta\left(V - R\frac{\partial I}{\partial x}\right), \tag{7.63}$$

$$\frac{\partial U_-(x)}{\partial x} = +j\kappa_{12}^* U_+(x) + j\theta_u U_-(x) - j\zeta^* \left(V - R\frac{\partial I}{\partial x}\right), \tag{7.64}$$

$$\frac{\partial I(x)}{\partial x} = -4j\zeta^* U_+(x) - 4j\zeta U_-(x) + j\omega C \left(V - R\frac{\partial I}{\partial x}\right). \tag{7.65}$$

Thus, by setting $\mu = (1 + j\omega CR)^{-1}$, these equations reduce to the same form as the original COM equation:

$$\frac{\partial U_+(x)}{\partial x} = -j(\theta_u - 4jR|\zeta|^2\mu)U_+(x) - j(\kappa_{12} - 4jR\zeta^2\mu)U_-(x)$$
$$+j\zeta\mu V, \tag{7.66}$$

$$\frac{\partial U_-(x)}{\partial x} = +j(\kappa_{12}^* - 4jR\zeta^{*2}\mu)U_+(x) + j(\theta_u - 4jR|\zeta|^2\mu)U_-(x)$$
$$-j\zeta^*\mu V, \tag{7.67}$$

$$\frac{\partial I(x)}{\partial x} = -4j\zeta^*\mu U_+(x) - 4j\zeta\mu U_-(x) + jV\omega C\mu. \tag{7.68}$$

When $R = 0$ and $V = 0$, the IDT coincides with a short-circuited grating with the grating periodicity of $p_I/2$. In this case, the COM equation reduces to

$$\frac{\partial U_+(x)}{\partial x} = -j\theta_u U_+(x) - j\kappa_{12} U_-(x), \tag{7.69}$$

$$\frac{\partial U_-(x)}{\partial x} = +j\kappa_{12}^* U_+(x) + j\theta_u U_-(x), \tag{7.70}$$

and we obtain the dispersion characteristics for the grating mode as

$$\theta_p = \beta - M\pi/p = \sqrt{\theta_u^2 - |\kappa_{12}|^2}. \tag{7.71}$$

By the way, (7.66)–(7.68) suggest that to set $R \to \infty$ is equivalent to setting $R = 0$ and $\delta I = 0$, and its corresponding structure is an open-circuited grating. In this case, the COM equations reduce to

$$\frac{\partial U_+(x)}{\partial x} = -j(\theta_u - 4|\zeta|^2/\omega C)U_+(x) - j(\kappa_{12} - 4\zeta^2/\omega C)U_-(x), \tag{7.72}$$

$$\frac{\partial U_-(x)}{\partial x} = +j(\kappa_{12}^* - 4\zeta^{*2}/\omega C)U_+(x) + j(\theta_u - 4|\zeta|^2/\omega C)U_-(x), \tag{7.73}$$

and we obtain the dispersion characteristics for the grating mode as

$$\theta_p = \beta - M\pi/p = \sqrt{(\theta_u - 4|\zeta|^2/\omega C)^2 - |\kappa_{12} - 4\zeta^2/\omega C|^2}. \tag{7.74}$$

Figure 7.18 shows the dispersion characteristics of SAWs on open- and short-circuited gratings. It is clear that the θ_us giving the stopband edges are uniquely determined from the COM parameters. This fact will be used for the theoretical determination of the COM parameters, which will be discussed in Sect. 7.3.2.

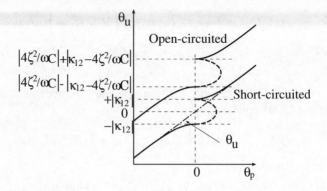

Fig. 7.18. SAW dispersion characteristics in short- and open-circuited gratings

7.2.4 Examples

Excitation by IDT. Let us discuss SAW excitation by an IDT without directivity as shown in Fig. 7.19.

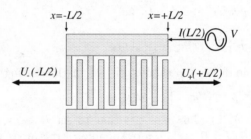

Fig. 7.19. SAW excitation by IDT

Since $U_+(-L/2) = 0$ and $U_-(+L/2) = 0$, substitution into (7.47) and (7.48) gives

$$c_\pm = -\frac{\xi V}{\Gamma_0 \exp(-j\theta_p L/2) + \exp(+j\theta_p L/2)}. \tag{7.75}$$

Added to this, $I(-L/2) = 0$. Then substitution of (7.75) into (7.49) gives

$$Y = I(+L/2)/V$$
$$= \frac{8j\zeta\xi L(\Gamma_0 + 1)\mathrm{sinc}(\theta_p L/2)}{\exp(j\theta_p L/2) + \Gamma_0 \exp(-j\theta_p L/2)} + j\omega CL - 8j\zeta\xi L, \tag{7.76}$$

If the internal reflection is negligible, i.e., $\kappa_{12} = 0$, (7.76) reduces to

$$\frac{Y}{L} = 4\zeta^2 L\mathrm{sinc}^2(\theta_u L/2) + j\omega C + 8j\zeta^2\theta_u^{-1}\mathrm{sinc}(\theta_u L) - 8j\zeta^2/\theta_u. \tag{7.77}$$

This is almost equivalent to that obtained by the delta-function model analysis described in Sect. 3.3.1.

Figure 7.20 shows the conductance $G = \Re(Y)$ of the IDT with $L = 20p_{\mathrm{I}}$ calculated from (7.76). In the figure, the horizontal axis is $\theta_{\mathrm{u}} p_{\mathrm{I}}/2\pi$ and the vertical axis is $G/4\zeta^2 L^2$. With an increase in $|\kappa_{12}|$, the curve is deformed from the $\sin x/x$ dependence shown in Fig. 3.15; it becomes asymmetric with respect to the center and its peak becomes steep due to internal reflection within the IDT. It is seen that the sign of κ_{12} affects the peak position. Note that when $|\kappa_{12}|L \to \infty$, the peak locates at $\theta_{\mathrm{u}} = -\kappa_{12}$.

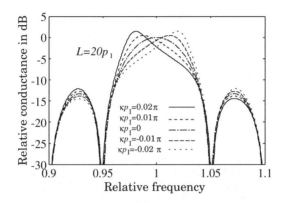

Fig. 7.20. Effect of internal reflection on IDT conductance characteristics

Detection. Let us discuss SAW detection by an IDT without directivity as shown in Fig. 7.21.

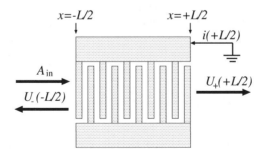

Fig. 7.21. SAW detection by IDT

Since $U_+(x) = A_{\mathrm{in}}$ at $x = -L/2$ and $U_-(x) = 0$ at $x = +L/2$, substitution into (7.47) and (7.48) gives

$$c_+ + c_- = A_{\mathrm{in}}/\{\exp(+j\theta_{\mathrm{p}}L/2) + \Gamma_0\exp(-j\theta_{\mathrm{p}}L/2)\}, \tag{7.78}$$

$$c_+ - c_- = A_{\mathrm{in}}/\{\exp(+j\theta_{\mathrm{p}}L/2) - \Gamma_0\exp(-j\theta_{\mathrm{p}}L/2)\}. \tag{7.79}$$

In addition, $I(x) = 0$ at $x = -L/2$ and $V = 0$. Then, substitution into (7.78) and (7.49) gives

$$\begin{aligned}
I(L/2) &= -4jL\zeta(\Gamma_0+1)(c_+ + c_-)\mathrm{sinc}(\theta_{\mathrm{p}}L/2)\\
&= -4jLA_{\mathrm{in}}\zeta\frac{(\Gamma_0+1)\mathrm{sinc}(\theta_{\mathrm{p}}L/2)}{\exp(+j\theta_{\mathrm{p}}L/2) - \Gamma_0\exp(-j\theta_{\mathrm{p}}L/2)}.
\end{aligned} \tag{7.80}$$

If internal reflection is negligible, i.e., $\kappa_{12} = 0$, (7.80) reduces to

$$I(L/2) = -4jLA_{\mathrm{in}}\zeta\,\mathrm{sinc}(\theta_{\mathrm{p}}L/2)\exp(-j\theta_{\mathrm{p}}L/2), \tag{7.81}$$

which is almost equivalent to that obtained by the delta-function model analysis described in Sect. 3.3.

One-Port SAW Resonator. Let us discuss the behavior of the one-port SAW resonator shown in Fig. 7.22 by using the COM equation. We define

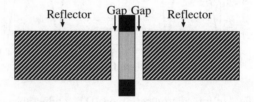

Fig. 7.22. One-port SAW resonator

the acoustic length L_{I} of the IDT as the IDT periodicity p_{I} times the number of finger-pairs. Also let us define the acoustic length of the reflector L_{r} and the gap length L_{g}. In the following analysis, identical COM parameters will be applied for both the IDT and reflectors for simplicity.

First, let us consider reflectors employing a short-circuited grating, where $V = 0$ and $U_-(L_{\mathrm{r}}) = 0$. Then (7.23) gives $c_- = -\Gamma_0 c_+\exp(-2j\theta_{\mathrm{p}}L_{\mathrm{r}})$, and substitution into (7.22) and (7.23) gives the reflection coefficient Γ at $x = 0$ of

$$\Gamma = \frac{U_-(0)}{U_+(0)} = \Gamma_0\frac{1 - \exp(-2j\theta_{\mathrm{p}}L_{\mathrm{r}})}{1 - \Gamma_0^2\exp(-2j\theta_{\mathrm{p}}L_{\mathrm{r}})}. \tag{7.82}$$

Figure 7.23 shows Γ calculated by using (7.82). The horizontal axis is $\theta_{\mathrm{u}}p_{\mathrm{I}}/2\pi + 1$. In the analysis, $\kappa_{12}p_{\mathrm{I}} = 0.02\pi$ and $L = 100p_{\mathrm{I}}$. The characteristics can be well explained by the discussion given in Sect. 2.3.2.

Next, let us consider the IDT. By taking the reflectors into account, the following relations hold;

$$u^-(+L_{\mathrm{I}}/2) = u^+(+L_{\mathrm{I}}/2) \times \Gamma_{\mathrm{t}}$$

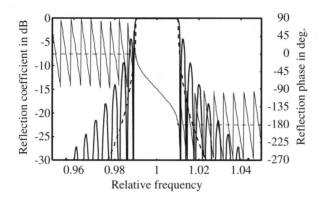

Fig. 7.23. Frequency response of reflector. Solid lines: Γ, broken lines: Γ_0. bold ones: amplitude, and thin ones: phase.

$$u^+(-L_\mathrm{I}/2) = u^-(-L_\mathrm{I}/2) \times \Gamma_\mathrm{t}$$
$$I(-L_\mathrm{I}/2) = 0,$$

where

$$\Gamma_\mathrm{t} = \Gamma \exp(-2j\beta L_\mathrm{g}). \tag{7.83}$$

Substitution of (7.47) and (7.48) into these relations gives

$$c_\pm = -\xi V \frac{1 - \Gamma_\mathrm{t}}{(\Gamma_0 - \Gamma_\mathrm{t})\exp(-j\theta_\mathrm{p}L_\mathrm{I}/2) + (1 - \Gamma_0\Gamma_\mathrm{t})\exp(+j\theta_\mathrm{p}L_\mathrm{I}/2)}. \tag{7.84}$$

Then substitution into (7.49) gives

$$I(L_\mathrm{I}/2) = -16jc_+\zeta(1 + \Gamma_0)\frac{\sin(\theta_\mathrm{p}L_\mathrm{I}/2)}{\theta_\mathrm{p}} - 8jL_\mathrm{I}\xi\zeta V + j\omega CL_\mathrm{I}V. \tag{7.85}$$

Then the input admittance Y of the IDT is given by

$$Y = \frac{j\zeta^2 L_\mathrm{I}}{\theta_\mathrm{u} + \kappa_{12}}\left(\frac{2}{\theta_\mathrm{p}L_\mathrm{I}}\left\{\cot\left(\frac{\theta_\mathrm{p}L_\mathrm{I}}{2}\right) + j\frac{(1+\Gamma_\mathrm{t})(1-\Gamma_0)}{(1-\Gamma_\mathrm{t})(1+\Gamma_0)}\right\}^{-1} - 1\right)$$
$$+ j\omega CL_\mathrm{I}. \tag{7.86}$$

First, let us discuss resonators where internal reflection within the IDT is negligible and reflection by the reflectors is dominant. Assuming $|\theta_\mathrm{p}L_\mathrm{I}| \ll 1$ gives

$$Y/L_\mathrm{I} \cong -8j\frac{\zeta^2}{\theta_\mathrm{u} + \kappa_{12}}\left(\frac{(1-\Gamma_\mathrm{t})(1+\Gamma_0)}{(1+\Gamma_\mathrm{t})(1-\Gamma_0)(+j\theta_\mathrm{p}L_\mathrm{I}/2)} + 1\right)^{-1} + j\omega C. \tag{7.87}$$

In addition, we also assume $|\theta_\mathrm{p} L_\mathrm{r}| \gg 1$ and $\Gamma_\mathrm{t} \cong \Gamma_0 \exp(-2j\beta L_\mathrm{g})$. Then (7.87) reduces to

$$Y/L_\mathrm{I} \cong -8j\frac{\zeta^2}{\theta_\mathrm{u}+\kappa_{12}}\left(1-\frac{\tan(\beta L_\mathrm{g}-\phi/2)}{\tan(\phi/2)(+j\theta_\mathrm{p}L_\mathrm{I}/2)}\right)^{-1}+j\omega C, \qquad (7.88)$$

where $\Gamma_0 = \exp(j\phi)$.

From this, the resonance frequency is given as a solution of

$$j\theta_\mathrm{p}L_\mathrm{I}/2 = \frac{\tan(\beta L_\mathrm{g}-\phi/2)}{\tan(\phi/2)}, \qquad (7.89)$$

whereas the antiresonance frequency is given as a solution of

$$j\theta_\mathrm{p}L_\mathrm{I}/2\left(1-\frac{8\zeta^2}{\omega C(\theta_\mathrm{u}+\kappa_{12})}\right) \cong \frac{\tan(\beta L_\mathrm{g}-\phi/2)}{\tan(\phi/2)}. \qquad (7.90)$$

If the IDT and reflectors have the same periodicity and L_g is chosen so that resonance occurs at $\theta_\mathrm{u} = 0$ giving maximum reflectivity, (7.89) reduces to

$$L_\mathrm{g}/p_\mathrm{I} = -\mathrm{sgn}(\kappa_{12})\{(2\pi)^{-1}\tan^{-1}(|\kappa_{12}|L_\mathrm{I}/2)+1/8\}, \qquad (7.91)$$

because $\theta_\mathrm{p} = -j|\kappa_{12}|$ at $\theta_\mathrm{u} = 0$.

Figure 7.24 shows the admittance $G + jB$ of the one-port SAW resonator calculated by using (7.76). The horizontal axis is $\theta_\mathrm{u} p_\mathrm{I}/2\pi+1$, and the vertical axis is $Y/\omega_\mathrm{r}CL_\mathrm{I}$. In the analysis, $L_\mathrm{I} = 10p_\mathrm{I}$, $L_\mathrm{r} = 50p_\mathrm{I}$, $\kappa_{12}p_\mathrm{I} = 0.02\pi$, and $L_\mathrm{g} = -0.17p_\mathrm{I}$ so that resonance occurs at $\theta_\mathrm{u} = 0$ where the reflection takes a maximum.

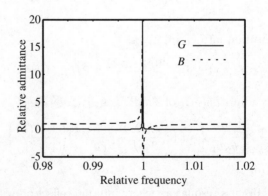

Fig. 7.24. Admittance characteristics of one-port SAW resonator ($L_\mathrm{I} = 10p_\mathrm{I}$, $L_\mathrm{g} = -0.17p_\mathrm{I}$)

If the number of finger-pairs is large enough, the IDT itself acts as a resonator without reflectors. Since this resonator can make its electrical impedance considerably lower than the structure shown in Fig. 7.22, it is desirable when materials with low permittivity such as quartz are employed as a substrate.

Figure 7.25 shows the calculated result when $L_{\mathrm{I}} = 100p_{\mathrm{I}}$, $L_{\mathrm{r}} = 50p_{\mathrm{I}}$, $L_{\mathrm{g}} = 0$ and $\kappa_{12}p_{\mathrm{I}} = 0.02\pi$. Resonance occurs at $\theta_{\mathrm{u}} = -\kappa_{12}$, and very large ripples appear at θ_{u} lower than the resonance. This is due to the ripple in the reflectivity of the gratings. This ripple is inherent for this type of resonator because they resonate near the stopband edge.

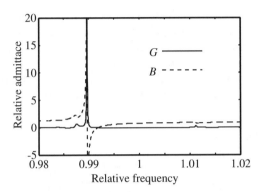

Fig. 7.25. Admittance characteristics of one-port SAW resonator ($L_{\mathrm{I}} = 100p_{\mathrm{I}}$, $L_{\mathrm{g}} = 0$)

Directional IDT. Even when the IDT possesses directivity, the procedure for the analysis is exactly the same. The following shows the p matrix element (see Sect. 3.5) derived from the COM equation for arbitrary IDTs:

$$p_{11} = \frac{-j\kappa_{12}^* \sin(\theta_{\mathrm{p}}L)}{\theta_{\mathrm{p}} \cos(\theta_{\mathrm{p}}L) + j\theta_{\mathrm{u}} \sin(\theta_{\mathrm{p}}L)}, \tag{7.92}$$

$$p_{12} = \frac{\theta_{\mathrm{p}}}{\theta_{\mathrm{p}} \cos(\theta_{\mathrm{p}}L) + j\theta_{\mathrm{u}} \sin(\theta_{\mathrm{p}}L)}, \tag{7.93}$$

$$p_{22} = \frac{-j\kappa_{12} \sin(\theta_{\mathrm{p}}L)}{\theta_{\mathrm{p}} \cos(\theta_{\mathrm{p}}L) + j\theta_{\mathrm{u}} \sin(\theta_{\mathrm{p}}L)}, \tag{7.94}$$

$$p_{31} = \frac{j\zeta \sin(\theta_{\mathrm{p}}L)\{1 + j\theta_{\mathrm{p}}^{-1}(\theta_{\mathrm{u}} - \eta^*) \tan(\theta_{\mathrm{p}}L/2)\}}{\theta_{\mathrm{p}} \cos(\theta_{\mathrm{p}}L) + j\theta_{\mathrm{u}} \sin(\theta_{\mathrm{p}}L)}, \tag{7.95}$$

$$p_{32} = \frac{j\zeta \sin(\theta_{\mathrm{p}}L)\{1 + j\theta_{\mathrm{p}}^{-1}(\theta_{\mathrm{u}} - \eta) \tan(\theta_{\mathrm{p}}L/2)\}}{\theta_{\mathrm{p}} \cos(\theta_{\mathrm{p}}L) + j\theta_{\mathrm{u}} \sin(\theta_{\mathrm{p}}L)}, \tag{7.96}$$

$$p_{33} = \frac{8j|\zeta|^2 \sin(\theta_{\mathrm{p}}L)\{\theta_{\mathrm{p}}(\theta_{\mathrm{u}} - \eta_{\mathrm{r}}) + j(\theta_{\mathrm{u}}^2 - 2\theta_{\mathrm{u}}\eta_{\mathrm{r}} + |\eta|^2) \tan(\theta_{\mathrm{p}}L/2)\}}{\theta_{\mathrm{p}}^3\{\theta_{\mathrm{p}} \cos(\theta_{\mathrm{p}}L) + j\theta_{\mathrm{u}} \sin(\theta_{\mathrm{p}}L)\}}$$
$$- 8j\theta_{\mathrm{p}}^{-3}|\zeta|^2(\theta_{\mathrm{u}} - \eta_{\mathrm{r}}) + j\omega CL, \tag{7.97}$$

where $\eta = \kappa_{12}\zeta^*/\zeta$ and $\eta = \eta_{\mathrm{r}} + j\eta_{\mathrm{i}}$. Based on the symmetry relation shown in (3.50), only independent components are shown.

From this, the directivity $D = |p_{31}/p_{32}|$ is given by

$$D = \left| \frac{1 + j\theta_{\mathrm{p}}^{-1}(\theta_{\mathrm{u}} - \eta^*)\tan(\theta_{\mathrm{p}}L/2)}{1 + j\theta_{\mathrm{p}}^{-1}(\theta_{\mathrm{u}} - \eta)\tan(\theta_{\mathrm{p}}L/2)} \right|, \tag{7.98}$$

which is governed by $\angle\eta$. If $|\kappa_{12}|L \ll 1$, D takes a maximum or a minimum at $\theta_{\mathrm{u}} \cong \eta_{\mathrm{r}}$, and the value is given by $D = |(1 - \eta_{\mathrm{i}}L/2)/(1 + \eta_{\mathrm{i}}L/2)|$.

Figure 7.26 shows the conductance $\Re(p_{33})$ of the IDT calculated by using (7.97). The horizontal axis is $\theta_{\mathrm{u}}p_{\mathrm{I}}/2\pi$, and the vertical axis is $G/4\zeta^2$. In the calculation, $L = 20p_{\mathrm{I}}$, $\kappa_{12}p_{\mathrm{I}} = 0.02\pi$, and the parameter ψ in the figure is $\angle\eta$. With an increase in ψ, the peak height at $\theta_{\mathrm{u}} \cong -\Re(\kappa_{12})$ decreases, and the other peak appears at $\theta_{\mathrm{u}} \cong \Re(\kappa_{12})$. When $\psi = 90°$, these two peaks have the same height, and $\Re(p_{33})$ becomes symmetric with respect to $\theta_{\mathrm{u}} = 0$.

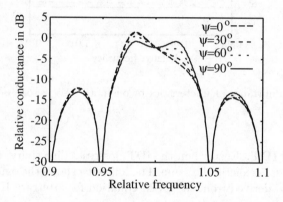

Fig. 7.26. ψ dependence of IDT conductance characteristics

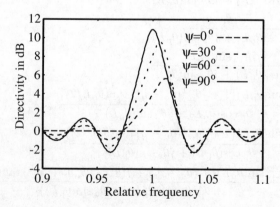

Fig. 7.27. ψ dependence of IDT directivity

Figure 7.27 shows the directivity D calculated by using (7.98). With an increase in ψ, D becomes large, and θ_u giving its maximum becomes close to $\theta_u = 0$; and then D takes a maximum at $\psi = 90°$. Then θ_u giving the peak decreases with a further increase in ψ; and when $\psi > 90°$, the θ_u dependence of D is the mirror image of that for $180° - \psi$. The sign of $\psi < 0$ does not affect the conductance characteristics, and the direction of the directivity is reversed.

Two-Port SAW Resonator. Two-port SAW Resonators are too complicate to perform the COM analysis analytically, and it should be carried out numerically by using the technique which will be described in Sect. 7.4.

Figure 7.28 shows the result for the two-port SAW resonator on 45°YZ-$Li_2B_4O_7$ with Al film of 0.08 μm thickness. In the analysis, the COM pa-

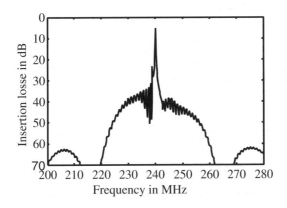

Fig. 7.28. Frequency response of two-port SAW resonator on 45°YZ-$Li_2B_4O_7$

rameters will be derived in Sect. 7.3.2 were employed, the number of IDT finger-pairs is 10, the acoustic length of the reflectors is $100p_I$, the aperture is $60p_I$, and the gap length between the IDT and reflector is $-\lambda/8$. In addition, the space L_g between the IDTs is chosen to be $1.85p_I$. Under this design, a single steep resonance peak appears in the middle of the reflector stopband.

Figure 7.29 shows the L_g dependence of the frequency response. As it increases, the resonance peak shifts toward the higher side of the reflector stopband, and the other peak appears. This result qualitatively agrees with the discussion given in Sect. 5.3.2 by using the Fabry–Perot model.

Note that, the optimal L_g is not an integer times p_I, which might seem to contradict the discussion. This is due to the fact that the SAW velocity in the gap region is different from that in the reflectors and/or IDTs. This must be taken into account in practical device design.

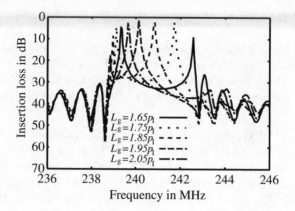

Fig. 7.29. Change in frequency response with L_g

Attention should be paid to the fact that an additional phase shift may occur at the gaps and spaces due to cancellation of the energy storing effect caused by lack of periodicity at the ends of the reflector and IDTs.

7.3 Determination of COM Parameters

7.3.1 Perturbation Theory

Since the COM theory is based upon the fact that the SAW excitation and reflection per unit period are relatively small, the perturbation method is readily applicable for the determination of the COM parameters.

Let us consider the following COM equation:

$$\frac{\partial U_+(x)}{\partial x} = -j\theta_\mathrm{u}U_+(x) - j\kappa_{12}U_-(x) + j\zeta V, \tag{7.99}$$

$$\frac{\partial U_-(x)}{\partial x} = +j\kappa_{12}^*U_+(x) + j\theta_\mathrm{u}U_-(x) - j\zeta^*V, \tag{7.100}$$

$$\frac{\partial I(x)}{\partial x} = -4j\zeta^*U_+(x) - 4j\zeta U_-(x) + j\omega CV. \tag{7.101}$$

In this section, let us express θ_u as

$$\theta_\mathrm{u} = \beta - M\pi/p = \omega/V_\mathrm{ref} - M\pi/p + \kappa_{11}, \tag{7.102}$$

where V_ref is the SAW velocity for reference, and that on the free surface is often employed. Note that the choice of V_ref affects the other parameters, and also influences the accuracy of the simulation. Thus, it must be close to the actual SAW velocity in the structure. κ_{11} is called the self-coupling coefficient. Equations (7.99) and (7.100) suggest that κ_{11} corresponds to the phase shift per period of the short-circuited grating without reflection, i.e.,

$V = 0$ and $\kappa_{12} = 0$, at the Bragg frequency. If we can derive $V_S = \omega/\beta$ for this condition, κ_{11} is given by

$$\kappa_{11}p = M\pi(1 - V_S/V_{\text{ref}}).\tag{7.103}$$

As an example, the equivalent circuit model [9] shown in Fig. 7.30 will be considered. In the figure, C_s is the static capacitance per period and C_e is a motional capacitance responsible for the energy storing effect [10]. When $m = 1$ denotes the electroded region and $m = 2$ the unelectroded region, β_m is the effective wavenumber in region m and R_m is the acoustic impedance given by $\pi/2\omega C_s K^2$, where K^2 is the electromechanical coupling factor. Note that the electrical port must be reversed alternately associating with the connection of the strips to the bus-bars.

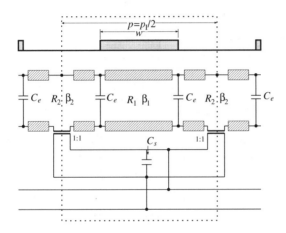

Fig. 7.30. Equivalent circuit under discussion

By setting $R_1 = R_2$ and $V = 0$, the circuit is equivalent to that for a short-circuited grating without reflection. The phase shift with propagation in this case has already been given in Sect. 2.3.1. From the result, θ_u is given by

$$\theta_u p + M\pi = \beta_1 w + \beta_2(p - w) + \omega C_e R_2.\tag{7.104}$$

If we choose $\beta_1 w + \beta_2(p - w) = M\pi$ as the reference, we obtain

$$\kappa_{11}p = \omega C_e R_2.\tag{7.105}$$

On the other hand, κ_{12} corresponds to the reflection coefficient per period for the short-circuited grating ($V = 0$ at its resonance frequency i.e., $\theta_u = 0$). For the equivalent circuit, this condition is equivalent to setting $\beta_1 w + \beta_2(p - w) = M\pi$, and $V = 0$ in the equivalent circuit. The reflection coefficient in this case can be derived using the procedure discussed in Sect. 2.1.2, and $-j\kappa_{12}p$ is given by

$$-j\kappa_{12}p = jT\sin(\beta_1 w - \psi/2 - \eta), \tag{7.106}$$

where $\Delta = R_1/R_2 - 1$, $\psi = \omega C_e R_2$, $T = \mathrm{sgn}(\Delta)\sqrt{\Delta^2 + \psi^2}$, and $\eta = \tan^{-1}(\psi/\Delta)$.

In addition, ζp corresponds to the excitation efficiency per period without reflection at the resonance frequency, and then the relation $\zeta p_I = \sqrt{\omega C p_I K^2/\pi}$ holds. This condition is equivalent to setting $R_1 = R_2 \equiv R$, $C_e = 0$ and $\beta_1 w + \beta_2(p - w) = M\pi$ in the equivalent circuit. In the COM model, since the acoustic power radiated from a unit section is given by $2\zeta^2$ when unit V is applied, we obtain

$$\zeta^2 = 1/2R. \tag{7.107}$$

Finally, C, the static capacitance per period, is given by

$$Cp_I = C_s. \tag{7.108}$$

This idea is not limited to the equivalent circuit model. Results of various well-established analytical techniques such as the perturbation method and variational analysis [1] are readily applicable.

7.3.2 Wave Theory Based Analysis

Theory. Various wave theories offer a dispersion relation for SAW propagation on open- and short-circuited gratings as shown in Fig. 7.31.

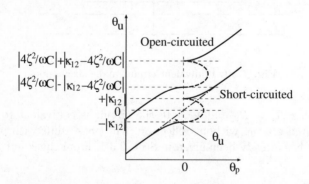

Fig. 7.31. SAW dispersion relation in open- and short-circuited gratings

Let us assume that the COM parameters are invariant against frequency. At frequencies $\omega_s^\pm$ and $\omega_o^\pm$ giving the stopband edges for the short- and open-circuited gratings, respectively, the following relations hold:

$$\omega_s^\pm/V_{\mathrm{ref}} - M\pi/p + \kappa_{11} = \pm|\kappa_{12}|, \tag{7.109}$$

$$\omega_o^\pm/V_{\mathrm{ref}} - M\pi/p + \kappa_{11} = |4\zeta^2/\omega C| \pm |\kappa_{12} - 4\zeta^2/\omega C| \tag{7.110}$$

from (7.71) and (7.74). Then we get

$$\kappa_{11} = M\pi/p - (\omega_s^+ + \omega_s^-)/2V_{\text{ref}}, \tag{7.111}$$

$$|\kappa_{12}| = (\omega_s^+ - \omega_s^-)/2V_{\text{ref}}, \tag{7.112}$$

$$|4\zeta^2/\kappa_{12}\omega C| = (\omega_o^+ + \omega_o^- - \omega_s^+ - \omega_s^-)/(\omega_s^+ - \omega_s^-), \tag{7.113}$$

$$|1 - 4\zeta^2/\kappa_{12}\omega C| = (\omega_o^+ - \omega_o^-)/(\omega_s^+ - \omega_s^-). \tag{7.114}$$

Note that $|\zeta^2/\kappa_{12}|$ and $\Re(\zeta^2/\kappa_{12})$ are determined from (7.112)-(7.114), but the sign of $\Im(\zeta^2/\kappa_{12})$ is not. Since $\angle(\zeta^2/\kappa_{12})$ determines the directivity of the IDT, the sign can usually be expected from the IDT configuration and material.

When ζ^2/κ_{12} is real, one of the stopband edges for the short-circuited grating coincides with that for the open-circuited grating as shown in Fig. 7.32. In other words, if one of the stopband edges coincides with each other, ζ^2/κ_{12} is real and no unidirectionality exists.

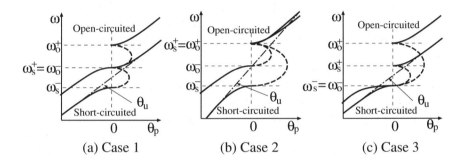

(a) Case 1   (b) Case 2   (c) Case 3

Fig. 7.32. SAW dispersion relation when COM parameters are real

In this case, (7.111)–(7.114) reduce to

$$\kappa_{11} = M\pi/p - (\omega_s^+ + \omega_s^-)/2V_{\text{ref}}, \tag{7.115}$$

$$\kappa_{12} = s(\omega_s^+ - \omega_s^-)/2V_{\text{ref}}, \tag{7.116}$$

$$4\zeta^2/\omega C = (\omega_o^+ + \omega_o^- - \omega_s^+ - \omega_s^-)/2V_{\text{ref}}, \tag{7.117}$$

where

$$s = \begin{cases} 1 & (\omega_s^+ = \omega_o^+ \text{ or } \omega_s^+ = \omega_o^-) \\ -1 & (\omega_s^- = \omega_o^-) \end{cases}. \tag{7.118}$$

Hence, (7.118) can be used to determine the sign of κ_{12} [11].

When mechanical reflection is negligible, κ_{12} is positive because $\omega_s^+ = \omega_o^-$ (see Sect. 2.4).

By the way, comparison of (7.77) with the delta-function model described in Sect. 3.3.1 gives

$$K_{\text{eff}}^2 = \frac{2\zeta^2 p_{\text{I}}}{\eta^2 \omega C}, \tag{7.119}$$

where η is the element factor. Substitution of (7.112) and (7.113) into (7.119) gives

$$K_{\text{eff}}^2 = \frac{p_{\text{I}}(\omega_o^+ + \omega_o^- - \omega_s^+ - \omega_s^-)}{4\eta^2 V_{\text{ref}}}. \tag{7.120}$$

When mechanical reflection is negligible, the approximate dispersion relation is already given in Sect. 2.4.3. Its application to (7.120) gives

$$\frac{K_{\text{eff}}^2}{K_V^2} = \frac{\pi(V|_{\omega_o^+} - V|_{\omega_s^-})}{\eta^2 V_{\text{ref}}} = \frac{\pi V_{\text{m}}}{4\eta^2 V_{\text{ref}}}\{F_{1/2}(\cos\theta) + F_{1/2}(-\cos\theta)\} \tag{7.121}$$

at $\beta p_{\text{I}} = 2\pi$, where $\theta = 2w/p_{\text{I}}$. The right-hand side of (7.121) is unity when $w/p_{\text{I}} = 0.25$, and its variation around $w/p_{\text{I}} \cong 0.25$ is gradual.

Numerical Analysis Based on FEMSDA. By using the free software FEMSDA [12] (see Sect. B.3.2), the dispersion relation of nonleaky SAWs propagating under the metallic gratings on 128°YX-LiNbO$_3$ was calculated. Figure 7.33 shows the result with h/p as a parameter, where h is the Al thickness and p is the grating period and $V_{\text{B}} = 4025$ m/sec is the slow-shear SSBW velocity. With an increase in h/p, the stopband width increases for the open-circuited grating whereas it decreases for the short-circuited grating. This is due to the fact that electrical reflection by the short-circuited grating has opposite sign to the mechanical reflection whereas that by the open-circuited grating including electrical regeneration has the same sign. Namely, $\kappa_{12} > 0$ at $h \cong 0$ and decreases monotonically with an increase in h/p. Note κ_{12} becomes zero at $h/p \cong 0.064$ and is negative when $h/p > 0.064$.

It is relatively hard to numerically determine stopband edges. On the other hand, κ_{11}, κ_{12} and V_{ref} can be determined with good accuracy by numerically fitting the dispersion relation for the short-circuited grating determined by the FEMSDA with that given by COM theory, i.e.,

$$\beta_{\text{s}}(\kappa_{11}, \kappa_{12}, V_{\text{ref}}) = M\pi/p + \sqrt{(\omega/V_{\text{ref}} - M\pi/p + \kappa_{11})^2 - \kappa_{12}^2}. \tag{7.122}$$

For numerical fitting, first, approximate values of $\omega_s^\pm$ are determined from the result of the FEMSDA. Second, initial values for the numerical fit are determined:

$$\kappa_{11} \leftarrow M\pi/p - (\omega_s^+ + \omega_s^-)/2V_{\text{B}}$$
$$|\kappa_{12}| \leftarrow (\omega_s^+ - \omega_s^-)/2V_{\text{B}}$$
$$V_{\text{ref}} \leftarrow V_{\text{B}}.$$

Third, κ_{11}, κ_{12} and V_{ref} are determined so that the wavenumbers given by (7.122) coincide with those at $\omega_s^\pm$ and the frequency giving the maximum $|\Im(\beta)|$. In addition, the stopband edges $\omega_o^\pm$ for the open-circuited grating are

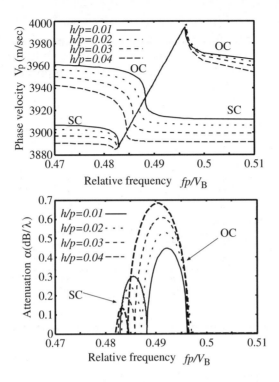

Fig. 7.33. Dispersion relation of nonleaky SAW on $128°$YX-LiNbO$_3$ derived by FEMSDA

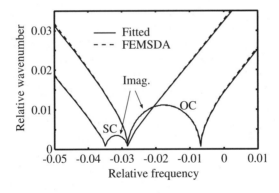

Fig. 7.34. Dispersion relation calculated by COM theory with that by FEMSDA

determined. Finally, substitution of $\omega_s^\pm$ and $\omega_o^\pm$ into (7.113) gives $|\zeta^2/\omega C|$, and (7.114) gives $\angle\zeta^2/\kappa_{12}\omega C$.

Figure 7.34 shows the dispersion relation of a nonleaky SAW calculated by using the determined COM parameters with that calculated by FEMSDA.

The horizontal axis is $2fp/V_B - 1$ and vertical axis is $\beta p/\pi - 1$. Agreement is quite good, and the effectiveness of this method is verified.

Figure 7.35 shows the h/p dependence of the COM parameters. In the figure, $c = V_B/V_{\text{ref}}$. All the parameters are mostly proportional to h/p, and their least square fit gives

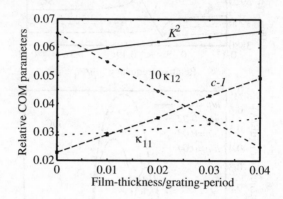

Fig. 7.35. h/p dependence of nonleaky SAW on $128°$YX-LiNbO$_3$

$$\kappa_{11}p = \pi\{0.028373 + 0.158314(h/p)\},$$
$$\kappa_{12}p = \pi\{0.006483 - 0.1008(h/p)\},$$
$$K_{\text{eff}}^2 = 0.057497 + 0.207871(h/p),$$
$$c = 1.022451 + 0.607222(h/p).$$

It is seen that K_{eff}^2, κ_{11} and c gradually increase with h/p, whereas κ_{12} decreases rapidly, and becomes zero at $h/p \cong 0.064$.

These expressions enable us to calculate the COM parameters for a specified film thickness without time-consuming numerical analysis by FEMSDA.

Figure 7.36 shows the dispersion relation of a nonleaky SAW under the short-circuited grating on $45°$YZ-Li$_2$B$_4$O$_7$. In the figure, $V_B = 3776$ m/sec is the slow-shear SSBW velocity. In this case, the stopband width for the short-circuited grating is larger than that for the open-circuited grating. Within this range of film thickness, both widths increase with the film thickness.

It is seen that the upper edges of these stopbands coincide with each other. This suggests that $\kappa_{12} > 0$ and the mechanical reflection and electrical reflection for the short-circuited grating are in phase. With an increase in the film thickness, the attenuation due to back-scattering of the BAWs increases at $f > 0.47V_B/p$.

Figure 7.37 shows the h/p dependence of the COM parameters. All parameters are mostly proportional to h/p, and their least square fit gives

$$\kappa_{11}p = \pi\{0.100278 + 0.263837(h/p)\},$$

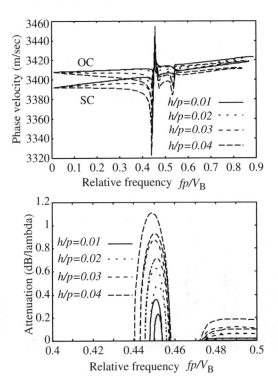

Fig. 7.36. Dispersion relation of nonleaky SAW on $45°$YZ-Li$_2$B$_4$O$_7$

$$\kappa_{12}p = \pi\{0.001975 + 0.464953(h/p)\},$$
$$K_{\mathrm{eff}}^2 = 0.009044 + 0.113987(h/p),$$
$$c = 1.108283 + 1.636644(h/p).$$

7.3.3 Analysis for Multi-Electrode IDTs

IDTs Composed of the Same Finger Widths and Gaps. Here let us consider the COM theory for IDTs which are composed of multiple fingers with the same widths and gaps.

As an example, let us consider the finger configuration shown in Fig. 7.38a. For the following discussion, the mutual connections among the fingers are separated, and the voltage V_i and current I_i for the i-th strip are chosen as variables.

The procedure described in Sect. 7.2 gives the COM equations for this structure:

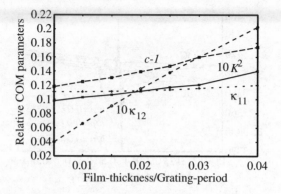

Fig. 7.37. h/p_I dependence of COM parameters for $45°$YZ-Li$_2$B$_4$O$_7$

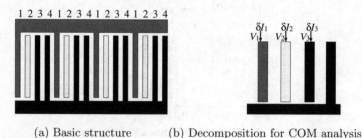

(a) Basic structure   (b) Decomposition for COM analysis

Fig. 7.38. IDTs composed of the same electrode widths and gaps

$$\frac{\partial U_+(x)}{\partial x} = -j\theta_u U_+(x) - j\kappa_{12} U_-(x) + j\zeta_1 V_1 + j\zeta_2 V_2 + j\zeta_3 V_3, \qquad (7.123)$$

$$\frac{\partial U_-(x)}{\partial x} = +j\kappa_{12}^* U_+(x) + j\theta_u U_-(x) - j\zeta_1^* V_1 - j\zeta_2^* V_2 - j\zeta_3^* V_3, \qquad (7.124)$$

$$\frac{\partial I_1(x)}{\partial x} = -4j\zeta_1^* U_+(x) - 4j\zeta_1 U_-(x) + j\omega C_{11} V_1 + j\omega C_{12} V_2$$
$$+ j\omega C_{13} V_3, \qquad (7.125)$$

$$\frac{\partial I_2(x)}{\partial x} = -4j\zeta_2^* U_+(x) - 4j\zeta_2 U_-(x) + j\omega C_{12} V_1 + j\omega C_{22} V_2$$
$$+ j\omega C_{23} V_3, \qquad (7.126)$$

$$\frac{\partial I_3(x)}{\partial x} = -4j\zeta_3^* U_+(x) - 4j\zeta_3 U_-(x) + j\omega C_{13} V_1 + j\omega C_{23} V_2$$
$$+ j\omega C_{33} V_3, \qquad (7.127)$$

where ζ_i is the transduction coefficient for the i-th finger. Since each strip is equivalent to each other for transduction, $|\zeta_1| = |\zeta_2| = |\zeta_3|$. It is clear that ζ_i can be expressed as

$$\zeta_1 = \zeta_0 \exp(-3j\beta p_I/8), \qquad (7.128)$$

$$\zeta_2 = \zeta_0 \exp(-j\beta p_{\mathrm{I}}/8), \tag{7.129}$$

$$\zeta_3 = \zeta_0 \exp(+j\beta p_{\mathrm{I}}/8), \tag{7.130}$$

by taking account of the spatial locations, where β is the wavenumber of the perturbed mode for the short-circuited grating realized by mutually connecting every finger of the IDT. On the other hand, C_{ij} is the static capacitance between the i-th and j-th fingers, and the following relations hold:

$$C_{11} = C_{22} = C_{33} \equiv C_0, \tag{7.131}$$

$$C_{12} = C_{23} \equiv C_1, \tag{7.132}$$

$$C_{13} \equiv C_2. \tag{7.133}$$

For the configuration shown in Fig. 7.38a, $I_2 = 0$ and $V_3 = 0$. Then (7.126) gives

$$V_2 = 4\{\zeta_2^* U_+(x) + \zeta_2 U_-(x)\}/\omega C_0 - C_1/C_0 \times V_1,$$

and substitution into (7.123)–(7.125) gives the following COM equations:

$$\frac{\partial U_+(x)}{\partial x} = -j\hat{\theta}_u U_+(x) - j\hat{\kappa}_{12} U_-(x) + j\hat{\zeta} V_1, \tag{7.134}$$

$$\frac{\partial U_-(x)}{\partial x} = +j\hat{\kappa}_{12}^* U_+(x) + j\hat{\theta}_u U_-(x) - j\hat{\zeta}^* V_1, \tag{7.135}$$

$$\frac{\partial I_1(x)}{\partial x} = -4j\hat{\zeta}^* U_+(x) - 4j\hat{\zeta} U_-(x) + j\omega \hat{C} V_1, \tag{7.136}$$

where

$$\hat{\zeta} = \zeta_1 - \zeta_2 C_1/C_0 = \zeta_0 \exp(-j\beta p_{\mathrm{I}}/8)$$
$$\times \{1 - \exp(-j\beta p_{\mathrm{I}}/4) C_1/C_0\}, \tag{7.137}$$

$$\hat{\theta}_u = \theta_u - 4|\zeta_2|^2/\omega C_0 = \theta_u - 4\zeta_0^2/\omega C_0, \tag{7.138}$$

$$\hat{\kappa}_{12} = \kappa_{12} - 4\zeta_2^2/\omega C_0 = \kappa_{12} - 4\exp(-j\beta p_{\mathrm{I}}/4)\zeta_0^2/\omega C_0, \tag{7.139}$$

$$\hat{C} = C_0 - C_1^2/C_0. \tag{7.140}$$

Similarly, where the double-electrode-type IDT is concerned, the electrical condition is as follows: $V_2 = V_3 \equiv V$, $V_1 = 0$ and $I_2 + I_3 \equiv I$. Then the effective transduction coefficient $\hat{\zeta}_d$ for the double-electrode-type IDT is given by

$$\hat{\zeta}_d = \zeta_2 + \zeta_3 = 2\zeta_0 \cos(\beta p_{\mathrm{I}}/8), \tag{7.141}$$

and the effective static capacitance $\hat{C}_d$ per unit length is given by

$$\hat{C}_d = 2(C_0 + C_1). \tag{7.142}$$

On the other hand, where the single-electrode-type IDT is concerned, the electrical condition is as follows: $V_3 = V_1 \equiv V$, $V_2 = 0$ and $I_1 + I_3 \equiv I$. Then the effective transduction coefficient $\hat{\zeta}_s$ for the single-electrode-type IDT is given by

$$\hat{\zeta}_s = \zeta_1 + \zeta_3 = 2\zeta_0 \cos(\beta p_{\mathrm{I}}/8) \exp(-j\beta p_{\mathrm{I}}/4), \tag{7.143}$$

and the effective static capacitance $\hat{C}_{\mathrm{s}}$ per unit length is given by

$$\hat{C}_{\mathrm{s}} = 2(C_0 + C_2). \tag{7.144}$$

Note θ_{u} and κ_{12} do not depend on the electrical connection.

In the previous section, we showed the procedure for the determination of the COM parameters for a single-electrode-type IDT by using FEMSDA. Thus, by using this result, ζ_0 can be determined from (7.143). In addition, θ_{u} and κ_{12} for the single-electrode-type IDT are readily applicable. Attention should be paid to the fact that these parameters should be determined at $\beta p = \pi/2$ corresponding to the resonance frequency for double-electrode-type IDTs instead of $\beta p = \pi$ corresponding to that for single-electrode-type IDTs.

For the determination of C_i, various methods, such as that described in Sect. 3.2.4, can be used. For example, when $w/p = 0.5$, it is known [3] that the capacitance coefficient $\hat{C}_{ij}$ between the m-th and n-th fingers for an infinitely long metallic grating is given by

$$\hat{C}_{ij} = \epsilon(\infty)W\frac{4/\pi}{1 - 4(i - j)^2}. \tag{7.145}$$

By using this, C_i is given by

$$C_0 p_{\mathrm{I}} = \epsilon(\infty)W \sum_{m=-\infty}^{+\infty} \frac{4/\pi}{1 - 4(4m)^2} \cong \epsilon(\infty)W \times 4/\pi, \tag{7.146}$$

$$C_1 p_{\mathrm{I}} = \epsilon(\infty)W \sum_{m=-\infty}^{+\infty} \frac{4/\pi}{1 - 4(4m + 1)^2} \cong -\epsilon(\infty)W \times 142/105\pi, \tag{7.147}$$

$$C_2 p_{\mathrm{I}} = \epsilon(\infty)W \sum_{m=-\infty}^{+\infty} \frac{4/\pi}{1 - 4(4m + 2)^2} \cong -\epsilon(\infty)W \times 4/15\pi, \tag{7.148}$$

where $p_{\mathrm{I}}(= 4p)$ is the periodicity of the IDT and W is the aperture.

Analysis Based on MULTI. The free software MULTI [13] (see Sect. B.3.2) is an extension of FEMSDA to analyze SAW propagation under metallic gratings with multiple fingers per period. This enables us to derive the COM parameters for complex finger geometry like unidirectional IDTs (UDTs).

Let us consider the finger geometry shown in Fig. 7.39. Due to symmetry, the reflection center for mechanical reflection is expected to be at the middle of finger "2". Since the electrical contribution of finger "3" is small, the excitation center is expected to be at the middle between fingers "1" and "2". Since the excitation center is displaced by about $\lambda/8$ from the reflection center, it is expected to act as a UDT.

To derive the COM parameters for the UDT, we must derive a dispersion relation for the short-circuited grating where all fingers connected to the

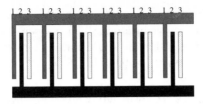

Fig. 7.39. SPUDT with three fingers per period

bus-bars are short-circuited as shown in Fig. 7.40a and that for the open-circuited grating where all fingers connected to the bus-bars are isolated as shown in Fig. 7.40b. Note that this isolation should not be applied to the mutual connection within a unit period.

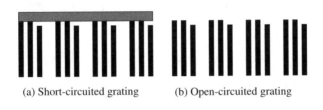

(a) Short-circuited grating   (b) Open-circuited grating

Fig. 7.40. Equivalent grating structure for COM analysis

Figure 7.41 shows the dispersion relation of a nonleaky SAW under the structure on 128°YX-LiNbO$_3$ calculated by MULTI. In the calculation, it is assumed that the Al thickness is $h/p_I = 0.04$, $w = p_I/8$, and the distances are $d_3 = -d_1 = p_I/4$ and $d_2/p_I = 0$. In the figure, $V_B = 4025$m/sec is the slow-shear SSBW velocity. It is seen that the stopband edges are separated from each other. This clearly exhibits the existence of unidirectionality.

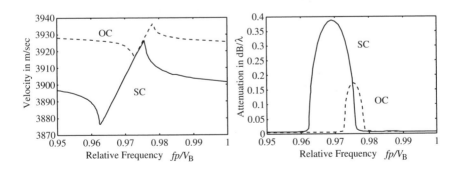

Fig. 7.41. SAW propagation under multi-finger grating near stopband

Even when unidirectionality exists, the COM parameters are readily obtained by applying the procedure described in Sect. 7.3.2. For example, for the result shown in Fig. 7.41, the COM parameters are determined as follows: $|\kappa_{11}p_{\mathrm{I}}| = 0.064\pi$, $|\kappa_{12}p_{\mathrm{I}}| = 0.0071\pi$, $|\xi|^2 p_{\mathrm{I}}/\omega C = 0.029$, $c = 1.047$ and $\angle(\zeta^2/\kappa_{12}) = 25.7°$.

Analysis Based on SYNC and MSYNC. The free software SYNC and MSYNC [14] are modified versions of FEMSDA and MULTI, respectively (see Sect. B.4.4). They are aimed at calculating the input admittance $\hat{Y}(\omega)$ per period p_{I} of infinitely long IDTs, and they run very much faster than FEMSDA and MULTI. As will be discussed below, these packages can also be used for the determination of the COM parameters, and are extremely useful for the analysis of IDTs with unidirectionality.

For numerical fitting, let us calculate $\hat{Y}(\omega)$ by using the COM equations (7.99)–(7.101). Since the IDT discussed here is infinitely long, the mode amplitude $U_{\pm}(x)$ is uniform, $\partial/\partial x = 0$. Then a solution of (7.99) and (7.100) is given by

$$\begin{pmatrix} U_+ \\ U_- \end{pmatrix} = \frac{V}{\theta_{\mathrm{u}}^2 - |\kappa_{12}|^2} \begin{pmatrix} \theta_{\mathrm{u}}\zeta - \kappa_{12}\zeta^* \\ \theta_{\mathrm{u}}\zeta^* - \kappa_{12}^*\zeta \end{pmatrix}. \tag{7.149}$$

Substitution into (7.150) gives the input admittance $\hat{Y}(\omega)$ per period:

$$\hat{Y}(\omega) = \frac{p_{\mathrm{I}}}{V}\frac{\partial I(x)}{\partial x} = j\omega C p_{\mathrm{I}} \left(1 - \frac{8|\zeta|^2}{\omega C}\frac{\theta_{\mathrm{u}} - |\kappa_{12}|\cos\psi}{\theta_{\mathrm{u}}^2 - |\kappa_{12}|^2} \right), \tag{7.150}$$

where $\psi = \angle(\kappa^*\zeta/\zeta^*)$.

Figure 7.42 shows $\hat{Y}(\omega)$ of a nonleaky SAW under the structure shown in Fig. 7.39 on $128°$YX-LiNbO$_3$ calculated by MSYNC. The horizontal axis

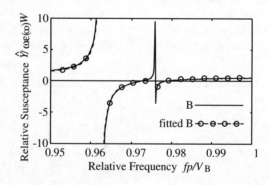

Fig. 7.42. Input admittance of infinite triple-finger IDT on $128°$YX-LiNbO$_3$

is the relative frequency $fp_{\mathrm{I}}/V_{\mathrm{B}}$ where $V_{\mathrm{B}} = 4025$ m/sec is the cut-off frequency for the slow-shear SSBW. The vertical axis is $\Im[\hat{Y}(\omega)]$ normalized by $\omega\epsilon(\infty)W$. Two resonances $\omega_{\mathrm{s}\pm}$ and antiresonances $\omega_{\mathrm{o}\pm}$ are clearly seen. In this frequency region, $\Re[\hat{Y}(\omega)] = 0$ because of the cut-off nature for BAW radiation.

From (7.150), it is seen that $\omega_{s\pm}$ corresponds to a solution of $\theta_u = \pm|\kappa_{12}|$. Assuming θ_u is linearly dependent upon ω, i. e., $\theta_u = \omega/V_{ref} - 2\pi/p_I$, one may obtain

$$V_{ref} = \frac{\omega_{sc}p_I}{2\pi}, \tag{7.151}$$

$$|\kappa_{12}|p_I = 2\pi\frac{\delta\omega_s}{\omega_{sc}}, \tag{7.152}$$

where $\omega_{sc} = (\omega_{s+} + \omega_{s-})/2$ and $\delta\omega_s = (\omega_{s+} - \omega_{s-})/2$.

In addition, $\omega_{o\pm}$ corresponds to a solution of

$$\theta_u = \frac{4|\zeta|^2}{\omega C} \pm \left|\kappa_{12} - \frac{4\zeta^2}{\omega C}\right|.$$

Then application of the above assumption results in

$$\frac{4|\zeta|^2 p_I}{\omega C} = 2\pi\left(\frac{\omega_{oc}}{\omega_{sc}} - 1\right) \tag{7.153}$$

$$\cos\psi = \frac{\delta\omega_s^2 - \delta\omega_o^2 - (\omega_{oc} - \omega_{sc})^2}{2\delta\omega_s(\omega_{oc} - \omega_{sc})}, \tag{7.154}$$

where $\omega_{oc} = (\omega_{o+} + \omega_{o-})/2$ and $\delta\omega_o = (\omega_{o+} - \omega_{o-})/2$.

Thus, by finding $\omega_{o\pm}$ and $\omega_{s\pm}$ from the calculated $\hat{Y}(\omega)$, all the COM parameters except C can readily be determined. Then C can be determined by fitting (7.150) to the numerically obtained $\hat{Y}(\omega)$.

Figure 7.42 also shows $\Im[\hat{Y}(\omega)]$ calculated by using the determined COM parameters. The agreement is excellent.

Note that this technique is applicable only in the case where directionality exists. When the IDT is bidirectional, one pole and one zero are degenerate.

The location of the degenerate zero and pole can be found by adding a tiny directionality. This is realized by giving a slight off-angle for the propagation direction from the crystal principal axis [14]. Note that application of this technique is difficult for leaky SAWs due to strong damping toward the added pole.

In addition, when the velocity dispersion is large, the relations (7.151)–(7.154) are no longer valid because they are based on the linear dependence of θ_u upon ω.

7.4 COM-Based Simulators

7.4.1 SAW Device Simulation

Let us represent each element such as an IDT and reflector by the six-port network shown in Fig. 7.43, and let $\mathbf{P}(x)$ be

$$\mathbf{P}(x) = (A_+, A_-, V_{\text{in}}, I_{\text{in}}, V_{\text{out}}, I_{\text{out}}), \tag{7.155}$$

where V_{in} and V_{out} are the input and output terminal voltages, and I_{in} and I_{out} the currents flowing into and out from the network, respectively. By

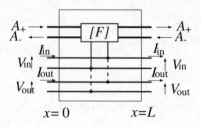

Fig. 7.43. Matrix representation of element.

assuming the initial value $\mathbf{P}(0)$, the solution of the COM equations (7.99)–(7.101) is given by

$$\mathbf{P}(L) = [H]\mathbf{P}(0), \tag{7.156}$$

where the hybrid matrix $[H]$ is given by

$$[H] = \begin{pmatrix} F_{11} & F_{12} & F_{13} & 0 & 0 & 0 \\ F_{21} & F_{22} & F_{23} & 0 & 0 & 0 \\ 0 & 0 & 1 & 0 & 0 & 0 \\ -F_{31} & -F_{32} & -F_{33} & 1 & 0 & 0 \\ 0 & 0 & 0 & 0 & 1 & 0 \\ 0 & 0 & 0 & 0 & 0 & 1 \end{pmatrix} \tag{7.157}$$

for input IDTs, and

$$[H] = \begin{pmatrix} F_{11} & F_{12} & 0 & 0 & F_{13} & 0 \\ F_{21} & F_{22} & 0 & 0 & F_{23} & 0 \\ 0 & 0 & 1 & 0 & 0 & 0 \\ 0 & 0 & 0 & 1 & 0 & 0 \\ 0 & 0 & 0 & 0 & 1 & 0 \\ -F_{31} & -F_{32} & 0 & 0 & -F_{33} & 1 \end{pmatrix} \tag{7.158}$$

for output IDTs. Their elements are

$$F_{11} = \cos(\theta_{\text{p}}L) - j\frac{\theta_{\text{u}}}{\theta_{\text{p}}}\sin(\theta_{\text{p}}L),$$

$$F_{12} = -j\frac{\kappa_{12}}{\theta_{\text{p}}}\sin(\theta_{\text{p}}L),$$

$$F_{13} = \frac{\theta_{\text{u}}\zeta - \kappa_{12}\zeta^*}{\theta_{\text{p}}^2}\{1 - \cos(\theta_{\text{p}}L)\} + j\frac{\zeta}{\theta_{\text{p}}}\sin(\theta_{\text{p}}L),$$

$$F_{21} = +j\frac{\kappa_{12}^*}{\theta_{\text{p}}}\sin(\theta_{\text{p}}L),$$

$$F_{22} = \cos(\theta_\mathrm{p}L) + j\frac{\theta_\mathrm{u}}{\theta_\mathrm{p}}\sin(\theta_\mathrm{p}L),$$

$$F_{23} = \frac{\theta_\mathrm{u}\zeta^* - \kappa_{12}^*\zeta}{\theta_\mathrm{p}^2}\{1 - \cos(\theta_\mathrm{p}L)\} - j\frac{\zeta^*}{\theta_\mathrm{p}}\sin(\theta_\mathrm{p}L),$$

$$F_{31} = -4\frac{\theta_\mathrm{u}\zeta^* - \kappa_{12}^*\zeta}{\theta_\mathrm{p}^2}\{1 - \cos(\theta_\mathrm{p}L)\} - 4j\frac{\zeta^*}{\theta_\mathrm{p}}\sin(\theta_\mathrm{p}L),$$

$$F_{32} = 4\frac{\theta_\mathrm{u}\zeta - \kappa_{12}\zeta^*}{\theta_\mathrm{p}^2}\{1 - \cos(\theta_\mathrm{p}L)\} - 4j\frac{\zeta}{\theta_\mathrm{p}}\sin(\theta_\mathrm{p}L),$$

$$F_{33} = -\frac{4j}{\theta_\mathrm{p}^3}(2\theta_\mathrm{u}|\zeta|^2 - \kappa_{12}\zeta^{*2} - \kappa_{12}^*\zeta^2)\{\theta_\mathrm{p}L - \sin(\theta_\mathrm{p}L)\} + j\omega CL.$$

In these equations,

$$\theta_\mathrm{p} = \begin{cases} -\sqrt{\theta_\mathrm{u}^2 - |\kappa_{12}|^2} & (\theta_\mathrm{u} < -|\kappa_{12}|) \\ -j\sqrt{|\kappa_{12}|^2 - \theta_\mathrm{u}^2} & (|\theta_\mathrm{u}| < |\kappa_{12}|) \\ \sqrt{\theta_\mathrm{u}^2 - |\kappa_{12}|^2} & (\theta_\mathrm{u} > |\kappa_{12}|) \end{cases} \tag{7.159}$$

is the wavenumber of the perturbed mode under a short-circuited grating.

On a uniformly metallized or free surface, the matrix elements are given by

$$F_{11} = \exp(-j\beta L),$$

$$F_{22} = \exp(+j\beta L),$$

and the other elements are zero.

Now consider the two-port SAW resonator shown in Fig. 7.44. Representing

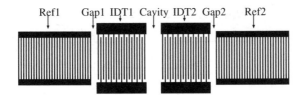

Fig. 7.44. Two-port SAW resonator

the IDTs and reflectors by the six-port networks in the form of Fig. 7.43, and cascade-connecting them as shown in Fig. 7.45, one obtains the total hybrid matrix $[H^\mathrm{t}]$ in the form

$$[H^\mathrm{t}] = [H_7][H_6][H_5][H_4][H_3][H_2][H_1]. \tag{7.160}$$

Since $A_+ = I_\mathrm{in} = I_\mathrm{out} = 0$ at $x = x_1$ and $A_- = 0$ at $x = x_2$, the transfer admittance matrix $[Y_\mathrm{t}]$ between the electrical input and output is given by

$$[Y_\mathrm{t}] = \begin{pmatrix} -H_{43}^\mathrm{t} + H_{23}^\mathrm{t}H_{42}^\mathrm{t}/H_{22}^\mathrm{t} & -H_{45}^\mathrm{t} + H_{25}^\mathrm{t}H_{42}^\mathrm{t}/H_{22}^\mathrm{t} \\ -H_{63}^\mathrm{t} + H_{23}^\mathrm{t}H_{62}^\mathrm{t}/H_{22}^\mathrm{t} & -H_{66}^\mathrm{t} + H_{25}^\mathrm{t}H_{62}^\mathrm{t}/H_{22}^\mathrm{t} \end{pmatrix}. \tag{7.161}$$

With this $[Y_t]$, various effects of parasitic and matching elements can be taken into account in the simulation.

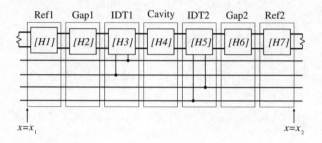

Fig. 7.45. Cascading matrix elements

7.4.2 Inclusion of Peripheral Circuit

In the UHF range, the effects of parasitic impedances, such as inductance of bonding wires and capacitance with the package, cannot be ignored, and they must be taken into account for the simulation. Here, an effective method of including parasitic elements in the simulation is described.

Figure 7.46 shows the circuit considered here. First, let us define the following vectors;

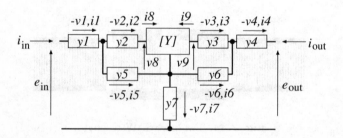

Fig. 7.46. Circuit considered here

$$\mathbf{v} = (v_1, v_2, v_3, v_4, v_5, v_6, v_7, v_8, v_9)^t, \tag{7.162}$$
$$\mathbf{i} = (i_1, i_2, i_3, i_4, i_5, i_6, i_7, i_8, i_9)^t. \tag{7.163}$$

Then they are related as

$$\mathbf{i} = \mathbf{Y}\mathbf{v}, \tag{7.164}$$

where

$$\mathbf{Y} = \begin{pmatrix} y_1 & 0 & 0 & 0 & 0 & 0 & 0 & 0 & 0 \\ 0 & y_2 & 0 & 0 & 0 & 0 & 0 & 0 & 0 \\ 0 & 0 & y_3 & 0 & 0 & 0 & 0 & 0 & 0 \\ 0 & 0 & 0 & y_4 & 0 & 0 & 0 & 0 & 0 \\ 0 & 0 & 0 & 0 & y_5 & 0 & 0 & 0 & 0 \\ 0 & 0 & 0 & 0 & 0 & y_6 & 0 & 0 & 0 \\ 0 & 0 & 0 & 0 & 0 & 0 & y_7 & 0 & 0 \\ 0 & 0 & 0 & 0 & 0 & 0 & 0 & Y_{11} & Y_{12} \\ 0 & 0 & 0 & 0 & 0 & 0 & 0 & Y_{12} & Y_{22} \end{pmatrix}. \tag{7.165}$$

Since the total current flow for each node is zero, we get

$$\begin{pmatrix} \mathbf{i}_t \\ 0 \end{pmatrix} = \begin{pmatrix} \mathbf{G}_1 \\ \mathbf{G}_2 \end{pmatrix} \mathbf{i}, \tag{7.166}$$

where $\mathbf{i}_t = (i_{\mathrm{in}}, i_{\mathrm{out}})^t$, and

$$\mathbf{G}_1 = \begin{pmatrix} 1 & 0 & 0 & 0 & 0 & 0 & 0 & 0 & 0 \\ 0 & 0 & 0 & -1 & 0 & 0 & 0 & 0 & 0 \end{pmatrix}, \tag{7.167}$$

$$\mathbf{G}_2 = \begin{pmatrix} 1 & -1 & 0 & 0 & -1 & 0 & 0 & 0 & 0 \\ 0 & 1 & 0 & 0 & 0 & 0 & 0 & -1 & 0 \\ 0 & 0 & 1 & 0 & 0 & 0 & 0 & 0 & 1 \\ 0 & 0 & 1 & -1 & 0 & 1 & 0 & 0 & 0 \\ 0 & 0 & 0 & 0 & 1 & -1 & -1 & 1 & 1 \end{pmatrix}. \tag{7.168}$$

Since the total voltage drop for each closed circuit is zero, we get

$$\begin{pmatrix} \mathbf{e}_t \\ 0 \end{pmatrix} = \begin{pmatrix} \mathbf{F}_1 \\ \mathbf{F}_2 \end{pmatrix} \mathbf{v}, \tag{7.169}$$

where $\mathbf{e}_t = (e_{\mathrm{in}}, e_{\mathrm{out}})^t$, and

$$\mathbf{F}_1 = \begin{pmatrix} 1 & 0 & 0 & 0 & 1 & 0 & 1 & 0 & 0 \\ 0 & 0 & 0 & -1 & 0 & -1 & 1 & 0 & 0 \end{pmatrix} \tag{7.170}$$

$$\mathbf{F}_2 = \begin{pmatrix} 0 & 1 & 0 & 0 & -1 & 0 & 0 & 1 & 0 \\ 0 & 0 & -1 & 0 & 0 & 1 & 0 & 0 & 1 \end{pmatrix}. \tag{7.171}$$

By rearranging (7.164), (7.166) and (7.169), we get the closed expression

$$\mathbf{i}_t = \mathbf{G}_1 \mathbf{Y} \begin{pmatrix} \mathbf{F}_1 \\ \mathbf{F}_2 \\ \mathbf{G}_2 \mathbf{Y} \end{pmatrix}^{-1} \begin{pmatrix} \mathbf{e}_t \\ 0 \\ 0 \end{pmatrix}. \tag{7.172}$$

Although the order of the matrix is equal to the number of elements, the computation time to solve (7.172) is negligible for present computers.

7.4.3 Results of Simulation

By using the COM parameters determined in Sect. 7.3.2, the one-port SAW resonator on $128°$YX-LiNbO$_3$ was analyzed. The result is shown in Fig. 7.47.

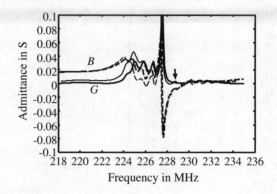

Fig. 7.47. Admittance characteristics of one-port SAW resonator on 128°YX-LiNbO₃

In the calculation, $h/p_{\mathrm{I}} = 0.88\%$. The simulation shown by the bold line agrees well with the experiment shown by the thin line, and the effectiveness of the method is demonstrated. Note the response indicated by the arrow in the figure may be due to coupling with the thickness vibration, which is not taken into account for the simulation. This experimental result was given by Mitsubishi Material Co. Ltd.

The measurements were performed under the situation where the device was mounted in the package, and its effect was taken into account. Figure 7.48 shows the equivalent circuit employed. In the figure, L_{p} is the inductance of the wires, and C_{p} is the parasitic capacitance. Since there remains a certain amount of error in material constants employed, adjustment for the COM parameters was also given.

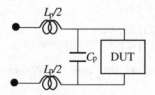

Fig. 7.48. Employed equivalent circuit

Referring to Fig. 7.49, the adjustment procedure is given as follows:

1. adjustment of the propagation loss to fit the half-width of the conductance at the resonance frequency (0.0035 dB/λ);
2. adjustment of κ_{12} to fit the stopband width ($\times 1.1$);
3. adjustment of κ_{11} to fit the stopband location; (SAW velocity reduction by 0.27%);

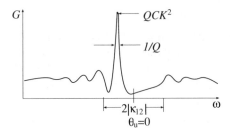

Fig. 7.49. Representative conductance characteristics

4. adjustment of L_p to fit the resonance frequency ($L_p =11$ nH);
5. adjustment of C_t to fit the susceptance at frequencies far from resonance ($C_p =3.5$ pF).

Figure 7.50 shows the result at intermediate stages of this procedure. Although the following adjustments may be required in general, they were not applied for this case:

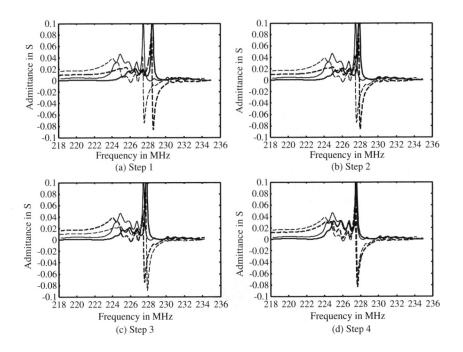

Fig. 7.50. Adjustment of COM parameters

1. adjustment of C to fit the overall amplitude;

2. adjustment of K^2 to fit the conductance height.

Next, a one-port SAW resonator on X-112°Y LiTaO₃ with $h/p_I = 0.88\%$ was analyzed. The result is shown in Fig. 7.51. The theoretical results given by the bold line agree well with the experimental results given by the thin line. The experimental results were given by Mitsubishi Material, Co. Ltd.

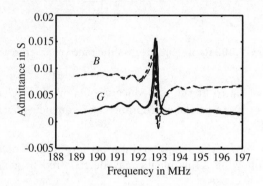

Fig. 7.51. Admittance characteristics of one-port SAW resonator on X-112°Y LiTaO₃

Here the following adjustments were given:

1. adjustment of the propagation loss to fit the half-width of the conductance peak (0.0028 dB/λ);
2. adjustment of L_p to fit the location of the resonance peak (L_p =5 nH);
3. adjustment of C_p to fit the susceptance (C_p =2.3 pF);
4. inclusion of the conductance parallel to the IDT (1.3 mS).

For precise determination of the COM parameters, one should employ the device configuration shown in Fig. 7.52. This enables us to use the microwave wafer-probe to remove parasitic impedances and to measure all COM parameters independently [15].

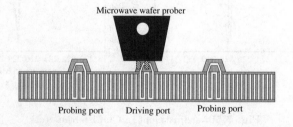

Fig. 7.52. Device structure fitted to microwave wafer-probe

References

1. B.A. Auld: Acoustic Waves and Fields in Solids, Vol. II, Chap. 11 Wiley, New York (1973) pp. 221–332.
2. F.G. Marshall and E.G.S. Paige: Novel Acoustic-Surface-Wave Directional Coupler with Diverse Applications, Electron. Lett., **7** (1971) pp. 460–464.
3. D.P. Morgan: Surface-Wave Devices for Signal Processing, Appendix D, Elsevier, Amsterdam (1985) pp. 363–373.
4. R. Ganss-Puchstein, C. Ruppel and H.R. Stocker: Spectrum Shaping SAW Filters for High-Bit-Rate Digital Radio, IEEE Trans. Ultrason., Ferroelec., and Freq. Contr., **UFFC-35**, 6 (1988) pp. 673–684.
5. Y. Kinoshita, M. Hikita, T. Tabuchi and H. Kojima: Broadband Resonant Filter Using Surface-Shear-Wave Mode and Twin-Turn Reflector, Electron. Lett., **15**, 4 (1979) pp. 130–131.
6. P.V. Wright: Natural Single-Phase Unidirectional Transducer, Proc. IEEE Ultrason. Symp. (1985) pp. 58–63.
7. C.S. Hartmann, P.V. Wright, R.J. Kansy and E.M. Garber: Analysis of SAW Interdigital Transducer with Internal Reflections and the Application to the Design of Single-Phase Unidirectional Transducers, Proc. IEEE Ultrason. Symp. (1982) pp. 40–45.
8. P.V. Wright: A New Generalized Modeling of SAW Transducers and Gratings, Proc. Freq. Contr. Symp. (1989) pp. 596–605.
9. T. Kojima and K. Shibayama: An Analysis of an Equivalent Circuit Model for an Interdigital Surface-Acoustic-Wave Transducer, Jpn. J. Appl. Phys., **27**, Suppl. 27-1 (1988) pp. 163–165.
10. R.C.M. Li and J. Melngailis: The Influence of Stored Energy at Step Discontinuities on the Behavior of Surface-Wave Gratings, IEEE Trans. Sonics and Ultrason., **SU-22** (1975) pp. 189–198.
11. Z.H. Chen and K. Yamanouchi: Theoretical Analysis of Relations Between Directivity of SAW Transducers and Its Dispersion Curves, Proc. IEEE Ultrason. Symp. (1989) pp. 71–74.
12. K. Hashimoto and M. Yamaguchi: Free Software Products for Simulation and Design of Surface Acoustic Wave and Surface Transverse Wave Devices, Proc. 1996 Freq. Contr. Symp. (1996) pp. 300–307.
13. K. Hashimoto, G.Q. Zheng and M. Yamaguchi: Fast Analysis of SAW Propagation Under Multi-Electrode-Type Grating, Proc. IEEE Ultrason. Symp. (1997) pp. 279–284.
14. K. Hashimoto, J. Koskela and M.M. Salomaa: Fast Determination of Coupling-of-Modes Parameters Based on Strip Admittance Approach, Proc. IEEE Ultrason. Symp. (1999) to be published.
15. C.S. Hartmann and B.P. Abott: Overview of Design Challenges for Single Phase Unidirectional SAW Filters, Proc. IEEE Ultrason. Symp. (1989) pp. 79–89.

8. Simulation of SH-type SAW Devices

This chapter describes SH-type SAWs that are now widely used. Various properties for the waves are discussed in the comparison with conventional Rayleigh-type SAWs. Detailed information is given about COM analysis modified for the SH-type SAWs, and simulations of the state-of-the-art SAW devices are illustrated.

8.1 Physics of SH-Type SAWs

8.1.1 Summary

As described in Chap. 1, SH-type SAWs such as the Bleustein–Gulyaev–Shimizu wave (BGS wave) [1, 2, 3] have several important features compared to conventional Rayleigh-type SAWs. For example, SH-type SAWs sometimes possess larger piezoelectricity than Rayleigh-type SAWs on the same substrate material. When the SV-type BAW is slow-shear, the SH-type SAW is usually leaky. However its propagation loss becomes negligible when its coupling with the SV component is small. In this case, since its velocity is close to the fast-shear BAW, the SH-type SAW is faster than the Rayleigh-type SAW. This feature is desirable for high-frequency applications.

The difference in the particle motion has several implications for device construction. In Rayleigh-type SAWs, reflection at the substrate edges can be suppressed effectively by placing absorbers. However, this technique is not so effective for SH-type SAWs. In addition, since the BAWs excited into the substrate are less sensitive to sand-blasting of the back surface, a relatively deep saw-cut must be applied for their suppression.

It is interesting to note that since the BGS wave is totally reflected at a straight edge, it can be used as a reflector for the resonator [4].

8.1.2 Propagation and Excitation on a Uniform Surface

Figure 8.1 shows the effective permittivity as a function of $V = S^{-1}$ for $36°$YX-LiTaO$_3$ [5], which is widely used in RF filters for mobile communication systems. In $\epsilon(S)$, there exists two poles: the one at 3150 m/sec is due to Rayleigh-type, and the other at 4100 m/sec is due to SH-type SAWs on the metallized surface. On the other hand, in $\epsilon(S)^{-1}$, the pole at 3150 m/sec is due to Rayleigh-type, and the other at 4200 m/sec is due to SH-type SAWs on the free surface. Note that the poles with larger velocities possess an imaginary part. This indicates that they are due to leaky SAWs.

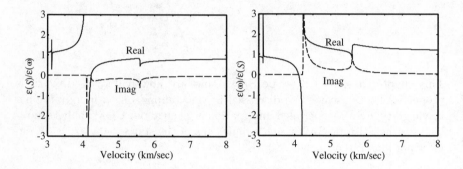

Fig. 8.1. Effective permittivity for 36°YX-LiTaO$_3$

It is interesting to note that the behavior of $\epsilon(S)$ near the complex pole is quite different from that of $\epsilon(S)^{-1}$. That is, $\Im[\epsilon(S)]$ is symmetric with respect to S whereas $\Im[\epsilon(S)^{-1}]$ is asymmetric. That is, the former behaves like a pole whereas latter behaves like a branch point. Hence, we can conclude that the SH-type leaky SAW is predominantly radiated on the metallized surface characterized by $\epsilon(S)$ whereas the SSBW [6] is predominantly excited on the free surface characterized by $\epsilon(S)^{-1}$.

In fact, the electromechanical coupling factor $K_{\text{Sm}}^2 = 7.75\%$ for the metallized surface evaluated from (6.1) is considerably different from $K_{\text{Sf}}^2 = 0.702\%$ for the free surface evaluated from (6.2). Note that these values are considerably different from $2\Delta V/V = 5.56\%$ evaluated by (6.3). This is because the behavior of $\epsilon(S)$ cannot be approximated by (6.40) when the SSBW and SAW velocities are in proximity, and estimation of the electromechanical coupling factor by $2\Delta V/V$ is no longer valid.

Figure 8.2 shows the effective permittivity of 41°YX-LiNbO$_3$ [7], where the propagation loss is negligible on the free surface. As with 36°YX-LiTaO$_3$, the behavior of $\epsilon(S)$ near the complex pole is quite different from that of $\epsilon(S)^{-1}$. This clearly shows that an SH-type leaky SAW is predominantly radiated on the metallized surface whereas an SSBW is predominantly excited on the free surface.

Figure 8.3 shows the effective permittivity of 64°YX-LiNbO$_3$ [7], where the propagation loss is negligible on the metallized surface. In this case, the behavior of $\epsilon(S)$ near the complex pole is similar to that of $\epsilon(S)^{-1}$, and it shows that the SH-type leaky SAW is predominantly radiated on both the free and metallized surfaces.

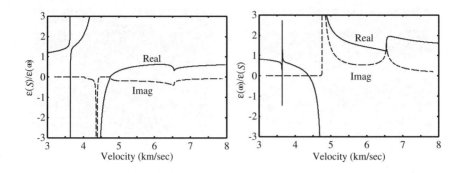

Fig. 8.2. Effective permittivity for 41°YX-LiNbO$_3$

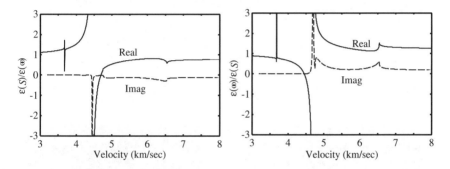

Fig. 8.3. Effective permittivity for 64°YX-LiNbO$_3$

Velocities and attenuations were determined by impulse response measurements between the launching and receiving IDTs. For the measurements, two filters of very broad bandwidth were fabricated on each substrate of 36°YX-LiTaO$_3$, 41°YX-LiNbO$_3$ and 64°YX-LiNbO$_3$ [8]. The surface between the launching and receiving IDTs of one filter was metallized with a thin Al film, while that of the other filter was kept free. The bottom surface of the substrates was not roughed, so that the BAWs radiated from the propagating leaky SAW may be reflected at the bottom surface and detected by the receiving IDT.

Figures 8.4, 8.5 and 8.6 show typical impulse responses for 36°YX-LiTaO$_3$, 41°YX-LiNbO$_3$ and 64°YX-LiNbO$_3$, respectively. Comparison between the theoretical and experimental velocities suggested that the dominant responses (Pulse L) for the metallized surface are those of leaky SAWs. The wave amplitude, the velocity of which corresponds to the SSBW velocity, was negligible on the metallized surface. Hence, it can be concluded that the leaky SAW is dominant for signal transfer on the metallized surface. On the other hand, the dominant responses (Pulse S/L) in Figs. 8.4a and 8.5a are those of the SSBW and/or leaky SAW propagating on the free surface of 36°YX-LiTaO$_3$

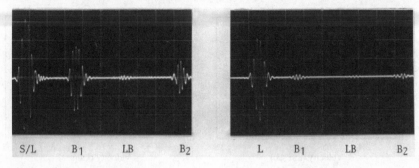

(a) Free surface (V: 10 mV/div.)   (b) Metallized surface (V: 100 mV/div.)

Fig. 8.4. Impulse response for $36°$YX-LiTaO$_3$. H: 100 ns/div., and $L = 150p_{\mathrm{I}}$

and $41°$YX-LiNbO$_3$, respectively. Since the SSBW and leaky SAW velocities on the free surface are in close proximity, it is difficult for a simple velocity measurement to distinguish between the SSBW and leaky SAW contributions to Pulse S/L.

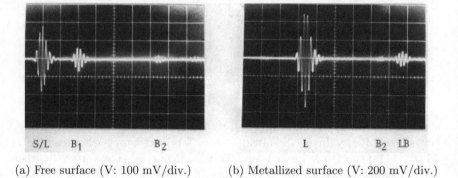

(a) Free surface (V: 100 mV/div.)   (b) Metallized surface (V: 200 mV/div.)

Fig. 8.5. Impulse response for $41°$YX-LiNbO$_3$. H: 100 ns/div., and $L = 175p_{\mathrm{I}}$

As for $64°$YX-LiNbO$_3$, the velocity measurement suggested that Pulses L for both the free and metallized surfaces are attributed to the leaky SAW.

There are responses which are independent of the surface electrical boundary condition. From time delay measurements, pulses B1 and B2 in Figs. 8.4–8.6 were identified as being attributed to fast-shear waves, which are radiated into the bulk and reflected once or twice at the bottom and top surfaces of the substrate as shown in Fig. 8.7a.

Figure 8.8 shows the propagation distance dependence of the delay time. This response is attributed to the slow-shear wave which is radiated from a propagating leaky SAW and reflected at the bottom surface as shown in Fig.

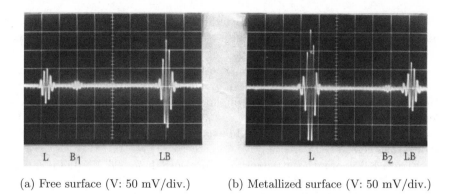

(a) Free surface (V: 50 mV/div.)   (b) Metallized surface (V: 50 mV/div.)

Fig. 8.6. Impulse response for $64°$YX-LiNbO$_3$. H: 100 ns/div., and $L = 100p_I$

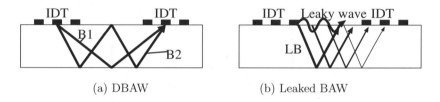

(a) DBAW   (b) Leaked BAW

Fig. 8.7. BAW received by IDT

8.7b. Note that this difference in the delay time is mainly due to a nonzero power flow angle arising from the anisotropy of the substrate. Note that, since the radiation angle θ_L of the leaked BAW is independent of frequency, its response cannot be detected by the IDT when the propagation distance is smaller than $2t \cot \theta_L$, where t is the substrate thickness.

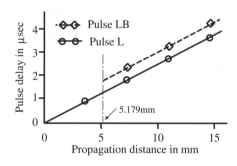

Fig. 8.8. Dependence of delay time for pulses LB and L on propagation pass length

Since the existence of Pulse LB indicates that a leaky SAW is excited and propagates, it is clear that Pulse S/L on 36°YX-LiTaO₃ includes the contribution of a leaky SAW propagating on the free surface.

Note that since the leaked BAW experiences reflection at the bottom surface, its response can be suppressed by sand-blasting the bottom surface. However, since it is SH-type, its roughness must be of the order of wavelengths. In devices in the GHz range, the response due to the leaked BAW is insignificant because the device dimension becomes smaller than the substrate thickness.

Figure 8.9 shows the attenuation of Pulse S/L for 36°YX-LiTaO₃ and 41°YX-LiNbO₃ as a function of the IDT separation x. The relative peak-to-peak value of Pulse S/L decreases monotonically with an increase in x. By a least square fit, the dependences were estimated to be $x^{-0.45}$ and $x^{-0.49}$, respectively, for 36°YX-LiTaO₃ and 41°YX-LiNbO₃.

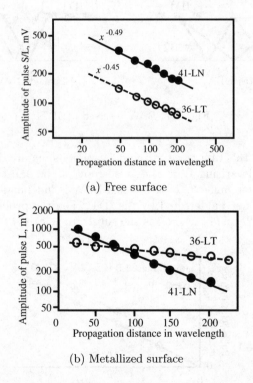

(a) Free surface

(b) Metallized surface

Fig. 8.9. Dependence of amplitude of pulses LB and L with propagation distance

By the way, the decay constants α for Pulse L (leaky SAW) on 36°YX-LiTaO₃ and 41°YX-LiNbO₃ were found to be about 2.4×10^{-2} dB/λ and 9.5×10^{-2} dB/λ, respectively.

Allowing for the effects of in-plane diffraction and electrical resistivity, these values for the free and metallized surfaces well reflect the distinctive attenuation characteristics associated with SSBWs and leaky SAWs, respectively. On the metallized surface, the attenuation characteristics again indicate that the leaky SAW is dominant, independent of x.

On the other hand, for the free surface, only the distinctive attenuation characteristics associated with the SSBW were observed throughout the experiments, up to a propagation distance of 200 λ.

8.1.3 Behavior on a Grating

Figure 8.10 shows the dispersion relation of the leaky SAWs propagating under the grating on 36°YX-LiTaO$_3$ with zero film thickness [9]. In the figure, p is the grating period, and V_B is fast-shear SSBW velocity (4226 m/sec). The stopband at $fp/V_B \cong 0.5$ is due to the coupling between forward and backward leaky SAWs, and at $fp/V_B \cong 0.42$ there is another stopband due to the coupling between the leaky SAW and Rayleigh-type SAW [10]. At $fp/V_B \cong 0.5 - 0.57$, strong attenuation occurs. This is due to back-scattering into the fast-shear BAW described in Sect. 2.2.1 [11]. Similarly, the attenuation at $fp/V_B \cong 0.57 - 0.65$ is mainly due to back-scattering into the L-type BAW.

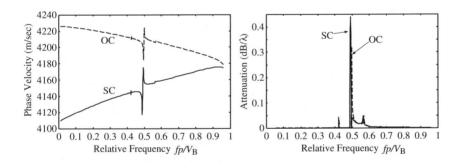

Fig. 8.10. Dispersion of leaky SAW under grating structure on 36°YX-LiTaO$_3$ with zero film thickness

Figure 8.11 shows the transmission response of a leaky SAW under the short-circuited grating on 36°YX-LiTaO$_3$ [11]. The theory agrees well with the experiment.

Figure 8.12 shows the change in the dispersion relation with Al film thickness h [12, 13]. With an increase in h, the stopband width increases, and at $h/p > 0.5$, its upper edge coincides with the cut-off frequency for the back-scattering of the fast-shear BAW. In addition, the amplitude of the back-scattered BAW also increases with an increase in h.

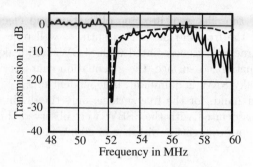

Fig. 8.11. Transmission response of leaky SAW under short-circuited grating on 36°YX-LiTaO₃ with zero film thickness. Solid line: experiment, and broken line: theory

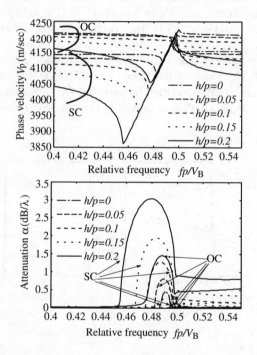

Fig. 8.12. Change in dispersion relation with Al film thickness (36°YX-LiTaO₃)

It should be noted that, although the propagation loss at frequencies lower than the stopband increases with an increase in h in this case, there is also a situation where the propagation loss decreases with an increase in h. This is due to the fact that the propagation loss is determined by the coupling strength with the slow-shear component, and is dependent upon the surface boundary condition.

Figure 8.13 shows the propagation loss near the stopband as a function of the angle θ for rotated Y-cut LiTaO$_3$ with h as a parameter [12, 13]. When $h = 0$, zero propagation loss is achieved at $\theta \cong 36°$. With an increase in h, θ giving zero propagation loss increases, and it attains $41°$ for $h = 0.1$ λ.

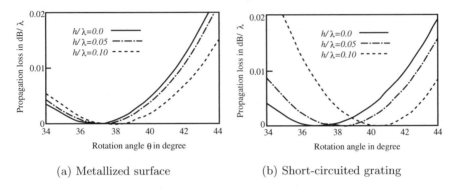

(a) Metallized surface   (b) Short-circuited grating

Fig. 8.13. Rotation angle dependence of propagation loss

One-port SAW resonators with Al thickness of 0.1 λ were fabricated by using the same photomask on several substrates with different θ. Figure 8.14 shows the change in the resonance Q as a function of θ [12, 13]. It is seen that Q takes a maximum at $\theta = 42°$, and comparing with that for $\theta = 36°$, it increases by 1.6. This increase in the resonant Q is significant for practical use. It should be noted that, since the change in θ scarcely affects other propagation characteristics of the leaky SAW and device performance, the design tools developed for 36°YX-LiTaO$_3$ are readily applicable to 42°YX-LiTaO$_3$.

Figure 8.15 shows the frequency response of ladder-type filters fabricated on 42°YX-LiTaO$_3$ and 36°YX-LiTaO$_3$ for the PCS-Rx application [13] developed by Fujitsu Ltd. The device was designed independently for each substrate to minimize the unwanted response in the Tx band (1850–1910 MHz) for a fixed passband width of 60 MHz (1930–1990 MHz). The results show that the out-of-band rejection level RL at the Tx band edge and the insertion loss IL are improved by 8 dB and 1 dB, respectively, in the filters based on 42°YX-LiTaO$_3$. Thanks to these improvements, it is only 42°YX-LiTaO$_3$ devices that can fulfill the tight specifications (RL >20 dB and IL < 3.5 dB).

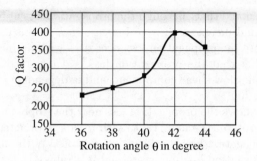

Fig. 8.14. Rotation angle dependence of resonance Q

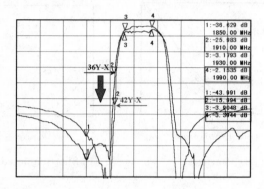

Fig. 8.15. Frequency response of ladder-type SAW filters for PCS-Rx application. H: 50 MHz/div., center: 1900 MHz, and V: 5dB/div.

8.1.4 Electrical Characteristics of IDTs

It was shown in Sect. 8.1.2 that the $\epsilon(S)$ and $\epsilon(S)^{-1}$ characteristics are considerably different from each other for 36°YX-LiTaO$_3$ and 41°YX-LiNbO$_3$. The discussion given in Sect. 6.2 suggests that the IDT is characterized by the charge distribution on it when the propagation surface is free whereas by the electric field when the propagation surface is metallized. One may expect that the charge distribution on the IDT is approximately given by the static one. Then the input admittance of the IDT with 10 finger-pairs was estimated by using (6.27) and (6.28) [9]. The result is shown in Fig. 8.16. In the calculation, 36°YX-LiTaO$_3$ was assumed as a substrate. It is seen that the characteristics reflect predominant excitation of the SSBW when the propagation surface is free, whereas it is that of the leaky SAW when the propagation surface is metallized.

Figure 8.17 shows the experimental results [9]. In contrast with the analysis, the result behaved as if the leaky SAW is predominantly excited independent of the propagation surface. This suggests that electrical properties

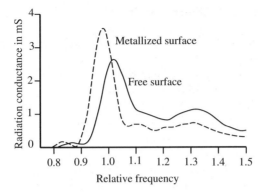

Fig. 8.16. Conductance of IDT on $36°$YX-LiTaO$_3$ calculated by electrostatic approximation

of the IDT are mostly determined by mutual coupling among fingers, and is scarcely affected by the propagation surface outside the IDT.

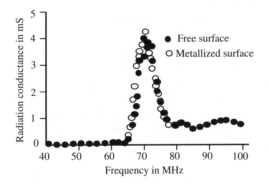

Fig. 8.17. Conductance of IDT on $36°$YX-LiTaO$_3$

This fact may be explained as follows: since the admittance of the IDT corresponds to the current flow when the IDT is driven by a constant voltage source, the admittance is characterized by excitation and propagation of the grating modes in the short-circuited grating, whose influence is similar to that of the metallized surface. The leaky SAW-like grating mode is converted into an SSBW when the propagation surface is free whereas it propagates as a leaky SAW [14] when the propagation surface is metallized.

8.1.5 Effects of Back-Scattered BAWs

From the early day of the development of RF filters employing leaky SAWs, it was known empirically that there exists some sort of dispersion near the passband, and its correction is required for the application of conventional simulation tools to leaky SAW devices [15].

By using FEMSDA (see Sect. B.3.2), the dispersion relation was calculated for the leaky SAW under a short-circuited grating (Al, $h/p = 0.1$) on $36°$YX-LiTaO$_3$, and the COM parameters were estimated by using the conventional procedure described in Sect. 7.3.2. Figure 8.18 shows the dispersion relation calculated from the obtained COM parameters with that obtained by FEMSDA. In the calculation, $h/p = 0.1$. The agreement is rather poor under comparison with the fitted result for Rayleigh-type SAW shown in Sect. 7.3.2, especially near the upper edge of the stopband. It is seen that the effective velocity at frequencies higher than the stopband is relatively larger than that at frequencies lower than the stopband.

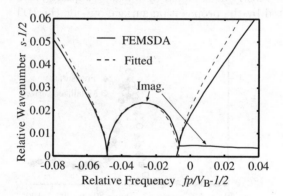

Fig. 8.18. Dispersion relation calculated by conventional COM theory with that obtained by FEMSDA

Let us consider this effect by using Fig. 8.19 [16]. Although the leaky SAW is scattered into BAWs by the grating structure, scattered BAWs cancel each other and are not radiated into the bulk under the SSBW cut-off frequency. Then the corresponding energy is stored and causes a reduction of the leaky SAW velocity. This phenomenon is the energy storing effect described in Sect. 2.2.2, and then at frequencies higher than the SSBW cut-off frequency, the energy storing effect diminishes due to BAW radiation instead of storage. Note that the scattering amplitude takes a maximum near the SSBW cut-off frequency, resulting in maximum velocity reduction.

It should be noted that the cut-off frequency is close to the leaky SAW resonance frequency. Then the variation in the energy storing effect consider-

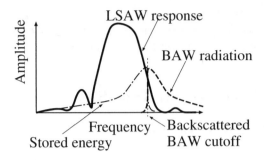

Fig. 8.19. Origin of dispersion near stopband

ably affects the resonant characteristics of grating structures such as reflectors and IDTs.

As for conventional Rayleigh-type SAWs, since they are considerably slower than SSBWs on the same substrate, the effects of the dispersion in the energy storing effect are not significant.

8.1.6 Influence of Grating Edge

The propagation characteristics of SH-type SAWs are quite sensitive to the surface boundary condition. In previous discussions, we were concerned with infinitely long grating structures. However, actual gratings are finite, and the effects of their edges must be considered.

That is, cancellation of the scattered BAWs does not occur at the grating edges as shown in Fig. 8.20. Thus, at the edge, the corresponding energy is radiated into the bulk as the BAWs. In addition, the energy storing effect disappears. This causes additional variation in the phase shift. This must be taken into account when designing resonators employing leaky SAWs because the phase shift directly affects the cavity length.

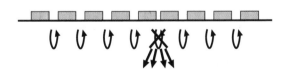

Fig. 8.20. Discontinuity at grating edge

Note that the penetration depth for SH-type SAWs varies with frequency, and this makes additional effects for propagation near the stopband. Since the penetration depth increases at the upper edge of the stopband, the grating mode cannot transfer its energy efficiently between adjacent gratings, and this will result in increased insertion loss [17].

From this reason, devices employing SH-type SAWs are often designed so that the resonance frequency coincides with the lower edge of the stopband without giving gaps between the IDT and reflector.

8.2 COM Theory for SH-Type SAWs

8.2.1 COM Parameter Derivation

As described in Sect. 8.1.5, relatively large frequency dispersion occurs near the resonance frequency of the gratings due to the coupling of the SAW with back-scattered BAWs.

Abbott suggested that the following expression is effective to approximate the dispersion relation of the STW [18] (see Sect. 8.3.2):

$$\theta_\mathrm{p} = c\sqrt{\{\Delta - \Delta_\mathrm{v} + |\kappa_\mathrm{B}|\nu(\Delta)\}^2 - \{\kappa + \kappa_\mathrm{B}\nu(\Delta)\}^2},\tag{8.1}$$

where $\Delta = \omega/V_\mathrm{B} - 2\pi/p_\mathrm{I}$ is the normalized frequency and V_B is the SSBW [6] velocity. The parameters κ_B and κ represent, respectively, the amplitude of the coupling between STWs and back-scattered BAWs, and that between forward and backward propagating STWs. Δ_v represents the STW velocity reduction, and

$$\nu(\Delta) = \frac{\eta_\mathrm{B}}{\sqrt{\Delta_\mathrm{B} - \Delta + \eta_\mathrm{B}}},\tag{8.2}$$

is a function giving the frequency dispersion due to the coupling of the STWs with the back-scattered BAWs, where Δ_B is its cut-off frequency and η_B is a parameter representing the effect of boundary conditions on SSBW propagation.

So as to apply Abbott's approximation to other kinds of SAWs, a new parameter c is introduced in (8.1), and c enables us to take coupling with other BAWs into consideration.

In Sect. 7.2, the following dispersion relation was obtained for the short-circuited grating:

$$\theta_\mathrm{p} = \sqrt{\theta_\mathrm{u}^2 - \kappa_{12}^2}.\tag{8.3}$$

Comparing this with (8.1), one can clearly see that these two equations are compatible with each other by setting

$$c\{\Delta - \Delta_\mathrm{v} + |\kappa_\mathrm{B}|\nu(\Delta)\} \Rightarrow \theta_\mathrm{u},\tag{8.4}$$

$$c\{\kappa + \kappa_\mathrm{B}\nu(\Delta)\} \Rightarrow \kappa_{12}.\tag{8.5}$$

Six parameters are needed to evaluate (8.1): the number of parameters is too many to be determined directly by a numerical fit with the result obtained by FEMSDA.

So, the first step is, by using the results of FEMSDA, to find three parameters for the approximate dispersion relation proposed by Plessky [19] (see Sect. 8.3.1):

$$\theta_p = c\sqrt{\Delta^2 - \frac{1}{4}\left(|\epsilon|^2 + \eta\sqrt{2|\epsilon|^2 - \eta^2 - 4\Delta}\right)^2},$$ (8.6)

where ϵ and η are the parameters determining the reflection and the penetration depth of the SAWs, respectively. Note again that the parameter c is also introduced in Plessky's approximation.

From (8.6), Δ_B is given as a function of ϵ and η by

$$\Delta_B = -\frac{\eta^2 - 2|\epsilon|^2}{4},$$ (8.7)

and the frequency Δ_+ at the upper edge of the stopband is given by

$$\Delta_+ = \begin{cases} -\dfrac{(\eta - |\epsilon|)^2}{2} & \text{if } \eta > 2|\epsilon| \\ \Delta_B & \text{otherwise} \end{cases},$$ (8.8)

while the frequency Δ_- at the lower edge is given by

$$\Delta_- = -\frac{(\eta + |\epsilon|)^2}{2}.$$ (8.9)

Although this approximate dispersion relation cannot include the piezoelectric contribution to the COM equations, only three parameters are required for the analysis.

From (8.8) and (8.9), η and ϵ are written as a function of $\Delta_\pm$ in the form

$$\eta = (\sqrt{-2\Delta_-} + \sqrt{-2\Delta_+})/2,$$ (8.10)
$$|\epsilon| = (\sqrt{-2\Delta_-} - \sqrt{-2\Delta_+})/2.$$ (8.11)

Since $\Delta_\pm$ can be estimated from the dispersion characteristics obtained by FEMSDA, the approximate values of η and ϵ are easily evaluated. It may also be reasonably assumed that $c \cong 1$. Using these three estimated parameters, c, η and ϵ, as initial values, their correct values can be determined by a numerical fit of (8.1) with the results of FEMSDA.

From ϵ and η, all the parameters needed in (8.1) are given by

$$\Delta_v \Leftarrow -\eta^2/2,$$ (8.12)
$$\kappa \Leftarrow |\epsilon|(\eta + |\epsilon|/2),$$ (8.13)
$$\kappa_B \Leftarrow -|\epsilon|^2\eta/(\eta + 2|\epsilon|)$$ (8.14)

with sufficient accuracy [18]. The two parameters c and Δ_B have already been determined in the previous step.

Figure 8.21 shows the normalized frequency $\Delta p_I/2\pi$ as a function of the normalized wavenumber $\theta_p p_I/2\pi - 1$ near the stopband calculated by FEMSDA. In the calculation, 36°YX-LiTaO$_3$ and Al were chosen as the substrate and electrode materials, respectively, and the electrode thickness h and width w were assumed to be $0.05p_I$ and $0.25p_I$, respectively. In the figure, the approximate dispersion curves by Plessky's and Abbott's approximations are also shown. Actually no difference is observed amongst the three curves, thereby confirming that the parameters needed for Plessky's and Abbott's approximations have been determined with sufficient accuracy.

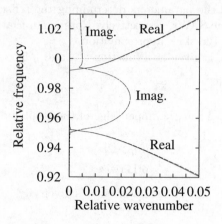

Fig. 8.21. Dispersion relation for leaky SAW on $36°$YX-LiTaO$_3$. Solid line: FEMSDA, broken line: Victor's relation (8.6), and dot-and-broken line: Abbott's relation (8.1)

It remains to consider the excitation efficiency. From the COM equations, the wavenumber β_o of the perturbed mode under the open-circuited grating is given by

$$\beta_o = 2\pi/p_I + \sqrt{(\theta_u - \xi)^2 - (\kappa_{12} - \xi)^2},\tag{8.15}$$

and that under the short-circuited grating is given by

$$\beta_p = 2\pi/p_I + \sqrt{\theta_u^2 - \kappa_{12}^2},\tag{8.16}$$

where $\xi = 4\zeta^2/\omega C$. Comparing (8.15) and (8.16), one may obtain

$$
\begin{aligned}
K^2 &= \frac{\pi p_I}{8} \frac{(\beta_p - 2\pi/p_I)^2 - (\beta_o - 2\pi/p_I)^2}{\theta_u - \kappa_{12}} \\
&= \frac{\pi p_I}{8c} \frac{(\beta_p - 2\pi/p_I)^2 - (\beta_o - 2\pi/p_I)^2}{\Delta - \Delta_v - \kappa + 2|\kappa_B|\eta(\Delta)}.
\end{aligned}
\tag{8.17}
$$

Since FEMSDA can calculate β_o as well as β_p as a function of ω, K^2 is also estimated from (8.17), where θ_u and κ have already been determined previously.

Figure 8.22 shows the estimated K^2 as a function of frequency, where $h = 0.05p_I$ and $w = 0.25p_I$. As substrate and electrode materials, $36°$YX-LiTaO$_3$ and Al were also assumed. Although an anomalous change in K^2 is observed in Fig. 8.22, the reason is not yet clearly understood. This may be due to coupling of the leaky SAW with the back-scattered BAWs and/or due to inadequate modeling for an unexpectedly strong coupling with the back-scattered BAWs. However, K^2 is assumed to be constant in the following analysis because the change near the leaky SAW resonance is not very large. Here K^2 at the lower edge of the stopband is employed as the constant.

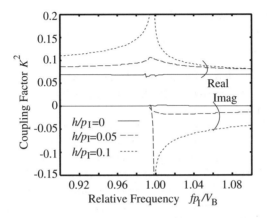

Fig. 8.22. Estimated K^2 on $36°$YX-LiTaO$_3$

Figure 8.23 shows the estimated COM parameters for leaky SAWs on $36°$YX-LiTaO$_3$ as a function of h/p_I, where α is the propagation loss in dB/λ. As can be seen in the figure, all the parameters change linearly with respect to h/p_I.

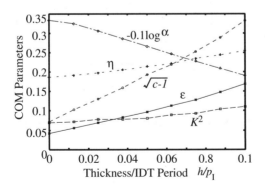

Fig. 8.23. h/p_I dependence of COM parameters

This suggests that the least square method could approximately express these changes;

$$\eta \cong \{0.182 + 0.349 * (2h/p_I)\} \sqrt{2\pi/p_I},$$
$$\epsilon \cong \{0.0388 + 0.618 * (2h/p_I)\} \sqrt{2\pi/p_I},$$
$$\alpha \cong 10^{-3.39+7.39*(2h/p_I)}(\text{ dB}/\lambda),$$
$$c \cong 1 + \{0.0678 + 1.27 * (2h/p_I)\}^2,$$
$$K^2 \cong 0.0655 + 0.206 * (2h/p_I),$$
$$\epsilon(\infty) = 50.255\epsilon_0,$$
$$V_B = 4226.54.$$

By including these dependences into the simulation software based on COM theory, device characteristics can be simulated by specifying only the film thickness and device structure.

Figure 8.24 shows the change in the estimated parameters with the metallization ratio $r = 2w/p_I$, where w is the strip width. It is seen that η representing the effective penetration depth increases monotonically with r. Associated with this, K^2 takes a maximum value when $r \cong 0.65$. These features are apparently different from those obtained for Rayleigh-type SAWs on 128°YX-LiNbO$_3$, where the change in η is small and K^2 takes a maximum value at $r \cong 0.5$ as will be shown in Sect. 8.2.3. Modeling for such r dependence of the COM parameters has not yet been completed.

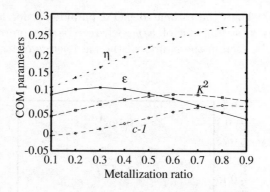

Fig. 8.24. w/p_I dependence of COM parameters

8.2.2 Simulation

Leaky SAW on 36°YX-LiTaO$_3$. Using the COM parameters approximately determined by the least square method, we have prepared a general-purpose simulator for 36°YX-LiTaO$_3$.

Figure 8.25 shows the simulated results for the US-ISM band ladder-type filter developed by Fujitsu Ltd. [20]. The simulation agreed very well with the measured results at both the passband and stopbands. Although a slight discrepancy is observed in the insertion loss, this may arise from ohmic losses of electrodes and measurement fixtures employing conductive rubber.

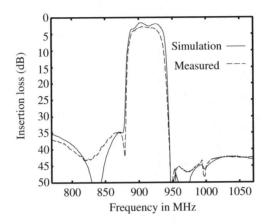

Fig. 8.25. Frequency response of ladder-type leaky SAW filter. Solid line: simulation, and and broken line: experiment.

Figure 8.26 shows the equivalent circuit used in the simulation. In the figure, L_{in}, L_{out}, and L_{ei} are inductances of the bonding wires, and L_c the inductance and C_{in}, C_{out} and C_{ei} the static capacitances of the SMD package.

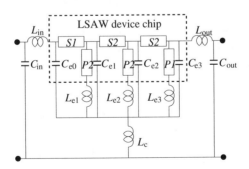

Fig. 8.26. Equivalent circuit used for the simulation

Figure 8.27 shows the frequency response calculated for a wide frequency range. The simulation agrees well with experiment except when $f > 2$ GHz where electrical feedthrough is dominant. Comparing with the results ignoring parasitic impedances, it is clear that their resonance with the static capacitances of the IDTs causes a null at 1.2 GHz, and it effectively enhances out-of-band rejection near the passband.

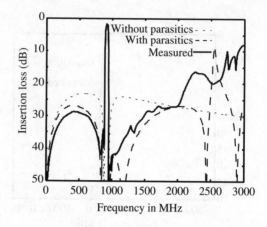

Fig. 8.27. Result of simulation for wide frequency range

In the simulation, the following modifications, which are mainly dependent on the material constants, were applied;

1. reduction of the leaky SAW velocity by 1%;
2. reduction of K^2 by 25%;
3. reduction of ϵ by 20%.

It should be noted that, although the value of L_c is small, it affects the rejection band characteristics significantly, and the device characteristics can be considerably improved by properly choosing L_c [21].

Figure 8.28 shows an equivalent circuit for the ladder-type filter within the rejection band.

For simplicity, the current flow i_{out} toward the load is ignored in the following analysis.

In the rejection band, the relation $|i_1| \gg |i_2| \gg |i_3|$ holds. Thus, we obtain

$$e_{in} = (Z_{s1} + j\omega L_c)(i_1 + i_2 + i_3) + Z_{p1}i_1 \cong (Z_{s1} + j\omega L_c + Z_{p1})i_1,$$
$$e_1 = Z_{p1}i_1 + j\omega L_c(i_1 + i_2 + i_3) \cong (j\omega L_c + Z_{p1})i_1,$$
$$e_2 = Z_{p2}i_2 + j\omega L_c(i_1 + i_2 + i_3) \cong (j\omega L_c(i_1/i_2) + Z_{p2})i_2,$$
$$e_{out} = j\omega L_c(i_1 + i_2 + i_3) + Z_{p3}i_3 \cong \{j\omega L_c(i_1/i_3) + Z_{p3}\}i_3.$$

If Z_{si} and Z_{pi} are capacitive, i_1, i_2 and i_3 are in phase with each other. Thus, for Z_{p1}, the existence of L_c is equivalent to increasing Z_{p1} by $j\omega L_c$, and its effect is negligible. On the other hand, L_c effectively increases Z_{p3} by

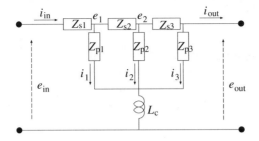

Fig. 8.28. Equivalent circuit of ladder-type filter within rejection band

$j\omega L_c(i_1/i_3)$. Even when L_c itself is small, its effect is not negligible if (i_1/i_3) is large. Although the same effect occurs for Z_{p2}, it is not significant because $(i_1/i_2) \ll (i_1/i_3)$. In the passband, since $|i_1| \cong |i_2| \cong |i_3|$, the effects of L_c are negligible.

SH-type Leaky SAW on 64°YX-LiNbO₃. The mathematical model developed for 36°YX-LiTaO₃ is readily applicable to a leaky SAW on 64°YX-LiNbO₃. Figure 8.29 shows the h/p_I dependence of the determined COM parameters. In this case, the COM parameters are almost proportional to

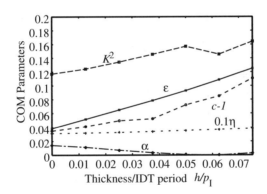

Fig. 8.29. h/p_I dependence of COM parameters for leaky SAW on 64°YX-LiNbO₃

h/p_I, and their least square fit gives

$$\eta = \{0.3069 + 0.4752 * (2h/p_I)\} \sqrt{2\pi/p_I},$$
$$\epsilon = \{0.03722 + 0.5714 * (2h/p_I)\} \sqrt{2\pi/p_I},$$
$$c = 1.027 + 0.4816 * (2h/p_I),$$
$$K^2 = 0.1191 + 0.2935 * (2h/p_I),$$
$$\alpha = \{0.1426 - 1.446 * (2h/p_I)\}^2,$$

$$\epsilon(\infty) = 51.73\epsilon_0,$$
$$V_B = 4752.36.$$

Note that although the present analysis indicates that α takes a minimum value at $h/p_I \cong 0.05$, experimental results did not show this tendency [22]. This might be due to insufficient accuracy of the material constants [23] employed for the analysis; α is very sensitive to variations in the material constants.

The performance of one-port resonators employing a leaky SAW on $64°$YX-LiNbO$_3$ was simulated. All the devices consist of an IDT and two Al grating reflectors with periodicity 16 μm, where the thickness of the electrodes is 0.4 μm. The following experimental results were given by Mitsubishi Material Co. Ltd.

Figure 8.30 shows both the experimental and simulated frequency responses of a one-port leaky SAW resonator without any discontinuity between the IDT and reflectors. Although the COM parameters were slightly modified for the best fit, it is seen that the simulated result is in good agreement with experiment.

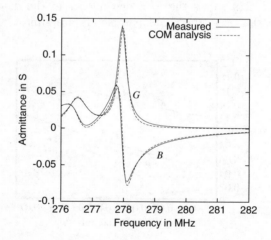

Fig. 8.30. Frequency response of one-port resonator without gap discontinuity

Figure 8.31 compares the measured response with two simulated results for a resonator having two discontinuous regions, i.e., each gap between the IDT and reflector is reduced by $\lambda/8 = p/4$. As can be seen from the figure, if the phase lag arising from the gap discontinuity is taken into consideration, the simulated result is in good agreement with the measured response. Here, the COM parameters used in the simulation are identical with those for Fig. 8.29, and a theoretically evaluated phase lag [24] was employed. This fact means that without taking account of the phase lag, the direct application of the conventional COM simulation technique results in a significant discrepancy in the vicinity of the resonance frequency.

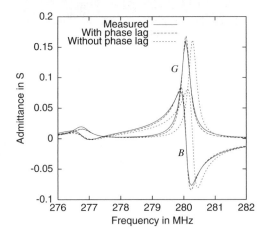

Fig. 8.31. Frequency response of one-port resonator with gap discontinuity

Note that the propagation loss in Fig. 8.31 was assumed to be 0.006 dB/λ, which is about twice as large as the value of 0.0032 dB/λ assumed in Fig. 8.30. Although this increased propagation loss might be due to the scattering of BAWs into the bulk of the substrate at the discontinuous regions, the reason for this is not yet completely understood.

STW on 90°off-X AT-Cut Quartz. On 90°off-X AT-cut quartz, there exists an SH-type SAW [25]. However, its electromechanical coupling factor is very small, and an SSBW is predominantly excited instead of the SAW [6]. Much attention was paid to the SSBW because of its large velocity ($\cong$ 5000 m/sec) and zero TCF [6]. However, the large attenuation associated with the SSBW propagation is quite a serious problem for practical use.

If the grating is placed on the propagation surface, the SSBW is trapped and is converted into an eigenmode by the perturbation. Thus, by using the resonator configuration, the device insertion loss can be improved dramatically without losing the above-mentioned merits [26, 27]. This eigenmode is called a surface transverse wave (STW). Figure 8.32 shows a typical configuration of STW resonators. Due to the guiding mechanism, the space between the IDTs is usually filled with the grating. Note that the center grating is usually designed to have a slightly smaller periodicity than the IDTs so as to operate as a waveguide instead of the reflector. Since the STW is a shear wave with very large velocity, its propagation loss is small due to the reason described in Sect. 1.1.1. Then very high Q resonators are realizable in the GHz range [28]. Added to this, it is known that STW devices exhibit large power handing capability [29].

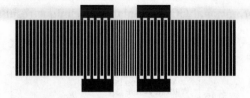

Fig. 8.32. Typical STW device configuration

Figure 8.33 shows the dispersion relation of the STW with Al film thickness h as a parameter. With an increase in h, the velocity decreases, and the dispersion and stopband width increase. At the upper stopband edge, the STW velocity coincides with that of the SSBW, and the STW disappears due to an increase in the penetration depth.

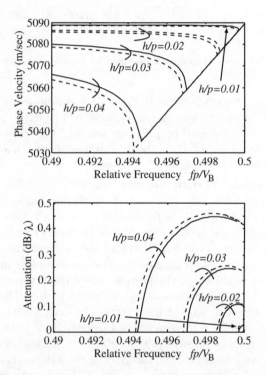

Fig. 8.33. Dispersion relation of STW (Al electrode). Solid line: open-circuited, and broken line: short-circuited grating

The mathematical model developed for $36°$YX-LiTaO$_3$ is also applicable to STWs. Figure 8.34 shows the h/p_I dependence of the determined COM parameters. In this case, the COM parameters are proportional to h/p_I, and

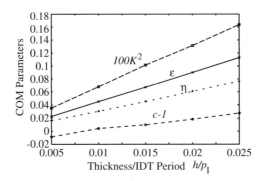

Fig. 8.34. h/p_I dependence of COM parameters for STW

their least square fit gives

$$\eta = \{-0.00011 + 1.524 * (2h/p_I)\}\sqrt{2\pi/p_I},$$
$$\epsilon = \{0.00902 + 2.209 * (2h/p_I)\}\sqrt{2\pi/p_I},$$
$$c = 0.995 + 0.5629 * (2h/p_I),$$
$$K^2 = 0.0000981 + 0.02679 * (2h/p_I),$$
$$\epsilon(\infty) = 5.5594\epsilon_0,$$
$$V_B = 5089.32.$$

It is interesting that K^2 also increases linearly with h/p_I.

The STW device developed by RF Monolithics, Inc., was simulated by using these parameters. Figure 8.35 shows its result [18]. It agrees very well with the experimental one. This clearly exhibits the effectiveness of this method and the accuracy of the material constants for quartz.

Longitudinal-Type Leaky SAW on (0, 47.3°, 90°) Li₂B₄O₇. Recently, the existence of low-loss longitudinal-type leaky SAWs on (0, 47.3°, 90°) $Li_2B_4O_7$ was reported [30], and much attention has been paid to them.

The COM parameters for a longitudinal-type leaky SAW were also estimated. Figure 8.36 shows the estimated COM parameters as a function of h/p_I. Since all parameters change quite linearly with respect to h/p_I, they are approximately expressed as

$$\eta = \{0.0759 + 1.10 * (2h/p_I)\}\sqrt{2\pi/p_I},$$
$$\epsilon = \{0.0226 + 2.24899 * (2h/p_I)\}\sqrt{2\pi/p_I},$$
$$\alpha = \{0.176 - 11.4 * (2h/p_I)\}^2 (\text{ dB}/\lambda),$$
$$c = 1 - \{0.199 - 13.2 * (2h/p_I)\}^2,$$

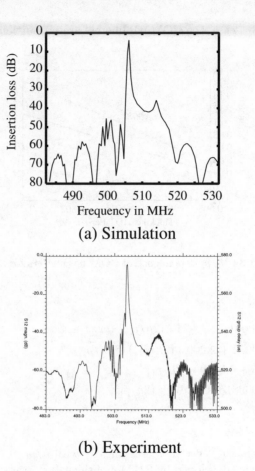

(a) Simulation

(b) Experiment

Fig. 8.35. Result of simulation for STW device

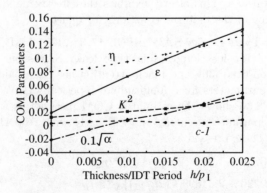

Fig. 8.36. h/p_I dependence of COM parameters

$$K^2 = 0.0108 + 0.5504 * (2h/p_{\mathrm{I}}),$$
$$\epsilon(\infty) = 10.503\epsilon_0,$$
$$V_{\mathrm{B}} = 6650.91.$$

An IIDT filter (see Sect. 4.4.1) on $(0, 47.3°, 90°)$ $\mathrm{Li_2B_4O_7}$ developed by Japan Energy Ltd. was simulated [20]. Figure 8.37 compares the simulated result with experiment. They are in good agreement. For this simulation, the leaky SAW velocity was increased by 2% and K^2 was reduced by 50%.

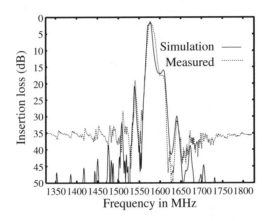

Fig. 8.37. Frequency response of IIDT filter on $(0, 47.3°, 90°)$ $\mathrm{Li_2B_4O_7}$. Solid line: experiment, and broken line: simulation

8.2.3 COM Parameters for Rayleigh-Type SAWs

128°YX-LiNbO₃. The mathematical model described in this section is also applicable to conventional Rayleigh-type SAWs. In fact, the simulations described in Sect. 7.4.3 were carried out by using the model with the following parameters.

Figure 8.38 shows the h/p_{I} dependence of the COM parameters for Rayleigh-type SAWs on 128°YX-LiNbO₃. In the following, the numerical fitted dependences of these parameters are shown:

$$\eta = \{0.2343 + 0.5737 * (2h/p_{\mathrm{I}})\}\sqrt{2\pi/p_{\mathrm{I}}},$$
$$\epsilon = \{0.02675 - 0.4474 * (2h/p_{\mathrm{I}})\}\sqrt{2\pi/p_{\mathrm{I}}},$$
$$c = 1.016 + 0.8085 * (2h/p_{\mathrm{I}}),$$
$$K^2 = 0.06253 + 0.2517 * (2h/p_{\mathrm{I}}),$$
$$\epsilon(\infty) = 56.68\epsilon_0,$$
$$V_{\mathrm{B}} = 4025.54.$$

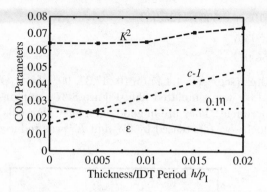

Fig. 8.38. h/p_I dependence of COM parameters on 128°YX-LiNbO$_3$.

Figure 8.39 shows the w/p_I dependence of the COM parameters for $h/p_I = 0.02$. It is seen that κ and K^2 takes a maximum at $w/p_I \cong 0.5$, and their behavior is somewhat different from that on 36°YX-LiTaO$_3$ shown in Fig. 8.24.

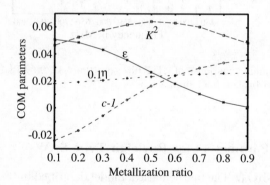

Fig. 8.39. w/p_I dependence of COM parameters on 128°YX-LiNbO$_3$ (Al electrode)

X-112°Y LiTaO$_3$. Figure 8.40 shows the h/p_I dependence of the COM parameters for Rayleigh-type SAWs on X-112°Y LiTaO$_3$. In the following, the numerical fitted dependences of these parameters are shown:

$$\eta = \{0.2206 + 0.1769 * (2h/p_I)\} \sqrt{2\pi/p_I},$$
$$\epsilon = \{0.005125 + 0.3023 * (2h/p_I)\} \sqrt{2\pi/p_I},$$
$$c = 1.24 + 0.2673 * (2h/p_I),$$
$$K^2 = 0.00993,$$
$$\epsilon(\infty) = 48.38\epsilon_0,$$

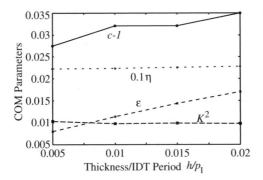

Fig. 8.40. h/p_I dependence of COM parameters on X-112°Y LiTaO$_3$

$V_B = 3374.16$.

COM Parameters on AT-cut Quartz. Figure 8.41 shows the h/p_I dependence of the COM parameters for Rayleigh-type SAWs on AT-cut quartz. In the following, the numerical fitted dependences of these parameters are shown:

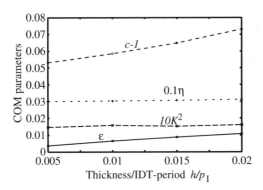

Fig. 8.41. h/p_I dependence of the COM parameters on AT-cut quartz

$$\eta = \{0.2957 + 0.3871 * (2h/p_I)\} \sqrt{2\pi/p_I},$$
$$\epsilon = \{0.001399 + 0.2397 * (2h/p_I)\} \sqrt{2\pi/p_I},$$
$$c = 1.046 + 0.6646 * (2h/p_I),$$
$$K^2 = 0.00155,$$
$$\epsilon(\infty) = 5.506\epsilon_0,$$
$$V_B = 3299.44.$$

8.3 Derivation of Approximate Dispersion Relations

8.3.1 Derivation of Plessky's Dispersion Relation

In this chapter, Plessky's dispersion equation [19] given by (8.6) was effectively used for the characterization of various kinds of SAWs. Although the dispersion equation might be extended so as to apply various kinds of problems, its derivation was not clearly written even in the original paper. Here we will show the derivation.

On a metallized surface of uniform 6mm substrates, the dispersion relation for the BGS wave [1, 2, 3] is given by

$$\Omega - K^2 = 0, \tag{8.18}$$

where $\Omega = \sqrt{1 - (\omega/\beta V_B)^2}$ is the decay constant toward the depth, β is the wavenumber of the BGS wave, and K^2 is the electromechanical coupling factor for the thickness vibration of the SH-type BAW.

When a grating with periodicity p is placed on the surface, a coupling arises between the forward propagating mode with wavenumber β_+ and the backward propagating mode with wavenumber β_-. In this case, the dispersion relation may be approximately given by

$$\left(\sqrt{1 - (\omega/\beta_+ V_B)^2} - K^2\right)\left(\sqrt{1 - (\omega/\beta_- V_B)^2} - K^2\right) = R^2, \tag{8.19}$$

where R is a parameter responsible for the coupling strength.

Let us consider the behavior of $\beta_+ \cong M\pi/p$ near the stopband. In this case, β_- generated by the reflection is given by $\beta_+ - 2M\pi/p$. Thus, by setting $\beta_+ = M\pi/p \times (1 + q)$ and $\omega = M\pi V_B/p \times (1 + w)$, we get

$$\sqrt{1 - \left(\frac{\omega}{\beta_\pm V_B}\right)^2} = \sqrt{1 - \left(\frac{1 + w}{1 \mp q}\right)^2} \cong \sqrt{2(\pm q - w)}. \tag{8.20}$$

Substitution into (8.19) gives

$$2\sqrt{w^2 - q^2} + K^4 - R^2 = K^2\left(\sqrt{2(q - w)} + \sqrt{2(-q - w)}\right). \tag{8.21}$$

Taking squares of both sides gives the following quadratic equation with respect to $\sqrt{w^2 - q^2}$:

$$\left(\sqrt{w^2 - q^2}\right)^2 - R^2\sqrt{w^2 - q^2} + (K^4 - R^2)^2/4 + K^4 w = 0. \tag{8.22}$$

Then the solution is

$$\sqrt{w^2 - q^2} = \frac{R^2 \pm K^2\sqrt{2R^2 - K^4 - 4w}}{2}. \tag{8.23}$$

Rearranging after taking squares of both sides gives

$$q = \sqrt{w^2 - \frac{1}{4}\left(R^2 \pm K^2\sqrt{2R^2 - K^4 - 4w}\right)^2}. \tag{8.24}$$

This is equivalent to (8.6) by setting $q \to \theta_p/2\pi$, $w \to \Delta/2\pi$, $R \to |\epsilon|/\sqrt{2\pi}$ and $K^2 \to \eta/\sqrt{2\pi}$.

8.3.2 Derivation of Abbott's Dispersion Relation

Abbott's dispersion equation [18] of (8.1) was also effectively used for the characterization of various kinds of SAWs. Although Abbott first derived the equation empirically through simulation, the basic form of the dispersion equation can be derived from extension of COM theory. Since the derivation might be interesting as an exercise for COM theory, here we will show the derivation.

Let us define the mode amplitudes of the SAW as $U_S^\pm(x)$, and those of the BAW as $U_B^\pm(x)$. The procedure described in Chap. 7 gives the COM equations including the coupling between all of these waves. When the system is unitary, the COM equations in the short-circuited grating are given by

$$\frac{\partial U_S^+(x)}{\partial x} = -j\theta_S U_S^+(x) - j\kappa_S U_S^-(x) \qquad\qquad - j\xi U_B^-(x), \qquad (8.25)$$

$$\frac{\partial \Phi_S^-(x)}{\partial x} = +j\kappa_S^* U_S^+(x) + j\theta_S U_S^-(x) + j\xi^* U_B^+(x), \qquad (8.26)$$

$$\frac{\partial U_B^+(x)}{\partial x} = \qquad\qquad -j\xi^* U_S^-(x) - j\theta_B U_B^+(x) - j\kappa_B U_B^-(x), \qquad (8.27)$$

$$\frac{\partial U_B^-(x)}{\partial x} = +j\xi U_S^+(x) \qquad\qquad + j\kappa_B^* U_B^+(x) + j\theta_B U_B^-(x), \qquad (8.28)$$

where $\theta_S = \omega/V_S - \pi/p$, $\theta_B = \omega/V_B - \pi/p$, and VS and V_B are the velocities of the unperturbed SAW and BAW, respectively.

Let us define the differential operator $\Lambda = j\partial/\partial x$. Then (8.25)–(8.28) can be written as

$$0 = (\theta_S - \Lambda)U_S^+(x) + \kappa_S U_S^-(x) \qquad\qquad + \xi U_B^-(x), \qquad (8.29)$$

$$0 = \kappa_S^* U_S^+(x) + (\theta_S + \Lambda)U_S^-(x) + \xi^* U_B^+(x), \qquad (8.30)$$

$$0 = \qquad\qquad \xi^* U_S^-(x) + (\theta_B - \Lambda)U_B^+(x) + \kappa_B U_B^-(x), \qquad (8.31)$$

$$0 = \xi U_S^+(x) \qquad\qquad + \kappa_B^* U_B^+(x) + (\theta_B + \Lambda)U_B^-(x). \qquad (8.32)$$

By solving (8.31) and (8.32) with respect to $U_B^\pm(x)$, we obtain

$$\begin{pmatrix} U_B^+(x) \\ U_B^-(x) \end{pmatrix} = \frac{1}{\Lambda^2 - \theta_B^2 + |\kappa_B|^2} \begin{pmatrix} \theta_B + \Lambda & -\kappa_B \\ -\kappa_B^* & \theta_B - \Lambda \end{pmatrix} \begin{pmatrix} 0 & \xi^* \\ \xi & 0 \end{pmatrix} \begin{pmatrix} U_S^+(x) \\ U_S^-(x) \end{pmatrix}. \qquad (8.33)$$

By substituting (8.33) into (8.29) and (8.30), we obtain the dispersion relation for the coupled mode;

$$\Lambda^4 - \Lambda^2(\theta_S^2 + \theta_B^2 - |\kappa_S|^2 - |\kappa_B|^2 - 2|\xi|^2) + \{(\theta_S^2 - |\kappa_S|^2)(\theta_B^2 - |\kappa_B|^2)$$
$$+ |\xi|^2(|\xi|^2 - 2\theta_S\theta_B - \kappa_B\kappa_S^* - \kappa_B^*\kappa_S)\} = 0. \qquad (8.34)$$

When the frequency is below the BAW cut-off, a portion of the acoustic power carried by the SAW is temporarily stored within the grating as the BAW. This is the energy storing effect described in 2.2.2. From the sense of the COM theory, it is expected that the "artificial" uncoupled BAW does not carry net acoustic power, but resonates within the grating structure. From (8.31) and (8.32), the condition requiring that multiple roots of $\Lambda = 0$ exist when the BAW is uncoupled, i.e., $\xi = 0$, is given by

$$\theta_B = -|\kappa_B|, \tag{8.35}$$

because $\theta_B < 0$.

In addition, let us assume that, in the presence of coupling but still below the cut-off, the coupled BAW also resonates within the grating and carries no net acoustic power. From (8.34) and (8.35), the condition requiring that multiple roots of $\Lambda = 0$ exist is

$$|\xi|^2 = 2\theta_S\theta_B + \kappa_B\kappa_S^* + \kappa_B^*\kappa_S. \tag{8.36}$$

From (8.34), the other two roots are given by $\Lambda = \sqrt{\theta_S^2 - |\kappa_S|^2 - 2|\xi|^2}$. Substitution of (8.35) and (8.36) gives

$$\Lambda = \sqrt{(\theta_S - 2\theta_B)^2 - (\kappa_S + 2\kappa_B)(\kappa_S^* + 2\kappa_B^*)}$$
$$= \sqrt{(\theta_S + 2|\kappa_B|)^2 - (\kappa_S + 2\kappa_B)(\kappa_S^* + 2\kappa_B^*)}. \tag{8.37}$$

Equation (8.37) has exactly the same form as (8.1).

References

1. J.L. Bleustein: A New Surface Wave in Piezoelectric Materials, Appl. Phys. Lett., **13** (1968) pp. 412.
2. Y.V. Gulyaev: Electroacoustic Surface Waves in Solids, Soviet Phys. JETP Lett., **9** (1969) pp. 63.
3. Y. Ohta, K. Nakamura and H. Shimizu: Surface Concentration of Shear Wave on Piezoelectric Materials with Conductor, Technical Report of IEICE, Japan **US69-3** (1969) in Japanese.
4. M. Kadota, K. Morozumi, T. Ikeda and T. Kasanami: Ceramic Resonators Using BGS Waves, Jpn. J. Appl. Phys., **31**, Suppl. 31-1 (1992) pp. 219–221.
5. K. Nakamura, M. Kazumi and H. Shimizu: SH-type and Rayleigh-Type Surface Waves on Rotated Y-cut LiTaO$_3$, Proc. IEEE Ultrason. Symp. (1977) pp. 819–822.
6. M. Lewis: Surface Skimming Bulk Waves, SSBW, Proc. IEEE Ultrason. Symp. (1977) pp. 744–752.
7. K. Yamanouchi and K. Shibayama: Propagation and Amplification of Rayleigh Waves and Piezoelectric Leaky Surface Waves in LiNbO$_3$, J. Appl. Phys., **43** (1972) pp. 856–862.
8. K. Hashimoto, M. Yamaguchi and H. Kogo: Experimental Verification of SSBW and Leaky SAW Propagating on Rotated Y-cuts of LiNbO$_3$ and LiTaO$_3$, Proc. IEEE Ultrason. Symp. (1983) pp. 345–349.
9. K. Hashimoto and M. Yamaguchi: Effects of Surface Electrical Boundary Condition on Excitation and Propagation of Highly Piezoelectric Leaky Surface Acoustic Waves, Proc. 7th European Time and Frequency Forum (1993) pp. 517–522.

10. V.P. Plessky and T. Thorvaldsson: Rayleigh Waves and Leaky SAW in Periodic Systems of Electrodes: Periodic Green's Function Analysis, Proc. IEEE Ultrason. Symp. (1992) pp. 461–464.
11. K. Hashimoto, M. Yamaguchi and H. Kogo: Interaction of High-Coupling Leaky SAW with Bulk Waves under Metallic-Grating Structure on $36°$YX-LiTaO$_3$, Proc. IEEE Ultrason. Symp. (1985) pp. 16–21.
12. O. Kawachi, G. Endoh, M. Ueda, O. Ikata, K. Hashimoto and M. Yamaguchi: Optimum Cut of LiTaO$_3$ for High Performance Leaky Surface Acoustic Wave Filters, Proc. IEEE Ultrason. Symp. (1986) pp. 71–76.
13. K. Hashimoto, M. Yamaguchi, S. Mineyoshi, O. Kawachi, M. Ueda, G. Endoh, and O. Ikata: Optimum Leaky-SAW Cut of LiTaO$_3$ for Minimised Insertion Loss Devices, Proc. Ultrason. Symp. (1997) pp. 245–254.
14. P.D. Bloch, N.G. Due, E.G.S. Paige and M. Yamaguchi: Observations on Surface Skimming Bulk Waves and Other Waves Launched from an IDT on Lithium Niobate, Proc. IEEE Ultrason. Symp. (1981) pp. 268–273.
15. V. Plessky and C.S. Hartmann: Characteristics of Leaky SAWs on $36°$-LiTaO$_3$ in Periodic Structures of Heavy Electrodes', Proc. IEEE Ultrason. Symp. (1993) pp. 1239–1246.
16. K. Hashimoto, G. Endoh and M. Yamaguchi: Coupling-of-Modes Modelling for Fast and Precise Simulation of Leaky Surface Acoustic Wave Devices, Proc. IEEE Ultrason. Symp. (1995) pp. 251–256.
17. E. Bigler, E. Gavignet, B.A. Auld, E. Ritz and E. Sang: Surface Transverse Wave (STW) Quartz Resonators in the GHz Range, Proc. 6th European Time and Frequency Forum (1992) pp. 219–222.
18. B.P. Abbott and K. Hashimoto: A Coupling-of Modes Formalism for Surface Transverse Wave Devices, Proc. IEEE Ultrason. Symp. (1995) pp. 239–245.
19. V. Plessky: Two Parameter Coupling-of-Modes Model for Shear Horizontal Type SAW Propagation in Periodic Gratings, Proc. IEEE Ultrason. Symp. (1993) pp. 195–200.
20. K. Hashimoto and M. Yamaguchi: General-Purpose Simulator for Leaky Surface Acoustic Wave Devices Based on Coupling-of-Modes Theory, Proc. IEEE Ultrason. Symp. (1996) pp. 117–122.
21. S. Mineyoshi, O. kawachi, M. Ueda and Y. Fujiwara: Analysis and Optimal SAW Ladder Filter Design Including Bonding Wire and Package Impedance, Proc. IEEE Ultrason. Symp. (1997) pp. 175–178.
22. C.S. Hartmann, V.P. Plessky: Experimental Measurements of Propagation, Attenuation, Reflection and Scattering of Leaky Waves in Al Electrode Gratings on $41°$, $54°$ and $64°$-LiNbO$_3$, Proc. IEEE Ultrason. Symp. (1993) pp. 1247–1250.
23. G. Kovacs, M. Anhorn, H.E. Engan, G. Visintini, and C.C.W. Ruppel: Improved Material Constants for LiNbO$_3$ and LiTaO$_3$, Proc. IEEE Ultrason. Symp. (1990) pp. 435–438.
24. Y. Sakamoto, K. Hashimoto and M. Yamaguchi: Behaviour of LSAW Propagation at Discontinuous Region of Periodic Grating, Jpn. J. Appl. Phys., **37**, 5B (1998) pp. 2905–2908.
25. M. Yamaguchi, K. Hashimoto and H. Kogo: Effects of Surface Metallisation on SSBW and BGW Propagation in Quartz, Electron. Lett., **17**, 17 (1981) pp. 602–603.
26. B.A. Auld, J.J. Gagnepain and M. Tan: Horizontal Shear Surface Waves on Corrugated Surface, Electron. Lett., **12** (1976) pp. 650–651.
27. Y.V. Gulyaev and V.P. Plessky: Slow Surface Acoustic Waves in Solids, Sov. Tech. Phys. Lett. **3** (1977) pp. 220–223.

28. I.D. Avramov, O. Ikata, T. Matsuda, T. Nishihara and Y. Satoh: Further Improvements of Surface Transverse Wave Resonator Performance in the 2.0 to 2.5 GHz Range, Proc. IEEE Freq. Contr. Symp. (1997) pp. 807–815.
29. I.D. Avramov: Microwave Oscillators Stabilized with Surface Transverse Wave Resonant Devices, Proc. IEEE Freq. Contr. Symp. (1992) pp. 391–408.
30. T. Sato and H. Abe: Longitudinal Leaky Surface Waves for High Frequency SAW Device Application, Proc. IEEE Ultrason. Symp. (1995) pp. 305–315.

A. Physics of Acoustic Waves

A.1 Elasticity of Solids

When an external force is applied to a solid, it deforms, and removal of the force may restore the solid into the initial state. This type of deformation is called elasticity. If the deformation remains after removal of the external force, we call this plasticity. Although usual materials exhibit elastic deformation when the force is relatively small, its further increase causes plastic deformation, and may finally result in fracture.

We define the displacement vector $\mathbf{u}$ characterizing the deformation. We denote by $\mathbf{L}$ and $\mathbf{L}'$ the position vectors for the initial and final states, respectively, for a specified point in the material. Then $\mathbf{u}$ is given by their difference (Fig. A.1).

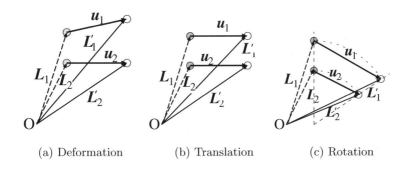

| (a) Deformation | (b) Translation | (c) Rotation |

Fig. A.1. Displacement. •: initial state, and ∘: final state

Note that although a displacement exists for the translation and rotation shown in Fig. A.1b and c, respectively, they do not cause deformation because the distance between the particles is unchanged. Hence, displacement cannot be used as a measure of deformation.

There are two kinds of deformation as shown in Fig. A.2. One is extension or compression shown in a and the other is shear shown in b.

In Fig. A.2, if the material is uniform, each part of the material deforms uniformly. Then, we can define a measure of the deformation by

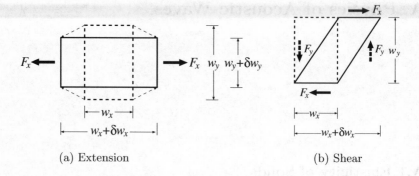

(a) Extension   (b) Shear

Fig. A.2. Deformation

$$S'_{ij} = \delta w_i / w_j. \tag{A.1}$$

By using the displacement vector, this can be rewritten as

$$S'_{ij} = \frac{\partial u_i}{\partial x_j}.$$

However S'_{ij} is still sensitive to the rotation shown in Fig. A.1c. So as to remove this effect, we define

$$S_{ij} = \frac{1}{2}\left(\frac{\partial u_i}{\partial x_j} + \frac{\partial u_j}{\partial x_i}\right), \tag{A.2}$$

and this is called the strain.

For the deformation shown in Fig. A.2a, positive S_{xx} indicates extension for the x direction whereas negative S_{yy} indicates compression in the y direction. On the other hand, positive S_{xy} indicates shear deformation as shown in Fig. A.2b.

Since $S_{ij} = S_{ji}$ from (A.2), there are six independent elements in S_{ij} which has nine elements. Then, the following abbreviated notation [1] is widely used instead of the double subscript;

$$\begin{pmatrix} S_1 & S_6 & S_5 \\ S_6 & S_2 & S_4 \\ S_5 & S_4 & S_3 \end{pmatrix} = \begin{pmatrix} S_{11} & 2S_{12} & 2S_{13} \\ 2S_{12} & S_{22} & 2S_{23} \\ 2S_{13} & 2S_{23} & S_{33} \end{pmatrix}, \tag{A.3}$$

or in matrix form

$$\mathbf{S} = \nabla \mathbf{u}, \tag{A.4}$$

where $\mathbf{S} = (S_1, S_2, S_3, S_4, S_5, S_6)^{\mathrm{t}}$, $\mathbf{u} = (u_1, u_2, u_3)^{\mathrm{t}}$, the superscript t indicates the transpose of the matrix, and

$$\nabla = \begin{pmatrix} \frac{\partial}{\partial x_1} & 0 & 0 \\ 0 & \frac{\partial}{\partial x_2} & 0 \\ 0 & 0 & \frac{\partial}{\partial x_3} \\ 0 & \frac{\partial}{\partial x_3} & \frac{\partial}{\partial x_2} \\ \frac{\partial}{\partial x_3} & 0 & \frac{\partial}{\partial x_1} \\ \frac{\partial}{\partial x_2} & \frac{\partial}{\partial x_1} & 0 \end{pmatrix}. \tag{A.5}$$

To specify a force applied to solids, we must indicate not only its direction but also the plane to which the force is applied. For example, although F_x is applied for both cases shown in Figs. A.2a and b, their reactions are completely different.

We define the stress T_{ij} as the force per unit area. The subscripts indicate the directions of the force and the plane to be applied. The direction of the plane is designated by that of the vector perpendicular to the plane (see Fig. A.3). Note there is the sign attached to the plane, and the plane on the $-x_i$

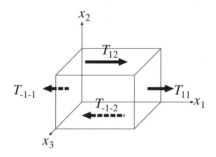

Fig. A.3. Force acting to planes

side of the volume is called $-x_i$ plane.

If the solid is stationary, the equilibrium condition for translation requires

$$T_{ij} = T_{-i-j}, \tag{A.6}$$

and that for rotation gives

$$T_{ij} = T_{ji}. \tag{A.7}$$

Equations (A.6) and (A.7) suggest that there are six independent elements in T_{ij} which has nine elements. Then, similar to (A.3), the following abbreviated notation is applicable;

$$\begin{pmatrix} T_1 & T_6 & T_5 \\ T_6 & T_2 & T_4 \\ T_5 & T_4 & T_3 \end{pmatrix} = \begin{pmatrix} T_{11} & T_{12} & T_{13} \\ T_{12} & T_{22} & T_{23} \\ T_{13} & T_{23} & T_{33} \end{pmatrix}. \tag{A.8}$$

Hooke's law indicates that the stress is proportional to the strain when they are not too large. Thus by using the abbreviated notation shown in (A.3) and (A.8), we can express their relation as

$$
\begin{pmatrix} T_1 \\ T_2 \\ T_3 \\ T_4 \\ T_5 \\ T_6 \end{pmatrix} = \begin{pmatrix} c_{11} & c_{12} & c_{13} & c_{14} & c_{15} & c_{16} \\ c_{12} & c_{22} & c_{23} & c_{24} & c_{25} & c_{26} \\ c_{13} & c_{23} & c_{33} & c_{34} & c_{35} & c_{36} \\ c_{14} & c_{24} & c_{34} & c_{44} & c_{45} & c_{46} \\ c_{15} & c_{25} & c_{35} & c_{45} & c_{55} & c_{56} \\ c_{16} & c_{26} & c_{36} & c_{46} & c_{56} & c_{66} \end{pmatrix} \begin{pmatrix} S_1 \\ S_2 \\ S_3 \\ S_4 \\ S_5 \\ S_6 \end{pmatrix},
\tag{A.9}
$$

where c_{ij} is called the stiffness constant or elastic constant.

We express the relation (A.9) in the matrix form

$$
\mathbf{T} = \mathbf{cS},
\tag{A.10}
$$

where $\mathbf{T} = (T_1, T_2, T_3, T_4, T_5, T_6)^t$.

The number of independent elements in c_{ij} reduces drastically by taking the crystallographic symmetry of solids into account. For example, when the coordinate system is chosen to coincide with that of crystal axes for 6mm materials, we get

$$
\begin{pmatrix} c_{11} & c_{12} & c_{13} & 0 & 0 & 0 \\ c_{12} & c_{11} & c_{13} & 0 & 0 & 0 \\ c_{13} & c_{13} & c_{33} & 0 & 0 & 0 \\ 0 & 0 & 0 & c_{44} & 0 & 0 \\ 0 & 0 & 0 & 0 & c_{44} & 0 \\ 0 & 0 & 0 & 0 & 0 & c_{66} \end{pmatrix}
\tag{A.11}
$$

and the relation $c_{11} - c_{12} = 2c_{66}$ holds. On the other hand, for 4mmm materials, we obtain

$$
\begin{pmatrix} c_{11} & c_{12} & c_{12} & 0 & 0 & 0 \\ c_{12} & c_{11} & c_{12} & 0 & 0 & 0 \\ c_{12} & c_{12} & c_{11} & 0 & 0 & 0 \\ 0 & 0 & 0 & c_{44} & 0 & 0 \\ 0 & 0 & 0 & 0 & c_{44} & 0 \\ 0 & 0 & 0 & 0 & 0 & c_{44} \end{pmatrix}.
\tag{A.12}
$$

For isotropic materials, the symmetry shown in (A.12) and the relation of $c_{11} - c_{12} = 2c_{44}$ hold. Detailed discussions are given in Ref. [2].

The motion of materials is governed by the Newton equation shown below:

$$
\rho \frac{\partial^2 u_i}{\partial t^2} = \sum_{j=1}^{3} \frac{\partial T_{ij}}{\partial x_j} \quad (i = 1, 2, 3),
\tag{A.13}
$$

or

$$
\rho \ddot{\mathbf{u}} = \nabla \cdot \mathbf{T}
\tag{A.14}
$$

in vector form, where ρ is the mass density. Note that the divergence operation for rank 2 tensors is defined by

$$
\nabla \cdot = \begin{pmatrix} \frac{\partial}{\partial x_1} & 0 & 0 & 0 & \frac{\partial}{\partial x_3} & \frac{\partial}{\partial x_2} \\ 0 & \frac{\partial}{\partial x_2} & 0 & \frac{\partial}{\partial x_3} & 0 & \frac{\partial}{\partial x_1} \\ 0 & 0 & \frac{\partial}{\partial x_3} & \frac{\partial}{\partial x_2} & \frac{\partial}{\partial x_1} & 0 \end{pmatrix}. \tag{A.15}
$$

Since T_{ij} can be expressed in terms of u_i by using the relations (A.2)–(A.10), (A.13) is regarded as a set of simultaneous differential equations with respect to u_i, and various acoustic properties can be determined by their solutions.

Let us consider plane waves propagating toward the x_3 direction. Since the wavefront is uniform, $\partial/\partial x_1 = \partial/\partial x_2 = 0$. Then (A.13) gives the wave equations of

$$
\rho \frac{\partial^2 u_1}{\partial t^2} = c_{44} \frac{\partial^2 u_1}{\partial x_3^2}, \tag{A.16}
$$

$$
\rho \frac{\partial^2 u_2}{\partial t^2} = c_{44} \frac{\partial^2 u_2}{\partial x_3^2}, \tag{A.17}
$$

$$
\rho \frac{\partial^2 u_3}{\partial t^2} = c_{11} \frac{\partial^2 u_3}{\partial x_3^2} \tag{A.18}
$$

for isotropic materials. It is well known that solutions of the wave equation are given by

$$
u_1 = A_{1+}(t - x_3/V_s) + A_{1-}(t + x_3/V_s), \tag{A.19}
$$
$$
u_2 = A_{2+}(t - x_3/V_s) + A_{2-}(t + x_3/V_s), \tag{A.20}
$$
$$
u_3 = A_{3+}(t - x_3/V_l) + A_{3-}(t + x_3/V_l), \tag{A.21}
$$

where $V_s = \sqrt{c_{44}/\rho}$ and $V_l = \sqrt{c_{11}/\rho}$ are the phase velocities for shear and longitudinal waves, respectively. In the equations, $A_{i\pm}$ is an arbitrary function and its sign designates the propagation direction, i.e., toward the $+x_3$ or $-x_3$ direction.

For problems including boundaries between different media, the following boundary conditions are applied:

- continuity of the stress components perpendicular to the boundary;
- continuity of the displacements.

A.2 Piezoelectricity

Piezoelectric materials are widely used for excitation and detection of acoustic waves.

Due to the piezoelectricity, the strain **S** induces an electric flux density **D** proportional to **S**. Added to this, since **D** is also induced by the electric field **E**, we get the following relation:

$$\begin{pmatrix} D_1 \\ D_2 \\ D_3 \end{pmatrix} = \begin{pmatrix} e_{11} & e_{12} & e_{13} & e_{14} & e_{15} & e_{16} \\ e_{21} & e_{22} & e_{23} & e_{24} & e_{25} & e_{26} \\ e_{31} & e_{32} & e_{33} & e_{34} & e_{35} & e_{36} \end{pmatrix} \begin{pmatrix} S_1 \\ S_2 \\ S_3 \\ S_4 \\ S_5 \\ S_6 \end{pmatrix}$$

$$+ \begin{pmatrix} \epsilon_{11}^S & \epsilon_{12}^S & \epsilon_{13}^S \\ \epsilon_{12}^S & \epsilon_{22}^S & \epsilon_{23}^S \\ \epsilon_{13}^S & \epsilon_{23}^S & \epsilon_{33}^S \end{pmatrix} \begin{pmatrix} E_1 \\ E_2 \\ E_3 \end{pmatrix}, \tag{A.22}$$

where e_{ij} is called the piezoelectric constant, and ϵ_{ij}^S is the dielectric constant, whose superscript S indicates the value measured under constant strain.

By taking the crystallographic symmetry into account, the number of independent elements for e_{ij} and ϵ_{ij}^S is dramatically reduced. For example, when the crystal axes for 6mm materials are chosen to coincide with the coordinate system, they are expressed as

$$\begin{pmatrix} 0 & 0 & 0 & 0 & e_{15} & 0 \\ 0 & 0 & 0 & e_{15} & 0 & 0 \\ e_{31} & e_{31} & e_{33} & 0 & 0 & 0 \end{pmatrix} \tag{A.23}$$

and

$$\begin{pmatrix} \epsilon_{11}^S & 0 & 0 \\ 0 & \epsilon_{11}^S & 0 \\ 0 & 0 & \epsilon_{33}^S \end{pmatrix}. \tag{A.24}$$

Similar to (A.10), we express the relation (A.22) in matrix form by

$$\mathbf{D} = \mathbf{e}\mathbf{S} + \epsilon^S \mathbf{E}. \tag{A.25}$$

As with the reaction, the stress $\mathbf{T}$ is induced by the applied electric field $\mathbf{E}$. By taking this into account, (A.10) relating the stress and strain is modified to

$$\mathbf{T} = \mathbf{c}^E \mathbf{S} - \mathbf{e}^t \mathbf{E}. \tag{A.26}$$

where the superscript E indicates the value measured under constant electric field.

Although the electric characteristics of materials are governed by the Maxwell equation, the quasi-electrostatic approximation [3] of

$$\nabla \times \mathbf{E} = 0 \tag{A.27}$$

can be applied for the discussion of acoustic wave propagation. Under this approximation, the electric field $\mathbf{E}$ can be expressed as

$$\mathbf{E} = -\nabla \phi, \tag{A.28}$$

where ϕ is the electric potential, and the Maxwell equation can be simplified to

$$\nabla \cdot \mathbf{D} = q, \tag{A.29}$$

where q is the charge density, whose contribution is negligible in usual piezo-electric materials.

Substitution of (A.25) and (A.26) into (A.13), (A.28) and (A.29) gives a set of simultaneous linear differential equations with respect to u_i and ϕ, and various acoustic properties in piezoelectric media can be derived from their solutions.

For example, since material constants for 6mm materials have symmetries given by (A.11), (A.23) and (A.24), plane wave solutions propagating toward the x_3 direction satisfy

$$\rho \frac{\partial^2 u_1}{\partial t^2} = c_{44}^E \frac{\partial^2 u_1}{\partial x_3^2} \tag{A.30}$$

$$\rho \frac{\partial^2 u_2}{\partial t^2} = c_{44}^E \frac{\partial^2 u_2}{\partial x_3^2}, \tag{A.31}$$

$$\rho \frac{\partial^2 u_3}{\partial t^2} = c_{33}^E \frac{\partial^2 u_3}{\partial x_3^2} + e_{33} \frac{\partial^2 \phi}{\partial x_3^2}, \tag{A.32}$$

$$0 = e_{33} \frac{\partial^2 u_3}{\partial x_3^2} - \epsilon_{33}^S \frac{\partial^2 \phi}{\partial x_3^2}. \tag{A.33}$$

These equations suggest that only the longitudinal wave couples with the piezoelectricity in this case. For plane waves propagating toward x_1 direction, we get

$$\rho \frac{\partial^2 u_1}{\partial t^2} = c_{11}^E \frac{\partial^2 u_1}{\partial x_1^2}, \tag{A.34}$$

$$\rho \frac{\partial^2 u_2}{\partial t^2} = c_{66}^E \frac{\partial^2 u_2}{\partial x_1^2}, \tag{A.35}$$

$$\rho \frac{\partial^2 u_3}{\partial t^2} = c_{44}^E \frac{\partial^2 u_3}{\partial x_1^2} + e_{15} \frac{\partial^2 \phi}{\partial x_1^2}, \tag{A.36}$$

$$0 = e_{15} \frac{\partial^2 u_3}{\partial x_1^2} - \epsilon_{11}^S \frac{\partial^2 \phi}{\partial x_1^2}. \tag{A.37}$$

In this case, only the shear wave polarized toward the x_3 direction couples with the piezoelectricity. It is interesting to note that the three acoustic waves have different velocities.

For specified problems, we often employ coordinate systems which are not aligned to the crystal axes. The derivation of material constants for arbitrary orientations is fully described in Ref. [1].

For problems including boundaries between different media, the following boundary conditions are applied;

- discontinuity of the electric flux density coincides with the charge distribution on the boundary;
- continuity of the electric potential.

A.3 Surface Acoustic Waves

Let us consider the propagation of a SAW toward the x_1 direction on the semi-infinite structure shown in Fig. A.4. For simplicity, the SAW is assumed to be a plane wave with $\partial/\partial x_2 = 0$.

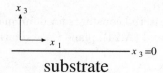

$$x_3 = 0$$

substrate

Fig. A.4. Semi-infinite structure

In this case, the Newton equation (A.13) is given by

$$-\rho\omega^2 u_n = \frac{\partial T_{n1}}{\partial x_1} + \frac{\partial T_{n3}}{\partial x_3} \quad (n = 1, 2, 3), \tag{A.38}$$

where $u_n \propto \exp(j\omega t)$ is assumed. The Maxwell equation (A.29) under the quasi-electrostatic approximation gives

$$0 = \frac{\partial D_1}{\partial x_1} + \frac{\partial D_3}{\partial x_3}. \tag{A.39}$$

Next let us assume that the solutions have an $\exp(-j\beta x_1)$ dependence. Although this assumption seems special, this approach will give the general solution to be described later. Under this assumption, substitution of (A.25) and (A.26) into (A.38) and (A.39) gives

$$\left(\begin{matrix}
\Omega^2 c_{55}^{\rm E} - 2j\Omega c_{15}^{\rm E} - c_{11}^{\rm E} + \rho V^2 & \Omega^2 c_{45}^{\rm E} - j\Omega(c_{14}^{\rm E} + c_{56}^{\rm E}) - c_{16}^{\rm E} \\
\Omega^2 c_{45}^{\rm E} - j\Omega(c_{14}^{\rm E} + c_{56}^{\rm E}) - c_{16}^{\rm E} & \Omega^2 c_{44}^{\rm E} - 2j\Omega c_{46}^{\rm E} - c_{66}^{\rm E} + \rho V^2 \\
\Omega^2 c_{35}^{\rm E} - j\Omega(c_{13}^{\rm E} + c_{55}^{\rm E}) - c_{15}^{\rm E} & \Omega^2 c_{34}^{\rm E} - j\Omega(c_{45}^{\rm E} + c_{36}^{\rm E}) - c_{56}^{\rm E} \\
\Omega^2 e_{35} - j\Omega(e_{15} + e_{31}) - e_{11} & \Omega^2 e_{34} - j\Omega(e_{14} + e_{36}) - e_{16}
\end{matrix}\right.$$

$$\left.\begin{matrix}
\Omega^2 c_{35}^{\rm E} - j\Omega(c_{13}^{\rm E} + c_{55}^{\rm E}) - c_{15}^{\rm E} & \Omega^2 e_{35} - j\Omega(e_{15} + e_{31}) - e_{11} \\
\Omega^2 c_{34}^{\rm E} - j\Omega(c_{45}^{\rm E} + c_{36}^{\rm E}) - c_{56}^{\rm E} & \Omega^2 e_{34} - j\Omega(e_{14} + e_{36}) - e_{16} \\
\Omega^2 c_{33}^{\rm E} - 2j\Omega c_{35}^{\rm E} - c_{55}^{\rm E} + \rho V^2 & \Omega^2 e_{33} - j\Omega(e_{13} + e_{35}) - e_{15} \\
\Omega^2 e_{33} - j\Omega(e_{13} + e_{35}) - e_{15} & -\Omega^2 \epsilon_{33}^{\rm S} + 2j\Omega \epsilon_{13}^{\rm S} + \epsilon_{11}^{\rm S}
\end{matrix}\right)
\left(\begin{matrix} u_1 \\ u_2 \\ u_3 \\ \phi \end{matrix}\right)$$

$$= \left(\begin{matrix} 0 \\ 0 \\ 0 \\ 0 \end{matrix}\right), \tag{A.40}$$

where $V = \omega/\beta$ is the phase velocity, and $\Omega \equiv \beta^{-1}\partial/\partial x_3$ is the differential operator.

For the existence of nontrivial solutions, the determinant of the matrix on the left hand side of (A.40) must be zero. The determinant will be an eighth-order polynomial with respect to Ω. Let us denote the eight solutions of the equation by Ω_n ($n = 1, 2, \cdots, 8$). Then the general solution of (A.40) is given by

$$u_i(x_1, x_3) = \sum_{n=1}^{8} A_{in} \exp(\beta \Omega_n x_3 - j\beta x_1), \tag{A.41}$$

where A_{in} are constants determined by the boundary conditions. Note that A_{in} are not independent of each other, and the ratios of A_{in} ($n = 1, 2, 3, 4$), i.e., $\alpha_n = A_{in}/A_{4n}$ are uniquely determined by solving the following linear equations obtained from (A.40);

$$\left(\begin{matrix} \Omega_n^2 c_{55}^{\mathrm{E}} - 2j\Omega c_{15}^{\mathrm{E}} - c_{11}^{\mathrm{E}} + \rho V^2 & \Omega_n^2 c_{45}^{\mathrm{E}} - j\Omega_n(c_{14}^{\mathrm{E}} + c_{56}^{\mathrm{E}}) - c_{16}^{\mathrm{E}} \\ \Omega_n^2 c_{45}^{\mathrm{E}} - j\Omega_n(c_{14}^{\mathrm{E}} + c_{56}^{\mathrm{E}}) - c_{16}^{\mathrm{E}} & \Omega_n^2 c_{44}^{\mathrm{E}} - 2j\Omega_n c_{46}^{\mathrm{E}} - c_{66}^{\mathrm{E}} + \rho V^2 \\ \Omega_n^2 c_{35}^{\mathrm{E}} - j\Omega_n(c_{13}^{\mathrm{E}} + c_{55}^{\mathrm{E}}) - c_{15}^{\mathrm{E}} & \Omega_n^2 c_{34}^{\mathrm{E}} - j\Omega_n(c_{45}^{\mathrm{E}} + c_{36}^{\mathrm{E}}) - c_{56}^{\mathrm{E}} \end{matrix}\right.$$

$$\left.\begin{matrix} \Omega_n^2 c_{35}^{\mathrm{E}} - j\Omega_n(c_{13}^{\mathrm{E}} + c_{55}^{\mathrm{E}}) - c_{15}^{\mathrm{E}} \\ \Omega_n^2 c_{34}^{\mathrm{E}} - j\Omega_n(c_{45}^{\mathrm{E}} + c_{36}^{\mathrm{E}}) - c_{56}^{\mathrm{E}} \\ \Omega_n^2 c_{33}^{\mathrm{E}} - j\Omega c_{35}^{\mathrm{E}} - c_{55}^{\mathrm{E}} + \rho V^2 \end{matrix}\right) \left(\begin{matrix} \alpha_{1n} \\ \alpha_{2n} \\ \alpha_{3n} \end{matrix}\right)$$

$$= - \left(\begin{matrix} \Omega_n^2 e_{35} - j\Omega_n(e_{15} + e_{31}) - e_{11} \\ \Omega_n^2 e_{34} - j\Omega_n(e_{14} + e_{36}) - e_{16} \\ \Omega_n^2 e_{33} - j\Omega_n(e_{13} + e_{35}) - e_{15} \end{matrix}\right). \tag{A.42}$$

Let us consider (A.41). If $\beta\Omega_n$ is purely real or complex, it indicates that the corresponding partial wave is evanescent toward the $\pm x_3$ direction. On the other hand, if $\beta\Omega_n$ is purely imaginary, it indicates that the partial wave is propagating obliquely in the substrate. That is, setting $\beta\Omega_n = -j\xi_n$ gives

$$A_{in} \exp(\beta\Omega_n x_3 - j\beta x_1) \quad \rightarrow \quad A_{in} \exp(-j\xi_n x_3 - j\beta x_1), \tag{A.43}$$

suggesting that the partial wave possesses a wavevector whose x_1 component is β and x_3 component is ξ_n.

In (A.41), six of the eight partial waves are due to longitudinal and two-kinds of shear waves propagating toward the $\pm x_3$ directions; and the other two are due to the electrostatic field. Note that since the electrostatic field does not contribute to the energy transfer, the corresponding partial wave is evanescent toward the $\pm x_3$ direction, and its Ω_n possesses a nonzero real part. Then the maximum number of Ω_n with zero real part is six over the whole range of β.

Note that we can draw the slowness surface described in Sect. 1.1 in the $x_1 - x_3$ plane, namely, the saggital plane, by plotting ξ_n/ω as a function of $\beta/\omega = V^{-1}$.

In the semi-infinite structure, no reflected wave arrives from the back surface at an infinite distance ($x_3 \rightarrow -\infty$). Then four corresponding partial waves which are evanescent or propagating toward the $+x_3$ direction are not involved in this case. So we must choose four Ω_n from eight under the following rul:

$$\Re(\beta\Omega_n) > 0 \qquad \text{(when } \Re(\beta\Omega_n) \neq 0)$$

$$\frac{\partial \xi_n}{\partial \beta} = -\frac{\partial(\beta\Im[\Omega_n])}{\partial \beta} > 0 \text{ (when } \Re(\beta\Omega_n) = 0) \qquad (A.44)$$

Equation (A.44) forces the partial waves to be evanescent or to have the group velocity directed into the depth. The rule for choosing proper partial waves is called the radiation condition.

When leaky SAWs are considered, the condition cannot be applied directly because β is complex and $\Re(\beta\Omega_n) \neq 0$ for all partial waves.

Here we will show an effective method to judge whether partial waves satisfy the radiation condition or not.

Firstly, we set $\Im(\beta) = 0$ temporally and calculate Ω_n. Then we count the number of Ω_n with pure imaginary part, and divide the number by two. This gives the number n_p of partial waves propagating into the depth. Second, we use complex β to calculate Ω_n. Then we choose n_p solutions from eight Ω_n in the order of smaller but positive $-\Re(\beta\Omega_n)$. The others are chosen in the order of larger $\Re(\beta\Omega_n)$. The reason why the leaky SAW field amplitude increases toward the depth has already been fully described in Sect. 1.2.3.

Note that, even for nonleaky SAWs, the same technique is applicable for the selection of Ω_n only by adding a very tiny imaginary part to β as well as the leaky SAWs. This technique is successfully implemented in the free software distributed by the author's group [4].

Next let us consider the air region ($x_3 > 0$). In the vacuum with permittivity ϵ_0, only the electrostatic field exists. Assuming $u_i \propto \exp(-j\beta x_1)$, the solution of (A.39) can be rewritten as

$$\phi(x_1, x_3) = \{B_+ \exp(-\beta x_3) + B_- \exp(+\beta x_3)\} \exp(-j\beta x_1) \qquad (A.45)$$
$$D_3(x_1, x_3) = \epsilon_0 \beta \{B_+ \exp(-\beta x_3) - B_- \exp(+\beta x_3)\} \exp(-j\beta x_1),$$

$$(A.46)$$

where $B_\pm$ are constants. Note that B_+ satisfies the radiation condition when $\Re(\beta) > 0$ whereas B_- does so when $\Re(\beta) < 0$. Thus the relation between D_3 and ϕ is written by

$$D_3(x_1, x_3) = s\epsilon_0 \beta \phi(x_1, x_3), \qquad (A.47)$$

where

$$s = \begin{cases} 1 & \Re(\beta) > 0 \\ -1 & \Re(\beta) < 0. \end{cases} \qquad (A.48)$$

The boundary conditions at the surface ($x_3 = 0$) are as follows;
(1) continuity of the stress normal to the boundary:

$$T_{3i}|_{x_3=0^-} = 0 \qquad (i = 1, 2, 3); \qquad (A.49)$$

(2) continuity of the surface electrical potential:

$$\phi|_{x_3=0+} = \phi|_{x_3=0-}, \tag{A.50}$$

and

(3) continuity of the electric displacement to the boundary:

$$D_3|_{x_3=0+} - D_3|_{x_3=0-} = q(x_1), \tag{A.51}$$

where $q(x_1)$ is the surface charge density.

From (A.25), (A.26) and (A.41), we get

$$\begin{pmatrix} T_1(\beta)/\beta \\ T_2(\beta)/\beta \\ T_3(\beta)/\beta \\ Q(\beta)/\beta \end{pmatrix} = \begin{pmatrix} f_{11} & f_{12} & f_{13} & f_{14} \\ f_{21} & f_{22} & f_{23} & f_{24} \\ f_{31} & f_{32} & f_{33} & f_{34} \\ f_{41} & f_{42} & _{43} & f_{44} \end{pmatrix} \begin{pmatrix} A_{41} \\ A_{42} \\ A_{43} \\ A_{44} \end{pmatrix} \tag{A.52}$$

where $T_i(\beta)$ and $Q(\beta)$ are the Fourier transforms of $T_{3i}(x_1)|_{x3=0-}$ and $q(x_1)$, respectively:

$$Q(\beta) = \frac{1}{2\pi} \int_{-\infty}^{+\infty} q(x_1) \exp(j\beta x_1) dx_1, \tag{A.53}$$

$$T_i(\beta) = \frac{1}{2\pi} \int_{-\infty}^{+\infty} T_{3i}(x_1)|_{x3=0-} \exp(j\beta x_1) dx_1, \tag{A.54}$$

and

$$\begin{pmatrix} f_{1n} \\ f_{2n} \\ f_{3n} \\ f_{4n} \end{pmatrix} = \begin{pmatrix} \Omega_n c_{55}^E - jc_{15} & \Omega_n c_{45}^E - jc_{56}^E & \Omega_n c_{35}^E - jc_{55}^E \\ \Omega_n c_{45}^E - jc_{14}^E & \Omega_n c_{44}^E - jc_{46}^E & \Omega_n c_{34}^E - jc_{45}^E \\ \Omega_n c_{35}^E - jc_{13}^E & \Omega_n c_{34}^E - jc_{36}^E & \Omega_n c_{33}^E - 2jc_{35}^E \\ \Omega_n e_{35} - je_{31} & \Omega_n e_{34} - je_{36} & \Omega_n e_{33} - je_{35} \end{pmatrix}$$

$$\begin{pmatrix} \Omega_n e_{35} - je_{15} \\ \Omega_n e_{34} - je_{14} \\ \Omega_n e_{33} - je_{13} \\ -\Omega_n \epsilon_{33}^S + j\epsilon_{13}^S + s\epsilon_0 \end{pmatrix} \begin{pmatrix} \alpha_{1n} \\ \alpha_{2n} \\ \alpha_{3n} \\ 1 \end{pmatrix}. \tag{A.55}$$

For the free surface, the vector on the left hand side of (A.52) is zero. Thus, for the existence of nontrivial solutions, the determinant D_f of the matrix on the right hand side of (A.52) must be zero. Namely, if we find β giving $D_f = 0$, it might correspond to the wavenumber β_f of the SAW on the free surface.

Note the condition $D_f = 0$ is not sufficient for the existence of SAW solutions. For example, it is clear from (A.52) that $D_f = 0$ at a particular β where $\Omega_m = \Omega_n$ for $m \neq n$. However this β is not relevant to the SAW solution because the general solution of (A.40) is different from that given in (A.41) for this case.

On the metallized surface, since $q(x_1)$ is nonzero and unknown, we must use the boundary condition of

$$\phi|_{x_3=0-} = 0 \tag{A.56}$$

instead of (A.50) and (A.51).

Let us define the Fourier transform of $\phi(x_1)|_{x3=0-}$:

$$\Phi(\beta) = \frac{1}{2\pi} \int_{-\infty}^{+\infty} \phi(x_1)|_{x3=0-} \exp(j\beta x_1)dx_1. \tag{A.57}$$

From (A.41), it is clear that

$$\Phi(\beta) = \sum_{n=1}^{4} A_{4n}. \tag{A.58}$$

Then we get

$$\begin{pmatrix} T_1(\beta)/\beta \\ T_2(\beta)/\beta \\ T_3(\beta)/\beta \\ \Phi(\beta) \end{pmatrix} = \begin{pmatrix} f_{11} & f_{12} & f_{13} & f_{14} \\ f_{21} & f_{22} & f_{23} & f_{24} \\ f_{31} & f_{32} & f_{33} & f_{34} \\ 1 & 1 & 1 & 1 \end{pmatrix} \begin{pmatrix} A_{41} \\ A_{42} \\ A_{43} \\ A_{44} \end{pmatrix}. \tag{A.59}$$

For the existence of nontrivial solutions, the determinant D_m of the matrix on the right hand side of (A.59) must be zero. Namely, if we find β giving $D_\mathrm{m} = 0$, it might correspond to the wavenumber β_m of the SAW on the metallized surface.

A.4 Effective Acoustic Admittance Matrix and Permittivity

The procedure described above is quite general, and is readily applicable to problems with various surface boundary conditions.

Let us define the Fourier transform $U_i(\beta)$ of $u_i(x_1)|_{x3=0-}$:

$$U_i(\beta) = \frac{1}{2\pi} \int_{-\infty}^{+\infty} u_i(x_1)|_{x3=0-} \exp(j\beta x_1)dx_1. \tag{A.60}$$

It is clear that

$$U_i(\beta) = \sum_{n=1}^{4} A_{in} = \sum_{n=1}^{4} \alpha_{in} A_{4n} \tag{A.61}$$

from (A.41). Thus from (A.57) and (A.61), we get

$$\begin{pmatrix} U_1(\beta) \\ U_2(\beta) \\ U_3(\beta) \\ \Phi(\beta) \end{pmatrix} = \begin{pmatrix} \alpha_{11} & \alpha_{12} & \alpha_{13} & \alpha_{14} \\ \alpha_{21} & \alpha_{22} & f_{23} & \alpha_{24} \\ \alpha_{31} & \alpha_{32} & \alpha_{33} & \alpha_{34} \\ 1 & 1 & 1 & 1 \end{pmatrix} \begin{pmatrix} A_{41} \\ A_{42} \\ A_{43} \\ A_{44} \end{pmatrix}. \tag{A.62}$$

By substituting A_{4n} derived from (A.52) into (A.62), we obtain the following relationship between $\mathbf{U} = (U_1, U_2, U_3)$, $\mathbf{T} = (T_1, T_2, T_3)$, Q and Φ:

$$\mathbf{U}(\beta) = s\beta^{-1}\{[\mathbf{R}_{11}(\beta)]\mathbf{T}(\beta) + [\mathbf{R}_{12}(\beta)]Q(\beta)\}, \tag{A.63}$$

$$\Phi(\beta) = s\beta^{-1}\{[\mathbf{R}_{21}(\beta)] \cdot \mathbf{T}(\beta) + R_{22}(\beta)Q(\beta)\}, \tag{A.64}$$

where $jsVR_{ij}(\beta) \equiv Y_{ij}(\beta)$ is called the effective acoustic admittance matrix [5] and is given by

$$\begin{pmatrix} [\mathbf{R}_{11}(\beta)] & [\mathbf{R}_{12}(\beta)] \\ [\mathbf{R}_{21}(\beta)] & R_{22}(\beta) \end{pmatrix} = s \begin{pmatrix} \alpha_{11} & \alpha_{12} & \alpha_{13} & \alpha_{14} \\ \alpha_{21} & \alpha_{22} & f_{23} & \alpha_{24} \\ \alpha_{31} & \alpha_{32} & \alpha_{33} & \alpha_{34} \\ 1 & 1 & 1 & 1 \end{pmatrix} \begin{pmatrix} f_{11} & f_{12} & f_{13} & f_{14} \\ f_{21} & f_{22} & f_{23} & f_{24} \\ f_{31} & f_{32} & f_{33} & f_{34} \\ f_{41} & f_{42} & f_{43} & f_{44} \end{pmatrix}^{-1}, \tag{A.65}$$

and $V = \omega/\beta$.

In the case where $T_i(\beta) = 0$, the relation between $Q(\beta)$ and $\Phi(\beta)$ simply reduces to

$$\frac{\Phi(\beta)}{Q(\beta)} = s\beta^{-1}R_{22}(\beta) \equiv s\beta^{-1}\epsilon(S)^{-1}, \tag{A.66}$$

where $S = \beta/\omega$ is the slowness, and

$$\epsilon(S) = -s\frac{D_{\mathrm{f}}(S)}{D_{\mathrm{m}}(S)} \tag{A.67}$$

is the effective permittivity [6] described in Sect. 6.2.

The inverse Fourier transform of (A.63) and (A.64) gives

$$\mathbf{u}(x_1,0) = \frac{1}{2\pi} \int_{-\infty}^{+\infty} s\beta^{-1}[\mathbf{R}_{11}(\beta)]\mathbf{T}(\beta)\exp(-j\beta x_1)d\beta$$

$$+ \frac{1}{2\pi} \int_{-\infty}^{+\infty} s\beta^{-1}[\mathbf{R}_{12}(\beta)]Q(\beta)\exp(-j\beta x_1)d\beta \tag{A.68}$$

$$\phi(x_1,0) = \frac{1}{2\pi} \int_{-\infty}^{+\infty} s\beta^{-1}[\mathbf{R}_{21}(\beta)] \cdot \mathbf{T}(\beta)\exp(-j\beta x_1)d\beta$$

$$+ \frac{1}{2\pi} \int_{-\infty}^{+\infty} s\beta^{-1}R_{22}(\beta)Q(\beta)\exp(-j\beta x_1)d\beta, \tag{A.69}$$

where $\mathbf{u} = (u_1, u_2, u_3)$.

Substitution of (A.53) and (A.54) into (A.68) and (A.69) gives the following convolution form:

$$\mathbf{u}(x_1,0) = \int_{-\infty}^{+\infty} [\mathbf{G}_{11}(x_1 - x_1')]\mathbf{T}(x_1',0)dx_1'$$

$$+ \int_{-\infty}^{+\infty} [\mathbf{G}_{12}(x_1 - x_1')]q(x_1')dx_1', \tag{A.70}$$

$$\phi(x_1,0) = \int_{-\infty}^{+\infty} [\mathbf{G}_{21}(x_1 - x_1')] \cdot \mathbf{T}(x_1',0)dx_1'$$

$$+ \int_{-\infty}^{+\infty} G_{22}(x_1 - x_1')q(x_1')dx_1', \tag{A.71}$$

where $\mathbf{T} = (T_{31}, T_{32}, T_{33})$, and $G_{mn}(x_1)$ is the Green function given by

$$[\mathbf{G}_{mn}(x_1)] = \frac{1}{2\pi} \int_{-\infty}^{+\infty} s\beta^{-1}[\mathbf{R}_{mn}(\beta)] \exp(-j\beta x_1) dx_1. \tag{A.72}$$

When $\mathbf{T}(x_1', 0) = 0$, (A.71) reduces to

$$\phi(x_1, 0) = \int_{-\infty}^{+\infty} G_{\mathrm{f}}(x_1 - x_1') q(x_1') dx_1', \tag{A.73}$$

which has already been given as (6.13), and $G_{\mathrm{f}}(x_1) = G_{22}(x_1)$ is the Green function for the free surface and has already been given as (6.14).

A.5 Acoustic Wave Properties in 6mm Materials

A.5.1 Rayleigh-Type SAWs

As an example, let us consider SAW propagation on 6mm materials as shown in Fig. A.5, where the IDT fingers are assumed to be aligned parallel to the crystal Z axis.

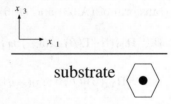

Fig. A.5. 6mm substrate and coordinate system

In this case, the crystal X, Y and Z axes coincide with x_1, x_3, and $-x_2$ axes, respectively, and this conversion corresponds to $(1, 2, 3, 4, 5, 6) \rightarrow (1, 3, 2, -4, -6, 5)$ in the abbreviated notation. Since material constants for 6mm materials are given in (A.11), (A.23) and (A.24), their substitution into (A.25) and (A.26) gives

$$T_{11} = c_{11}S_1 + c_{12}S_3 = c_{11}\frac{\partial u_1}{\partial x_1} + c_{12}\frac{\partial u_3}{\partial x_3}, \tag{A.74}$$

$$T_{12} = c_{44}S_6 - e_{15}E_1 = c_{44}\frac{\partial u_2}{\partial x_1} + e_{15}\frac{\partial \phi}{\partial x_3}, \tag{A.75}$$

$$T_{13} = c_{66}S_5 = c_{66}\frac{\partial u_1}{\partial x_3} + c_{66}\frac{\partial u_3}{\partial x_1}, \tag{A.76}$$

$$T_{23} = c_{44}S_4 - e_{15}E_3 = c_{44}\frac{\partial u_2}{\partial x_3} + e_{15}\frac{\partial \phi}{\partial x_3}, \tag{A.77}$$

$$T_{33} = c_{12}S_1 + c_{11}S_3 = c_{12}\frac{\partial u_1}{\partial x_1} + c_{11}\frac{\partial u_3}{\partial x_3}, \tag{A.78}$$

$$D_1 = \epsilon_{11}E_1 + e_{15}S_6 = e_{15}\frac{\partial u_2}{\partial x_1} - \epsilon_{11}\frac{\partial \phi}{\partial x_1}, \tag{A.79}$$

$$D_3 = \epsilon_{11}E_3 + e_{15}S_4 = e_{15}\frac{\partial u_2}{\partial x_3} - \epsilon_{11}\frac{\partial \phi}{\partial x_3}. \tag{A.80}$$

It is seen that u_1 and u_3 are isolated from u_2 and ϕ, and the former composes the Rayleigh-type SAW and the latter composes the SH-type SAW, called a BGS wave [7, 8, 9] (see Sect. 8.1.1).

Substitution of (A.74), (A.76) and (A.78) into (A.38) gives

$$\begin{pmatrix} \Omega^2 c_{66} - c_{11} + \rho V^2 & -j\Omega(c_{12} + c_{66}) \\ -j\Omega(c_{12} + c_{66}) & c_{11}\Omega^2 - c_{66} + \rho V^2 \end{pmatrix} \begin{pmatrix} u_1 \\ u_3 \end{pmatrix} = \begin{pmatrix} 0 \\ 0 \end{pmatrix}. \tag{A.81}$$

Then the general solution is expressed as follows;

$$u_i(x_1, x_3) = \sum_{n=1}^{2} A_{in} \exp(\beta\Omega_n x_3 - j\beta x_1), \tag{A.82}$$

where

$$\Omega_1 = s\sqrt{1 - \rho V^2/c_{11}}, \tag{A.83}$$

$$\Omega_2 = s\sqrt{1 - \rho V^2/c_{66}}. \tag{A.84}$$

Note that we choose the square root with positive imaginary part when their argument is negative, and

$$r_1 = A_{31}/A_{11} = j\Omega_1, \tag{A.85}$$

$$r_2 = A_{32}/A_{12} = j\Omega_2^{-1}. \tag{A.86}$$

Substitution of (A.82) into (A.76) and (A.78) gives

$$T_{31}(x_1, 0) = \beta c_{44} \sum_{n=1}^{2} (\Omega_n - jr_n)A_{1n} \exp(-j\beta x_1), \tag{A.87}$$

$$T_{33}(x_1, 0) = \beta \sum_{n=1}^{2} (-jc_{12} + c_{11}\Omega_n r_n)A_{1n} \exp(-j\beta x_1), \tag{A.88}$$

and they are zero from the boundary condition. Then the dispersion relation for the Rayleigh-type SAW is given by

$$4\Omega_1\Omega_2 = (\Omega_2^2 + 1)^2. \tag{A.89}$$

A.5.2 Effective Permittivity for BGS Waves

Next let us consider the field composed by ϕ and u_2. Substitution of (A.75), (A.77), (A.79) and (A.80) into (A.38) and (A.39) gives

$$\begin{pmatrix} \Omega^2 c_{44} - c_{44} + \rho V^2 & e_{15}\Omega^2 - e_{15} \\ e_{15}\Omega^2 - e_{15} & -\epsilon_{11}\Omega^2 + \epsilon_{11} \end{pmatrix} \begin{pmatrix} u_2 \\ \phi \end{pmatrix} = \begin{pmatrix} 0 \\ 0 \end{pmatrix}. \tag{A.90}$$

Then the general solution is given by

$$u_2(x_1, x_3) = \sum_{n=1}^{2} A_{2n} \exp(\beta\Omega_n x_3 - j\beta x_1), \tag{A.91}$$

$$\phi(x_1, x_3) = \sum_{n=1}^{2} A_{4n} \exp(\beta\Omega_n x_3 - j\beta x_1), \tag{A.92}$$

where

$$\Omega_1 = s, \tag{A.93}$$

$$\Omega_2 = s\sqrt{1 - (V/V_B)^2}, \tag{A.94}$$

$V_B \left(= \sqrt{c_{44}^D/\rho} \right)$ is the SSBW velocity propagating parallel to the surface, $c_{44}^D = c_{44} + e_{15}^2/\epsilon_{11}$, and

$$\alpha_1 = A_{21}/A_{41} = 0, \tag{A.95}$$

$$\alpha_2 = A_{22}/A_{42} = \epsilon_{11}/e_{15}. \tag{A.96}$$

Substitution of (A.91) and (A.92) into (A.77) and (A.80) gives

$$T_{32}(x_1, 0) = \beta \sum_{n=1}^{2} \Omega_n A_{4n}(c_{44}\alpha_n + e_{15}) \exp(-j\beta x_1), \tag{A.97}$$

$$D_3(x_1, 0^-) = \beta \sum_{n=1}^{2} \Omega_n A_{4n}(e_{15}\alpha_n - \epsilon_{11}) \exp(-j\beta x_1). \tag{A.98}$$

In the vacuum, the relation between D_3 and ϕ is given by (A.47). Since

$$\Phi(\beta) = \phi(x_1, 0) = A_{41} + A_{42} = (1 - \Omega_1\Omega_2^{-1} K_{15m}^2) A_{41} \tag{A.99}$$

and

$$D_3(x_1, 0^+) = s\epsilon_0\beta(A_{41} + A_{42}), \tag{A.100}$$

we get

$$Q(\beta) = D_3(x_1, 0^+) - D_3(x_1, 0^-) = s\beta[\epsilon(0)A_{41} + \epsilon_0 A_{42}], \tag{A.101}$$

where

$$\epsilon(0) = \epsilon_0 + \epsilon_{11}. \tag{A.102}$$

Next, let us apply the boundary conditions. Since $T_{32}(x_1, 0) = 0$, we get

$$A_{42} = -\Omega_1\Omega_2^{-1} K_{15m}^2 A_{41}, \tag{A.103}$$

where

$$K_{15m}^2 = e_{15}^2/c_{44}^D \epsilon_{11} \tag{A.104}$$

is the electromechanical coupling factor for the shear wave in 6mm materials.

On the free surface, since $Q(\beta) = 0$, substitution of (A.103) into (A.101) gives the dispersion relation for the BGS wave on the free surface:

$$s\Omega_2 = K_{15f}^2, \tag{A.105}$$

where

$$K_{15f}^2 = K_{15m}^2/(1 + \epsilon_{11}/\epsilon_0). \tag{A.106}$$

On the metallized surface, since $\Phi(\beta) = 0$, $A_{41} + A_{42} = 0$. Then substitution of (A.103) into this relation gives the dispersion relation for the BGS wave on the metallized surface:

$$s\Omega_2 = K_{15m}^2. \tag{A.107}$$

A.5.3 Effective Acoustic Admittance Matrix

From (A.82), (A.87), (A.88), (A.91), (A.92), (A.97), and (A.101), the effective acoustic admittance matrix $Y_{ij}(\beta) = jsV R_{ij}(\beta)$ defined in (A.63) and (A.64) is given by

$$R_{11}(\beta) = \frac{s\Omega_2(1 - \Omega_2^2)}{c_{44}\{4\Omega_1\Omega_2 - (1 + \Omega_2^2)^2\}}, \tag{A.108}$$

$$R_{13}(\beta) = -R_{31}(\beta) = \frac{-js\{2\Omega_1\Omega_2 - (1 + \Omega_2^2)\}}{c_{44}\{4\Omega_1\Omega_2 - (1 + \Omega_2^2)^2\}}, \tag{A.109}$$

$$R_{22}(\beta) = \frac{1}{c_{44}^{D}(s\Omega_2 - K_{15f}^2)}, \tag{A.110}$$

$$R_{24}(\beta) = -R_{42}(\beta) = -\frac{e_{15}}{c_{44}^{D}(s\Omega_2 - K_{15f}^2)}, \tag{A.111}$$

$$R_{33}(\beta) = \frac{s\Omega_1(1 - \Omega_2^2)}{c_{44}\{4\Omega_1\Omega_2 - (1 + \Omega_2^2)^2\}}, \tag{A.112}$$

$$R_{44}(\beta) = \frac{s\Omega_2 - K_{15m}^2}{\epsilon(0)(s\Omega_2 - K_{15f}^2)}, \tag{A.113}$$

where Ω_1 and Ω_2 are those defined by (A.83) and (A.84), respectively, and the other components of $R_{ij}(\beta)$ are zero.

Thus the effective permittivity is given by

$$\epsilon(S) = \epsilon(0)\frac{s\Omega_2 - K_{15f}^2}{s\Omega_2 - K_{15m}^2}. \tag{A.114}$$

A.6 Wave Excitation

A.6.1 Integration Path

By using the effective permittivity given by (A.114), we will estimate

$$G_f(x_1) = \frac{1}{2\pi} \int_{-\infty}^{+\infty} \frac{1}{s\beta\epsilon(S,\omega)} \exp(-j\beta x_1) d\beta \tag{A.115}$$

and

$$G_m(x_1) = \frac{1}{2\pi} \int_{-\infty}^{+\infty} \frac{\epsilon(S,\omega)}{s\beta} \exp(-j\beta x_1) d\beta \tag{A.116}$$

given in (6.14) and (6.20), respectively.

For the estimation, we employ the Cauchy–Riemann theorem [10]. Figure A.6 shows the integration path used for the case where $x_1 > 0$. That is, the

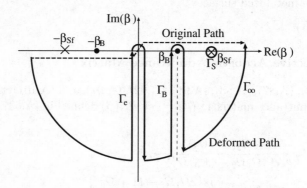

Fig. A.6. Integration path.

integration path for (A.115) and (A.116) is modified as follows;

$$\int_{-\infty}^{+\infty} d\beta = \left(\oint_{\Gamma_e} + \oint_{\Gamma_S} + \oint_{\Gamma_B} + \int_{\Gamma_\infty} \right) d\beta. \tag{A.117}$$

In the equation, Γ_e is the integral path clockwise around the imaginary axis. This originates from the discontinuity in s at $\beta = 0$, and gives the electrostatic coupling. On the other hand, Γ_S is the integral path around the pole $\beta = \beta_{Sf}$ where $\epsilon(\beta/\omega) = 0$, and this gives the BGS wave excitation. In addition, Γ_B is the branch cut around $\beta_B(= \omega/V_B)$, and this gives the SSBW excitation. The remaining Γ_∞ is the integration path as $|\beta| \to \infty$, and the integral is zero since the integrand is always zero.

A.6.2 Electrostatic Coupling

The contribution $G_e(x)$ of the electrostatic coupling is given analytically by

$$G_e(x) = \frac{1}{2\pi} \left(\int_{0^- - j\infty}^{0^-} + \int_{0^+}^{0^+ - j\infty} \right) \frac{\exp(-j\beta x)}{s\beta\epsilon(\beta/\omega)} d\beta$$

$$= \frac{1}{\pi} \int_0^{+\infty} \frac{\exp(-yx)}{y\epsilon(-jy/\omega)} dy$$

$$\cong \frac{1}{\pi\epsilon(\infty)} \int_0^{+\infty} \frac{\exp(-yx)}{y} dy = -\frac{1}{\pi\epsilon(\infty)} \log|x|. \tag{A.118}$$

For the derivation, the fact that the total charge on the surface is zero is used.

A.6.3 BGS Wave Excitation

The BGS wave velocity V_{Sf} on the free surface is given as solutions of $\epsilon(S) = 0$. On the other hand, the BGS wave velocity V_{Sm} on the metallized surface is given as solutions of $\epsilon(S)^{-1} = 0$. From (A.105) and (A.107), they are given by

$$V_{Sf} = V_B \sqrt{1 - K_{15f}^4}, \tag{A.119}$$

$$V_{Sm} = V_B \sqrt{1 - K_{15m}^4}. \tag{A.120}$$

The contribution $G_{Sf}(x)$ of the BGS wave excitation in $G_f(x)$ is given by applying the residue theorem to (A.115). The solution has already been given by (6.22), i.e.,

$$G_{Sf}(x_1) = \frac{1}{2\pi} \oint_{\Gamma_S} \frac{\exp(-j\beta x_1)}{s\beta\epsilon(\beta/\omega)} d\beta = -j \operatorname*{Res}_{\beta \to \beta_{Sf}} \left\{ \frac{\exp(-j\beta x_1)}{s\beta\epsilon(\beta/\omega)} \right\}$$

$$= j\frac{K_{Sf}^2 \exp(-j\beta_{Sf}|x_1|)}{2\epsilon(\infty)}, \tag{A.121}$$

where K_{Sf}^2 is the electromechanical coupling factor for the BGS wave on the free surface given by (6.1), i.e.,

$$K_{Sf}^2 = -2\epsilon(\infty) \left[S_{Sf} \frac{\partial\epsilon(S)}{\partial S}|_{S=S_{Sf}} \right]^{-1}. \tag{A.122}$$

On the other hand, the contribution $G_{Sm}(x)$ of the BGS wave excitation in $G_m(x)$ is given by applying the residue theorem to (A.116). The solution has already been given as (6.23), i.e.,

$$G_{Sm}(x_1) = \frac{1}{2\pi} \oint_{\Gamma_S} \frac{\epsilon(\beta/\omega)\exp(-j\beta x_1)}{s\beta} d\beta = -j \operatorname*{Res}_{\beta \to \beta_{Sm}} \left\{ \frac{\exp(-j\beta x_1)}{s\beta\epsilon(\beta/\omega)} \right\}$$

$$= j\epsilon(\infty)\frac{K_{Sm}^2 \exp(-j\beta_{Sm}|x_1|)}{2}, \tag{A.123}$$

where K_{Sm}^2 is the electromechanical coupling factor for the BGS wave on the metallized surface given by (6.2), i.e.,

$$K_{Sm}^2 = 2\epsilon(\infty)^{-1} \left[S_{Sm} \frac{\partial\epsilon(S)^{-1}}{\partial S}|_{S=S_{Sm}} \right]^{-1}. \tag{A.124}$$

Substitution of (A.114) into (A.122) and (A.124) gives

$$K_{Sf}^2 = 2K_{15f}^2 \frac{(1 - K_{15m}^2)(K_{15m}^2 - K_{15f}^2)}{(1 - K_{15f}^2)(1 + K_{15f}^4)}, \tag{A.125}$$

$$K_{Sm}^2 = 2K_{15m}^2 \frac{(1 - K_{15f}^2)(K_{15m}^2 - K_{15f}^2)}{(1 - K_{15m}^2)(1 + K_{15m}^4)}. \tag{A.126}$$

Since $\epsilon_{11} \gg \epsilon_0$ in the usual piezoelectric materials employed in SAW devices, (A.106) indicates $K_{15m}^2 \gg K_{15f}^2$ and $V_{Sf} \cong V_B$. Then we get

$$\begin{aligned}
K_{Sf}^2 &\cong 2K_{15f}^2 K_{15m}^2 (1 - K_{15m}^2) & (K_{15f}^2 \ll K_{15m}^2) \\
&\cong 2K_{15f}^2 K_{15m}^2 & (K_{15m}^2 \ll 1) & \tag{A.127} \\
K_{Sm}^2 &\cong \frac{2K_{15m}^4}{(1 - K_{15m}^2)(1 + K_{15m}^4)} & (K_{15f}^2 \ll K_{15m}^2) \\
&\cong 2K_{15m}^4 & (K_{15m}^2 \ll 1). & \tag{A.128}
\end{aligned}$$

This suggests that the BGS wave excitation is critically influenced by the surface electrical boundary condition.

A.6.4 SSBW Excitation

Since Ω changes its sign at the steepest descent path Γ_B in Fig. A.6, we get a contribution $G_{Bf}(x_1)$ of the SSBW excitation in $G_f(x_1)$ as

$$\begin{aligned}
G_{Bf}(x_1) &= \frac{1}{2\pi\epsilon(0)} \left(\int_{\beta_B^- - j\infty}^{\beta_B^-} + \int_{\beta_B^+}^{\beta_B^+ - j\infty} \right) \frac{\Omega - K_{15m}^2}{\beta(\Omega - K_{15f}^2)} \\
&\quad \times \exp(-j\beta x_1) d\beta \\
&= \frac{K_{15f}^2 - K_{15m}^2}{\pi\epsilon(0)} \int_{S_B^+}^{S_B^+ - j\infty} \frac{\Omega \exp(-j\beta x_1)}{\beta(\Omega^4 - K_{15m}^4)} d\beta, \tag{A.129}
\end{aligned}$$

where Ω is chosen so that its real part is positive.

When $\beta_B x_1 \gg 1$, setting $\beta = \beta_B - j\alpha$ gives

$$\begin{aligned}
G_{Bf}(x_1) &\cong (K_{15f}^2 - K_{15m}^2) \frac{\exp(-j\beta_B x_1 + j\pi/4)}{\pi\epsilon(0)} \\
&\quad \times \int_0^{+\infty} \frac{\sqrt{2\alpha/\beta_B} \exp(-\alpha x_1)}{2j\alpha(1 - K_{15f}^2) + K_{15f}^4 \beta_B} d\alpha. \tag{A.130}
\end{aligned}$$

Then setting $t^2 = \alpha x_1$, $x_{cf} = 2(1 - K_{15f}^2)/K_{15f}^4 \beta_B$ and $K_{Bf}^2 = (K_{15m}^2 - K_{15f}^2)/\{2\epsilon(0)(1 - K_{15f}^2)\}$ gives the same form as (4.58):

$$G_{Bf}(x_1) \cong -K_{Bf}^2 H_0^{(2)}(\beta_B x_1) U(x_1/x_{cf}), \tag{A.131}$$

where $H_0^{(2)}(\theta)$ and $U(r)$ are the functions which have already been given in (4.59) and (4.60), respectively.

References

1. B.A. Auld: Acoustic Waves and Fields in Solids, Vol. I, Chap. 3, Wieley, New York (1973) pp. 57–99.
2. B.A. Auld: Acoustic Waves and Fields in Solids, Vol. I, Chap. 7, Wieley, New York (1973) pp. 191–264.
3. B.A. Auld: Acoustic Waves and Fields in Solids, Vol. I, Chap. 4, Wieley, New York (1973) pp. 101–134.
4. K. Hashimoto and M. Yamaguchi: Free Software Products for Simulation and Design of Surface Acoustic Wave and Surface Transverse Wave Devices, Proc. 1996 Freq. Contr. Symp. (1996) pp. 300–307.
5. K. Hashimoto, Y. Watanabe, M. Akahane and M. Yamaguchi: Analysis of Acoustic Properties of Multi-Layered Structures by Means of Effective Acoustic Impedance Matrix, Proc. IEEE Ultrasonics Symp. (1990) pp. 937–942.
6. R.F. Milsom, N.H.C. Reilly and M. Redwood: Analysis of Generation and Detection of Surface and Bulk Acoustic Waves by Interdigital Transducers, IEEE Trans. Sonics and Ultrason., **SU-24** (1977) pp. 147–166.
7. J.L. Bleustein: A New Surface Wave in Piezoelectric Materials, Appl. Phys. Lett., **13** (1968) pp. 412.
8. Y.V. Gulyaev: Electroacoustic Surface Waves in Solids, Soviet Phys. JETP Lett., **9** (1969) pp. 63.
9. Y. Ohta, K. Nakamura and H. Shimizu: Surface Concentration of Shear Wave on Piezoelectric Materials with Conductor, Technical Report of IEICE, Japan **US69-3** (1969) in Japanese.
10. K. Yashiro and N. Goto: Analysis of Generation of Acoustic Waves on the Surface of a Semi-Infinite Piezoelectric Solid, IEEE Trans. Sonics and Ultrason., **SU-25**, 3 (1978) pp. 146–153.

B. Analysis of Wave Propagation on Grating Structures

B.1 Summary

In Chaps. 7 and 8, it was shown that SAW devices are simulated well by using COM theory, and the parameters required for the analysis are determined by the SAW propagation characteristics on infinite metallic grating structures.

For the analysis, the charge concentration on electrode edges must be skillfully taken into account so as to reduce the required computational time (see Sect. 3.2.4). In Sect. 2.4, it was shown that Bløtekjær's theory [1, 2] is quite powerful for the characterization of SAW propagation on these structures. The method can be modified for the case where the SAW is propagating obliquely on a metallic grating [3].

Zhang, et al. extended the theory to characterize acoustic wave excitation on the grating structure [4]. Since the method employs numerical techniques, it requires a considerable amount of computational time to achieve sufficient accuracy. Then the authors proposed use of the complex integral for the analytical evaluation of excitation and propagation characteristics of acoustic waves on periodic metallic grating structures [5].

Note that the applicability of Bløtekjær's theory was limited because the original theory does not take account of the mass loading effect due to the finite thickness of grating electrodes.

By the way, Reichinger and Baghai-Wadji proposed a method of analyzing the dispersion relation of SAWs propagating on periodic metallic grating structures with finite thickness [6]. The method employs the finite element method (FEM) for the analysis of the grating electrode region, whereas the boundary element method (BEM) is used for the analysis of the substrate region; FEM is effective to analyze electrodes of arbitrary shape, and BEM, though, markedly reduces computational time compared with FEM. In the theory, no particular consideration was given to the charge concentration at electrode edges.

Ventura, et al. discussed the use of spectral domain analysis (SDA) for the analysis of the substrate region, by which the mathematical treatment required for the analysis can be considerably simplified compared to BEM [7, 8]. So as to take the charge concentration at electrode edges into account, their method employed the technique described in Sect. 3.2.4, which is effective for the analysis of gratings with finite length.

The authors proposed another method to calculate SAW properties under periodic metallic grating structures with finite thickness [9]. The method employs both FEM and SDA for the analysis of the grating electrode and substrate grating electrodes is taken into consideration as part of the electrical quantities in a substrate. This makes it possible to employ Bløtekjær's theory, which enables rapid analysis of various wave properties on infinite metallic grating structures. Of course, the technique is also applicable for the case of oblique propagation [10].

The authors also extended the method, aiming at analyzing SAW excitation and propagation on metallic grating structures consisting of electrodes having unequal width, pitch and/or thickness [11]. Instead of Bløtekjær's theory, Aoki's theory [12, 13] and its extension [14] were employed for structures with two fingers and three fingers per period, respectively, to take account of the charge distribution for multiple electrode grating structures.

Note that the free software, FEMSDA, OBLIQ and MULTI based on these methods, has been developed and is now widely used amongst SAW researchers [15, 16].

Here useful techniques employed in this software is discussed in detail. In Sect. B.2, Bløtekjær's original theory and its extensions such as Aoki's theory are fully described. Section B.3 describes the inclusion of the mass loading effect into these theories. In Sect. B.4, it is shown how excitation properties are retrieved from these extended Bløtekjær theories.

B.2 Metallic Gratings

B.2.1 Bløtekjær's Theory for Single-Electrode Gratings

Here let us explain Bløtekjær's original theory [1, 2] in detail. Consider the periodic metallic grating shown in Fig. B.1 with an infinite acoustic length, the electrode periodicity and width of which are p and w. An acoustic plane wave propagates toward the x_1 direction. The electrode length is assumed to be much larger than the wavelength, and the effects of electrode thickness h and resistivity are neglected.

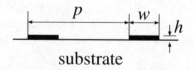

Fig. B.1. Periodic metallic grating

Let us express the electric charge $q(x_1)$ on the electrodes and electric field $e(x_1)$ within the electrode gaps as follows

$$q(x_1) = \sum_{m=-\infty}^{+\infty} \frac{A_m \exp(-j\beta_{m-1/2}x_1)}{\sqrt{\cos(\beta_g x_1) - \cos\Delta}}, \tag{B.1}$$

$$e(x_1) = \sum_{m=-\infty}^{+\infty} \frac{B_m \mathrm{sgn}(x_1) \exp(-j\beta_{m-1/2}x_1)}{\sqrt{-\cos(\beta_g x_1) + \cos\Delta}}, \tag{B.2}$$

where $\Delta = \pi w/p$, $\beta_g = 2\pi/p$, $\beta_m = \beta_g(m+s)$ and $\beta_0(= s\beta_g)$ is the wavenumber of the grating mode. The square root terms in these equations express the divergence of $q(x_1)$ and $e(x_1)$ at the electrode edges.

Using the effective permittivity $\epsilon(\beta/\omega)$ explained in Sect. 6.2, one can relate $q(x_1)$ to $e(x_1)$ as

$$E(\beta_n) = jS_n\epsilon(\beta_n/\omega)^{-1}Q(\beta_n). \tag{B.3}$$

In this equation, $S_n = \mathrm{sgn}(\beta_n)$, and $Q(\beta_n)$ and $E(\beta_n)$ are the Fourier expansion coefficients of $q(x_1)$ and $e(x_1)$ given by

$$Q(\beta_n) = p^{-1} \int_{-p/2}^{+p/2} q(x_1) \exp(+j\beta_n x_1) dx_1$$

$$= 2^{-0.5} \sum_{m=-\infty}^{+\infty} A_m P_{n-m}(\cos\Delta), \tag{B.4}$$

$$E(\beta_n) = p^{-1} \int_{-p/2}^{+p/2} e(x_1) \exp(+j\beta_n x_1) dx_1$$

$$= -2^{-0.5} \sum_{m=-\infty}^{+\infty} S_{n-m} B_m P_{n-m}(\cos\Delta), \tag{B.5}$$

where $P_m(\theta)$ is the m-th order Legendre functions.

Substitution of (B.4) and (B.5) into (B.3) gives

$$\sum_{m=-\infty}^{+\infty} \left(\frac{jS_n\epsilon(\infty)}{\epsilon(\beta_n/\omega)} A_m + S_{n-m} B_m \right) P_{n-m}(\cos\Delta) = 0. \tag{B.6}$$

For numerical calculation, assume that the range of summation over m in (B.6) is limited to $M_1 \le m \le M_2$. When $\epsilon(\beta_n/\omega)$ for a specific substrate material can approximately be described by $\epsilon(\infty)$ for $n \ge N_2$ or $n \le N_1$, the following relation should be satisfied so that (B.6) holds for $n \ge N_2 - M_1$ or $n \le N_1 - M_2$:

$$B_m = -j\epsilon(\infty)^{-1}A_m, \tag{B.7}$$

for all m. Hence, one obtains

$$\sum_{m=M_1}^{M_2} A_m \left(S_{n-m} - S_n \frac{\epsilon(\infty)}{\epsilon(\beta_n/\omega)} \right) P_{n-m}(\cos\Delta) = 0. \tag{B.8}$$

If $M_2 = N_2$ and $M_1 = N_1 + 1$, (B.8) is automatically satisfied for $n \ge N_2$ and $n \le N_1$. In this case, since $n = [M_1, M_2 - 1]$, the total number of unknowns

A_m is greater than that of the equations by one (see (B.8)). So the ratio of A_m can be determined by solving the simultaneous linear equations with $(M_2 - M_1)$ unknowns.

Then A_m gives the total charge Q on an electrode;

$$Q = \int_{-w/2}^{+w/2} q(x_1)dx = 2^{-0.5}p \sum_{m=M_1}^{M_2} A_m P_{m+s-1}(\cos \Delta), \qquad (B.9)$$

and the potential Φ [1];

$$\Phi = -\int_{-\infty}^{-w/2} e(x_1)dx_1 = -\int_{-p+w/2}^{-w/2} \frac{e(x_1)dx_1}{1 - \exp(2\pi js)}$$

$$= \frac{2^{-1.5}p}{\epsilon(\infty)\sin(s\pi)} \sum_{m=M_1}^{M_2} (-)^m A_m P_{m+s-1}(-\cos \Delta). \qquad (B.10)$$

As with the ratio between Q and Φ, the strip admittance $Y(s,\omega)$ is defined [1];

$$Y(s,\omega) = \frac{j\omega Q}{\Phi} = 2j\omega \sin(s\pi)\epsilon_g(s,\omega), \qquad (B.11)$$

where $\epsilon_g(s,\omega)$ is the effective permittivity for the grating structure [5] defined by

$$\epsilon_g(s,\omega) = \epsilon(\infty) \frac{\displaystyle\sum_{m=M_1}^{M_2} A_m P_{m+s-1}(\cos \Delta)}{\displaystyle\sum_{m=M_1}^{M_2} (-1)^m A_m P_{m+s-1}(-\cos \Delta)}. \qquad (B.12)$$

Note that $\epsilon_g(-s,\omega) = \epsilon_g(s,\omega)$ from reciprocity (see Sect. B.4.1).

Note that $Q = 0$ and $\Phi = 0$ hold for open-circuited (OC) and short-circuited (SC) gratings, respectively. Hence, the dispersion relation of the grating modes propagating on an OC grating is obtained by substituting A_m into (B.9) and setting $Q = 0$ (or $\epsilon_g = 0$). For an SC grating, on the other hand, the dispersion relation is determined by (B.10) and $\Phi = 0$ (or $\epsilon_g^{-1} = 0$). Further discussions on this subject were given in Sect. 2.4.

B.2.2 Wagner's Theory for Oblique Propagation

Here an extension of Bløtekjær's theory made by Wagner and Männer [3] is described. It aims at the analysis of SAW oblique propagation on a massless, metallic grating.

Consider plane SAWs obliquely propagating on the metallic grating shown in Fig. B.2. Assuming the wavenumbers of SAWs propagating toward the x_1 and x_2 directions to be $\beta^{(1)}$ and $\beta^{(2)}$, respectively, one can relate these wavenumbers to the propagation angle θ with respect to the x_1 axis as

$$\beta^{(2)} = \beta^{(1)} \tan \theta. \tag{B.13}$$

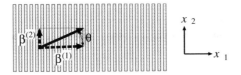

Fig. B.2. Obliquely propagating SAWs

Since the metallic grating is uniform in the x_2 direction, the SAW field $\phi(x_1, x_2)$ on the surface of the substrate varies according to

$$\phi(x_1, x_2) \propto \exp(-j\beta^{(2)} x_2). \tag{B.14}$$

Because of the periodicity of the grating structure in the x_1 direction, $\phi(x_1, x_2)$ also satisfies the Floquet theorem (see Sect. 2.2.1), i.e.,

$$\phi(x_1 + p, x_2) = \phi(x_1, x_2) \exp(-j\beta^{(1)} p). \tag{B.15}$$

From (B.14) and (B.15), $\phi(x_1, x_2)$ is expressed in the form

$$\phi(x_1, x_2) = \sum_{n=-\infty}^{+\infty} \Phi(\beta_n) \exp(-j\beta_n x_1 - j\beta^{(2)} x_2) \tag{B.16}$$

where $\beta_n = \beta^{(1)} + 2\pi n/p$, and $\Phi(\beta_n)$ is the amplitude. The equation suggests that $\phi(x_1, x_2)$ on the grating structure could be expressed as a sum of various plane waves having wavenumbers β_n and $\beta^{(2)}$ toward the x_1 and x_2 directions, respectively. Hence, although the SAW field generally has three-dimensional variation, it is reduced to a two-dimensional problem, and $\phi(x_1, x_2)$ can be analyzed by specifying the angular frequency ω and $\beta^{(2)}$ as a parameter and applying the method discussed in Sect. B.2.1.

Note, however, that since the grating electrodes are assumed to be of infinite length in the x_2 direction, the electric field associated with the obliquely propagating SAWs is always short-circuited, independent of the electrical connection; the OC grating also behaves like the SC grating.

B.2.3 Aoki's Theory for Double-Electrode Gratings

Here another extension of Bløtekjær's theory is given, and was done by Aoki and Ingebrigtsen [12, 13] so as to discuss SAW propagation in periodic metallic grating structures with multiple fingers per period.

Figure B.3 shows the periodic metallic grating of infinite acoustic length, where two types of electrodes with widths w_1 and w_2, respectively, are aligned within one periodic length p. The analysis assumes that SAWs propagate toward x_1. The effects of electrode thickness and resistivity are ignored.

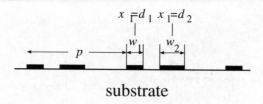

substrate

Fig. B.3. Configuration used for analysis

Define the following functions;

$$f_i(x_1) = \frac{\sqrt{2}}{2} \sum_{\ell=-\infty}^{+\infty} P_\ell(\Omega_i) \exp\{-j\beta_{\mathrm{g}}(\ell + \frac{1}{2})(x_1 - d_i)\}$$

$$= \begin{cases} \dfrac{1}{\sqrt{\cos\{\beta_{\mathrm{g}}(x_1 - d_i)\} - \Omega_i}} & (|x_1 - d_i| < w_i/2) \\ 0 & (|x_1 - d_i| > w_i/2) \end{cases}, \tag{B.17}$$

$$g_i(x_1) = \frac{j\sqrt{2}}{2} \sum_{\ell=-\infty}^{+\infty} S_\ell P_\ell(\Omega_i) \exp\{-j\beta_{\mathrm{g}}(\ell + \frac{1}{2})(x_1 - d_i)\}$$

$$= \begin{cases} 0 & (|x_1 - d_i| < w_i/2) \\ \dfrac{\mathrm{sgn}(x_1 - d_i)}{\sqrt{\Omega_i - \cos\{\beta_{\mathrm{g}}(x_1 - d_i)\}}} & (|x_1 - d_i| > w_i/2) \end{cases}, \tag{B.18}$$

where $\Omega_i = \cos(\pi w_i/p)$.

Then we represent $q(x_1)$ and $e(x_1)$ in the form of a Fourier expansion with weighting factors $f_i(x_1)$ and $g_i(x_1)$ (see (B.17) and (B.18)):

$$q(x_1) = \{f_1(x_1)g_2(x_1) + g_1(x_1)f_2(x_1)\} \sum_{m=M_1}^{M_2} A_m \exp(-j\beta_m x_1)$$

$$= \sum_{n=-\infty}^{+\infty} \sum_{m=M_1}^{M_2} A_m F_{n-m} \exp(-j\beta_n x_1), \tag{B.19}$$

$$e(x_1) = g_1(x_1)g_2(x_1) \sum_{m=M_1}^{M_2} B_m \exp(-j\beta_m x_1)$$

$$= \sum_{n=-\infty}^{+\infty} \sum_{m=M_1}^{M_2} B_m G_{n-m} \exp(-j\beta_n x_1). \tag{B.20}$$

In (B.19) and (B.20), A_m is the unknown coefficient,

$$F_n = \frac{j}{2} \sum_{\ell=-\infty}^{+\infty} P_{n-\ell-1}(\Omega_1) P_\ell(\Omega_2)(S_\ell + S_{n-\ell-1})$$

$$\times \exp(j\eta^{(1)}_{n-\ell-1/2} + j\eta^{(2)}_{\ell+1/2}), \tag{B.21}$$

$$G_n = -\frac{1}{2} \sum_{\ell=-\infty}^{+\infty} P_{n-\ell-1}(\Omega_1)P_\ell(\Omega_2)S_{n-\ell-1}S_\ell$$

$$\times \exp(j\eta^{(1)}_{n-\ell-1/2} + j\eta^{(2)}_{\ell+1/2}), \tag{B.22}$$

where $\eta^{(i)}_n = n\beta_g d_i = 2\pi n d_i/p$.

Because the electrodes do not overlap each other, $f_1(x_1)f_2(x_1) = 0$. Since

$$f_1(x_1)f_2(x_1) = \sum_{n=-\infty}^{+\infty} \exp(j\beta_n x_1) \sum_{\ell=-\infty}^{+\infty} P_{n-\ell-1}(\Omega_1)P_\ell(\Omega_2)$$

$$\times \exp(j\eta^{(1)}_{n-\ell-1/2} + j\eta^{(2)}_{\ell+1/2}) \tag{B.23}$$

from (B.17), the following relation hold for arbitrary n:

$$\sum_{\ell=-\infty}^{+\infty} P_{n-\ell-1}(\Omega_1)P_\ell(\Omega_2)\exp(j\eta^{(1)}_{n-\ell-1/2} + j\eta^{(2)}_{\ell+1/2}) = 0. \tag{B.24}$$

This relation simplifies (B.21) and (B.22) to

$$G_n = -\sum_{\ell=0}^{n-1} P_{n-\ell-1}(\Omega_1)P_\ell(\Omega_2)\exp(j\eta^{(1)}_{n-\ell-1/2} + j\eta^{(2)}_{\ell+1/2}) \tag{B.25}$$

for positive n, $G_0 = 0$, $G_{-n} = G_n^*$, and

$$F_n = -jS_nG_n. \tag{B.26}$$

Since the summation has only to be done over finite ℓ, (B.25) and (B.26) could calculate F_n and G_n much faster than (B.21) and (B.22).

From (B.19) and (B.20), the Fourier transforms $Q(\beta_n)$ and $E(\beta_n)$ of $q(x_1)$ and $e(x_1)$, respectively, are given by

$$Q(\beta_n) = -j \sum_{m=M_1}^{M_2} A_m S_{n-m} G_{n-m}, \tag{B.27}$$

$$E(\beta_n) = \sum_{m=M_1}^{M_2} B_m G_{n-m}. \tag{B.28}$$

Substitution of (B.27) and (B.28) into (B.3) gives

$$\sum_{m=M_1}^{M_2} G_{n-m}\{B_m - A_m S_{n-m} S_n/\epsilon(\beta_n/\omega)\} = 0. \tag{B.29}$$

Note that $\epsilon(\beta_n/\omega)$ approaches $\epsilon(\infty)$ when $|n| \to \infty$, and that the relation $\epsilon(\beta_n/\omega) \cong \epsilon(\infty)$ holds if $|\omega/\beta_n|$ is not very close to the SAW velocity V_S. So if it is assumed that $\epsilon(\beta_n/\omega) = \epsilon(\infty)$ for $n < N_1$ or $N > N_2$, $B_m = A_m/\epsilon(\infty)$ so that (B.29) holds for $n < M_1$ or $n \geq M_2$. Thus (B.29) becomes

$$\sum_{m=M_1}^{M_2} A_m G_{n-m}\{1 - S_{n-m}S_n \epsilon(\infty)/\epsilon(\beta_n/\omega)\} = 0 \tag{B.30}$$

for $n = [M_1 + 1, M_2 - 1]$.

By applying the Floquet theorem, the potential $\Phi^{(\ell)}(s)$ of the ℓ-th electrode is given by

$$\Phi^{(\ell)}(s) = -\int_{-\infty}^{d_\ell - w_\ell/2} e(x_1)dx_1 = -\int_{-p+d_\ell+w_\ell/2}^{d_\ell - w_\ell/2} \frac{e(x_1)dx_1}{1 - \exp(2\pi js)}. \tag{B.31}$$

Substitution of (B.20) into (B.31) gives

$$\Phi^{(1)}(s) = -j\frac{\exp(-\pi js)}{2\sin(2\pi s)} \sum_{m=M_1}^{M_2} A_m\{V_m^{(1)} + V_m^{(2)}\exp(2\pi js)\}, \tag{B.32}$$

$$\Phi^{(2)}(s) = -j\frac{\exp(-\pi js)}{2\sin(2\pi s)} \sum_{m=M_1}^{M_2} A_m\{V_m^{(1)} + V_m^{(2)}\}, \tag{B.33}$$

where

$$V_m^{(1)} = \int_{-p+d_2+w_2/2}^{d_1-w_1/2} g_1(x_1)g_2(x_1)\exp(-j\beta_m x_1)dx_1, \tag{B.34}$$

$$V_m^{(2)} = \int_{d_1+w_1/2}^{d_2-w_2/2} g_1(x_1)g_2(x_1)\exp(-j\beta_m x_1)dx_1. \tag{B.35}$$

The current $I^{(k)}(s)$ flowing into the k-th electrode is given by

$$I^{(k)}(s) = j\omega \int_{d_k-w_k/2}^{d_k+w_k/2} q(x_1)dx_1 = \sum_{m=M_1}^{M_2} A_m W_m^{(k)}(s), \tag{B.36}$$

where

$$W_m^{(k)}(s) = j\omega \int_{d_k-w_k/2}^{d_k+w_k/2} g_k(x_1)f_{3-k}(x_1)\exp(-j\beta_m x_1)dx_1. \tag{B.37}$$

Numerical analysis is carried out by the following procedure. After calculating $V_m^{(i)}$ and $W_m^{(i)}$ by numerical integration for given values of s and ω, the linear equations composed of (B.30), (B.32) and (B.33) are solved with respect to A_m. This gives A_m in the form

$$A_m = p_{m1}(s)\Phi^{(1)}(s) + p_{m2}(s)\Phi^{(2)}(s). \tag{B.38}$$

Substitution of (B.38) into (B.36) gives

$$I^{(k)}(s) = \sum_{\ell=1}^{2} Y_{k\ell}(s,\omega)\Phi^{(\ell)}(s), \tag{B.39}$$

where

$$Y_{k\ell}(s,\omega) = \sum_{m=M_1}^{M_2} W_m^{(k)}(s)p_{m\ell}(s), \tag{B.40}$$

and $k = 1$ or 2. In this equation, $Y_{k\ell}(s,\omega)$ is the transfer admittance matrix for the grating structure.

Various grating modes can be characterized by the transfer admittance matrix $Y_{k\ell}(s,\omega)$. For example, if all electrodes are electrically shorted, the dispersion relation of the grating modes is obtained by $Y_{k\ell}(s,\omega)^{-1} = 0$, because $I^{(k)}(s) \neq 0$ even when $\Phi^{(\ell)}(s) = 0$. When the electrode "2" is electrically open,

$$0 = Y_{21}(s,\omega)\Phi^{(1)}(s) + Y_{22}(s,\omega)\Phi^{(2)}(s).$$

Substitution of this relation into (B.39) gives

$$I^{(1)}(s) = \tilde{Y}_{11}\Phi^{(1)}(s) \tag{B.41}$$

where

$$\tilde{Y}_{11}(s,\omega) = Y_{11}(s,\omega) - \frac{Y_{12}(s,\omega)Y_{21}(s,\omega)}{Y_{22}(s,\omega)}. \tag{B.42}$$

Then when the electrode "2" is open-circuited, the dispersion relation of the grating mode is given by $\tilde{Y}_{11}(s,\omega) = 0$ when the electrode "1" is open-circuited whereas $\tilde{Y}_{11}(s,\omega)^{-1} = 0$ when the electrode "1" is short-circuited.

B.2.4 Extension to Triple-Electrode Gratings

Next, let us discuss another metallic grating, where three types of electrodes with widths w_1, w_2 and w_3 are aligned within one periodic length p. The following derivation was done by Hashimoto and Yamaguchi [14].

Let $q(x_1)$ and $e(x_1)$ be

$$q(x_1) = \{f_1(x_1)g_2(x_1)g_3(x_1) + g_1(x_1)f_2(x_1)g_3(x_1)$$

$$+ g_1(x_1)g_2(x_1)f_3(x_1)\} \sum_{m=M_1}^{M_2} A_m \exp(-j\beta_{m-\frac{1}{2}}x_1)$$

$$= \sum_{n=-\infty}^{+\infty} \sum_{m=M_1}^{M_2} A_m F_{n-m} \exp(-j\beta_n x_1), \tag{B.43}$$

$$e(x) = g_1(x_1)g_2(x_1)g_3(x_1) \sum_{m=M_1}^{M_2} B_m \exp(-j\beta_{m-\frac{1}{2}}x_1)$$

$$= \sum_{n=-\infty}^{+\infty} \sum_{m=M_1}^{M_2} B_m G_{n-m} \exp(-j\beta_n x_1), \tag{B.44}$$

where

$$F_n = \frac{-\sqrt{2}}{4} \sum_{k=-\infty}^{+\infty} \sum_{\ell=-\infty}^{+\infty} P_k(\Omega_1) P_\ell(\Omega_2) P_{n-k-\ell-1}(\Omega_3)$$

$$\times \exp(j\eta_{k+1/2}^{(1)} + j\eta_{\ell+1/2}^{(2)} + j\eta_{n-k-\ell-1/2}^{(3)})$$

$$\times (S_\ell S_k + S_k S_{n-k-\ell-1} + S_{n-k-\ell-1} S_\ell), \tag{B.45}$$

$$G_n = \frac{-j\sqrt{2}}{4} \sum_{k=-\infty}^{+\infty} \sum_{\ell=-\infty}^{+\infty} P_k(\Omega_1) P_\ell(\Omega_2) P_{n-k-\ell-1}(\Omega_3)$$

$$\times \exp(j\eta_{k+1/2}^{(1)} + j\eta_{\ell+1/2}^{(2)} + j\eta_{n-k-\ell-1/2}^{(3)}) S_\ell S_k S_{n-k-\ell-1}. \tag{B.46}$$

Using the relations $f_1(x_1)f_2(x_1)f_3(x_1) = 0$, $g_1(x_1)f_2(x_1)f_3(x_1) = 0$, $f_1(x_1)g_2(x_1)f_3(x_1) = 0$ and $f_1(x_1)f_2(x_1)g_3(x_1) = 0$, one can simplify (B.45) and (B.46) to

$$G_n = -j\sqrt{2} \sum_{\ell=1}^{n} \sum_{k=0}^{n-\ell} P_k(\Omega_1) P_{\ell-1}(\Omega_2) P_{n-k-\ell}(\Omega_3)$$

$$\times \exp(j\eta_{k+1/2}^{(1)} + j\eta_{\ell+1/2}^{(2)} + j\eta_{n-k-\ell-1/2}^{(3)}), \tag{B.47}$$

for positive n, $G_0 = 0$, $G_{-n-1} = G_n^*$, and

$$F_n = -jS_n G_n. \tag{B.48}$$

As can be seen, (B.48) is the same as (B.26). This implies that (B.27)–(B.29) developed for double-electrode gratings also hold for triple-electrode gratings. Note, however, that (B.30) should be modified to make the number of linear equations equal to the number of unknowns A_m, i.e.,

$$\sum_{m=M_1}^{M_2} A_m G_{n-m} \{1 - S_{n-m} S_n \epsilon(\infty)/\epsilon(\beta_n/\omega)\} = 0 \tag{B.49}$$

for $n = [M_1 + 1, M_2 - 2]$.

The potential $\Phi^{(\ell)}(s)$ of the ℓ-th electrode is given by

$$\Phi^{(1)}(s) = -j\frac{\exp(-j\pi s)}{2\sin(2\pi s)} \sum_{m=M_1}^{M_2} A_m \{V_m^{(1)} + (V_m^{(2)} + V_m^{(3)})\exp(2j\pi s)\},$$

$$\tag{B.50}$$

$$\Phi^{(2)}(s) = -j\frac{\exp(-j\pi s)}{2\sin(2\pi s)} \sum_{m=M_1}^{M_2} A_m \{V_m^{(1)} + V_m^{(2)} + V_m^{(3)}\exp(2j\pi s)\},$$

$$\tag{B.51}$$

$$\Phi^{(3)}(s) = -j\frac{\exp(-j\pi s)}{2\sin(2\pi s)} \sum_{m=M_1}^{M_2} A_m \{V_m^{(1)} + V_m^{(2)} + V_m^{(3)}\}, \tag{B.52}$$

where

$$V_m^{(1)} = \int_{-p+d_3+w_3/2}^{d_1-w_1/2} g_1(x_1)g_2(x_1)g_3(x_1)\exp(-j\beta_{m-1/2}x_1)dx_1, \qquad (B.53)$$

$$V_m^{(2)} = \int_{d_1+w_1/2}^{d_2-w_2/2} g_1(x_1)g_2(x_1)g_3(x_1)\exp(-j\beta_{m-1/2}x_1)dx_1, \qquad (B.54)$$

$$V_m^{(3)} = \int_{d_2+w_2/2}^{d_3-w_3/2} g_1(x_1)g_2(x_1)g_3(x_1)\exp(-j\beta_{m-1/2}x_1)dx_1. \qquad (B.55)$$

The current $I^{(k)}(s)$ flowing into the k-th electrode is given by

$$I^{(k)}(s) = \sum_{m=M_1}^{M_2} A_m W_m^{(k)}(s), \qquad (B.56)$$

where

$$W_m^{(1)}(s) = j\omega \int_{d_1-w_1/2}^{d_1+w_1/2} f_1(x_1)g_2(x_1)g_3(x_1)\exp(-j\beta_{m-1/2}x_1)dx_1, \qquad (B.57)$$

$$W_m^{(2)}(s) = j\omega \int_{d_2-w_2/2}^{d_2+w_2/2} g_1(x_1)f_2(x_1)g_3(x_1)\exp(-j\beta_{m-1/2}x_1)dx_1, \qquad (B.58)$$

$$W_m^{(3)}(s) = j\omega \int_{d_3-w_3/2}^{d_3+w_3/2} g_1(x_1)g_2(x_1)f_3(x_1)\exp(-j\beta_{m-1/2}x_1)dx_1. \qquad (B.59)$$

From the linear equations (with $(M_2 - M_1 + 1)$ unknowns) composed of (B.49) and (B.50)–(B.52), A_m can be determined in the form

$$A_m = p_{m1}(s)\Phi^{(1)}(s) + p_{m2}(s)\Phi^{(2)}(s) + p_{m3}(s)\Phi^{(3)}(s). \qquad (B.60)$$

Substitution of A_m into (B.56) gives

$$I^{(k)}(s) = \sum_{\ell=1}^{3} Y_{k\ell}(s,\omega)\Phi^{(\ell)}(s), \qquad (B.61)$$

where

$$Y_{k\ell}(s,\omega) = \sum_{m=M_1}^{M_2} W_m^{(k)}(s)p_{m\ell}(s), \qquad (B.62)$$

and $k = 1, 2, 3$. Equations (B.61) and (B.62) are equivalent to (B.39) and (B.40) developed for double-electrode gratings.

B.3 Analysis of Metallic Gratings with Finite Thickness

B.3.1 Combination with Finite Element Method

Here a method developed by Endoh, et al. [9] is given for the analysis of metallic grating structures with finite film thickness. This method is based upon including the effects of finite electrode thickness into the original effective permittivity, and the results are directly applicable to Bløtekjær's theory and its extensions described above.

Let us consider plane acoustic waves propagating toward the x_1 direction in the periodic metallic grating structure shown in Fig. B.4. The periodicity, width and height of the grating electrodes are p, w and h, respectively. The electrodes are of infinite length in the x_2 direction, and the effects of electrical resistivity are neglected.

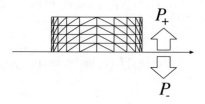

Fig. B.4. FEM analysis of electrode region and power flow at the boundary

From the Poynting theorem, the acoustic complex power $P^\pm$ supplied from the boundary at $x_3 = 0^\pm$ is given by

$$P^\pm = \mp j\omega \int_{-p/2}^{+p/2} \mathbf{u}(x_1) \cdot \mathbf{T}(x_1)^*|_{x_3=0^\pm} dx_1, \tag{B.63}$$

where $\mathbf{u}(x_1)$ and $\mathbf{T}(x_1)$ are vectors composed of particle displacement $u_i(x_1)$ and the stress $T_{3i}(x_1)$ as

$$\mathbf{u}(x_1) = \{u_1(x_1), u_2(x_1), u_3(x_1)\},$$

$$\mathbf{T}(x_1) = \{T_{31}(x_1), T_{32}(x_1), T_{33}(x_1)\}.$$

For the acoustic wave field at $x_3 = 0^+$, define the vectors $\hat{\mathbf{u}}$ and $\hat{\mathbf{T}}$ composed of the particle displacements and the integration of the stresses at the nodal points of $x_3 = 0^+$. If the driving frequency ω is specified, the application of the FEM to the grating electrode region relates the vectors to each other in the form

$$\hat{\mathbf{T}} = -[\mathbf{F}]\,\hat{\mathbf{u}}, \tag{B.64}$$

where $[\mathbf{F}]$ is the matrix derived from the FEM analysis. Substitution of (B.64) into (B.63) gives

$$P^+ = j\omega \hat{\mathbf{u}}^* \cdot [\mathbf{F}]^{*t} \, \hat{\mathbf{u}}, \tag{B.65}$$

where t denotes the transpose of a matrix.

Next, consider the acoustic wave field at $x_3 = 0^-$. Since the field variables satisfy the Floquet theorem, $\mathbf{T}(x_1)$ of a grating mode with wavenumber β_0 can be expressed in the form

$$\mathbf{T}(x_1) = \sum_{n=-\infty}^{+\infty} \mathbf{T}(\beta_n) \exp(-j\beta_n x_1), \tag{B.66}$$

where $\beta_n = \beta_0 + 2n\pi/p$, and

$$\mathbf{T}(\beta_n) = \frac{1}{p} \int_{-p/2}^{+p/2} \exp(j\beta_n x_1)\mathbf{T}(x_1)dx_1. \tag{B.67}$$

Carrying out the numerical integration and taking account of (B.64), one may write (B.67) in the form

$$\mathbf{T}(\beta_n) = [\mathbf{G}(\beta_n)] \, \hat{\mathbf{T}} = -[\mathbf{G}(\beta_n)] \, [\mathbf{F}] \, \hat{\mathbf{u}}, \tag{B.68}$$

because $\mathbf{T}(x_1) = \mathbf{0}$ in the unelectroded region. In (B.68), $[\mathbf{G}(\beta_n)]$ is a matrix giving the transform in (B.67).

Using the definition shown in (B.67), one may transform $\mathbf{u}(x_1)$, the surface potential $\phi(x_1)$ and the charge $q(x_1)$ into $\mathbf{U}(\beta_n)$, $\varPhi(\beta_n)$ and $Q(\beta_n)$, respectively. Hence, P^- in (B.63) is rewritten as

$$P^- = j\omega p \sum_{n=-\infty}^{+\infty} \mathbf{T}(\beta_n)^* \cdot \mathbf{U}(\beta_n)|_{x_3=0^-}. \tag{B.69}$$

As described by (A.63) and (A.64), it was shown that these variables are related to each other by

$$\mathbf{U}(\beta_n) = S_n\beta_n^{-1}\{[\mathbf{R}_{11}(\beta_n)]\mathbf{T}(\beta_n) + [\mathbf{R}_{12}(\beta_n)]Q(\beta_n)\}, \tag{B.70}$$

$$\varPhi(\beta_n) = S_n\beta_n^{-1}\{[\mathbf{R}_{21}(\beta_n)] \cdot \mathbf{T}(\beta_n) + R_{22}(\beta_n)Q(\beta_n)\}, \tag{B.71}$$

where R_{ij} is the effective acoustic admittance matrix and $S_n = \mathrm{sgn}(\beta_n)$.

Since $\mathbf{T}(\beta_n)|_{x_3=0^+} = \mathbf{T}(\beta_n)|_{x_3=0^-}$ from the boundary condition, substitution of (B.68) and (B.70) into (B.69) gives

$$P^- = j\omega p \sum_{n=-\infty}^{+\infty} S_n\beta_n^{-1}\hat{\mathbf{u}}^* \cdot [\mathbf{F}]^{*t} \, [\mathbf{G}(\beta_n)]^{*t}$$

$$\times\{[\mathbf{R}_{11}(\beta_n)] \, [\mathbf{G}(\beta_n)] \, [\mathbf{F}] \, \hat{\mathbf{u}} - [\mathbf{R}_{12}(\beta_n)]Q(\beta_n)\}. \tag{B.72}$$

The total power P supplied from the boundary is $P^+ + P^-$. Since $\mathbf{u}(x_1)$ is continuous at the boundary, P should be zero if the solution is rigorous. Although P generally takes a nonzero value because of numerical evaluation, it must satisfy the following stationary condition;

$$\frac{\partial P}{\partial u(x_\ell)^*} = 0 \tag{B.73}$$

for each component $u(x_\ell)$ of $\hat{\mathbf{u}}$. Substitution of (B.65) and (B.72) into (B.73) gives

$$\hat{\mathbf{u}} = \sum_{n=-\infty}^{+\infty} [\mathbf{L}(\beta_n)]Q(\beta_n), \tag{B.74}$$

where

$$\mathbf{L}(\beta_n) = [\mathbf{A}]^{-1} [\mathbf{G}(\beta_n)]^{*t} [\mathbf{R}_{12}(\beta_n)],$$

and

$$[\mathbf{A}] = \frac{S_n\beta_n}{p}[\mathbf{I}] + \sum_{n=-\infty}^{+\infty} [\mathbf{G}(\beta_n)]^{*t}[\mathbf{R}_{11}(\beta_n)][\mathbf{G}(\beta_n)][\mathbf{F}].$$

Substituting (B.74) into (B.68) and (B.71), one finally obtains the following relation

$$\Phi(\beta_n) = S_n\beta_n^{-1} \sum_{\ell=-\infty}^{+\infty} H_{n\ell}Q(\beta_n), \tag{B.75}$$

where

$$H_{n\ell} = -[\mathbf{R}_{21}(\beta_n)] \cdot [\mathbf{G}(\beta_n)] [\mathbf{F}] [\mathbf{L}(\beta_\ell)] + R_{22}(\beta_n)\delta_{n\ell}. \tag{B.76}$$

Equation (B.75) represents the relationship between the surface potential and charge, where the mass loading effect has already been taken into account.

B.3.2 Application to Extended Bløtekjær Theories

FEMSDA. By using the relation (B.75) instead of (B.3), we can include the effects of finite film thickness into Bløtekjær's theory. Namely, substitution of (B.4) and (B.5) into (B.75) gives the following simultaneous linear equations instead of (B.8):

$$\sum_{m=M_1}^{M_2} A_m \left[\epsilon_\infty S_n \sum_{\ell=-\infty}^{+\infty} H_{n\ell}P_{\ell-m}(r) - S_{n-m}P_{n-m}(r) \right] = 0 \tag{B.77}$$

for $n = [M_1, M_2 - 1]$. By solving them with respect to A_m and substituting into (B.9) and (B.10), the effects of finite film thickness can be taken into account in the effective permittivity for the grating structure, and various SAW characteristics can be evaluated.

The software FEMSDA was developed based on this algorithm, and it is now freely distributed through the Internet from the author's lab [15, 16].

Figure B.5 shows, as an example, the velocity dispersion of Rayleigh-type SAWs on $128°$YX-LiNbO$_3$ calculated by using FEMSDA, where the shape of the Al electrodes was assumed to be rectangular and $w/p = 0.5$. In the figure, $V_B = 4025$ m/sec is the slow-shear SSBW velocity, and OC and SC indicate the dispersion relations for open- and short-circuited gratings, respectively. It is clearly seen that the SAW velocity decreases with an increase in h/p

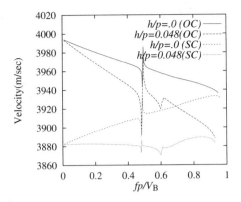

Fig. B.5. Velocity dispersion of SAWs on a metallic grating on 128°YX-LiNbO₃

due to the mass loading effect. For the OC grating, the stopband width at $fp/V_B \cong 0.5$ increases with an increase in h/p, whereas the stopband of the SC grating becomes narrow. The stopband of the SC grating nearly disappears when $h/p \cong 0.048$. This is because the electrical perturbation is canceled by the mechanical perturbation at this electrode thickness [17]. Note that the discontinuous change at $fp/V_B \cong 0.6$ is due to the cut-off nature of the back-scattered longitudinal bulk waves.

Figure B.6 shows the conversion of the calculation error as a function of M_2. The error was estimated from the fractional change in the SAW velocity from the value when M_2 is sufficiently large. In the calculation, $M_1 = -M_2$ and $fp/V_B = 0.45$. Although the error becomes worse with an increase in h/p,

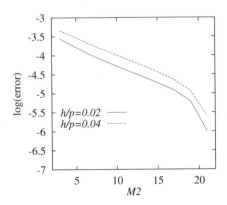

Fig. B.6. Calculation error in SAW velocity as a function of M_2

its decrease with M_2 is still steep and monotonic. From the figure, $M_2 = 10$ seems sufficient for most cases.

Figure B.7 shows the conversion of the calculation error as a function of the numbers N_x and N_z of FEM subdivisions toward the width and thickness directions, respectively. Note that the discretization was performed as shown in Fig. B.4, where equidistance sampling was done for the thickness direction whereas the sampling was made dense at electrode edges so as to take the stress concentration into account. It is seen that the error decreases monotonically with an increase in N_x, and $N_x \cong 10$ seems enough. Note that N_z should be increased with an increase in h/p.

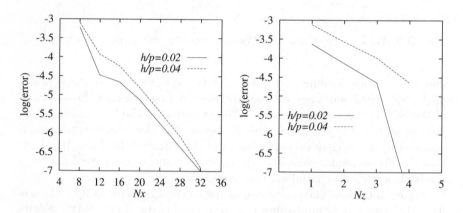

Fig. B.7. Calculation error in SAW velocity as a function of N_x and N_z

OBLIQ. The method described above is also applicable to the oblique incident case described in Sect. B.2.2. This idea was implemented in the software OBLIQ [15, 16].

Figure B.8 shows the frequency dispersion of the velocity of the Rayleigh-type SAW with propagation angle θ (see Fig. B.2) as a parameter. As a substrate 128°YX-LiNbO$_3$ was chosen, and for the metallic grating, Al electrodes are assumed to be rectangular with $h/p = 0.024$ and $w/p = 0.5$.

For 128°YX-LiNbO$_3$, the influence of the mass loading effect on the SAW slowness curve is relatively small, and the dispersion arising from the effect of the short-circuited electric field is dominant. The SAW velocity increases sightly with θ, and is almost independent of h/p.

MULTI. For Aoki's theory, application of the above method is simple. The only thing that needed to be done was to modify the simultaneous linear equations of (B.30) into

$$\sum_{m=M_1}^{M_2} A_m \left[G_{n-m} - S_n\epsilon(\infty) \sum_{\ell=-\infty}^{+\infty} S_{\ell-m} H_{n\ell} G_{\ell-m} \right] = 0 \qquad (B.78)$$

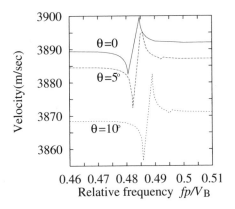

Fig. B.8. Frequency dispersion of Rayleigh-type SAW velocity on 128°YX-LiNbO$_3$

for $n = [M_1 + 1, M_2 - 1]$.

For the three-electrode grating described in Sect. B.2.4, the range of n should be changed to $[M_1 + 1, M_2 - 2]$ (see (B.49)).

As a demonstration, the SAW propagation characteristics on the double-electrode grating shown in Fig. B.3 were analyzed. Al and 128°YX-LiNbO$_3$ are assumed as the electrode and substrate materials. The parameters of the electrodes are $h_1/p = 0$, $h_2/p = 0.02$, $w_1 = w_2 = p/4$, $d_1 = -p/4$, and $d_2 = p/4$ (see Fig. B.3 for d_1 and d_2).

Figure B.9 shows the calculated result for the first stopband occurring near $fp/V_B \cong 0.5$. The electrical connection of the grating electrodes with a bus-bar is represented by OO, OS, SO, and SS, where O and S indicate, respectively, that the electrodes are open- and short-circuited with a bus-bar. In addition, IS indicates that the two electrodes are electrically short-circuited with respect to each other but isolated from the bus-bar.

It is clear that the SAW propagation characteristics markedly depend on the electrical connection of the grating electrodes. The stopband widths for the OO and SS connections are very narrow: reflection by an electrode cancels out reflection by its adjacent electrode, because reflection coefficients of the two electrodes, which are separated by the spatial distance of $\lambda/4$, are not very much different from each other. On the other hand, the stopbands for the OS and SO connections are relatively large. In these cases the reflection coefficients of the two electrodes are opposite in sign, whilst the spatial distance is $\lambda/4$: reflections from the two electrodes are added.

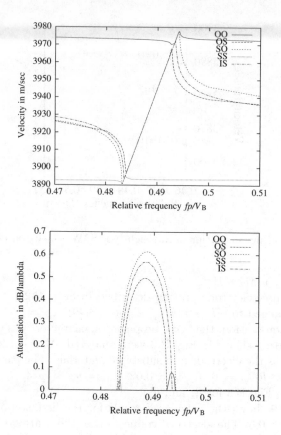

Fig. B.9. SAW dispersion relation near the first stopband in double-electrode grating

B.4 Wave Excitation and Propagation in Grating Structures

B.4.1 Effective Permittivity for Grating Structures

Let us consider the infinite grating structure with a single strip per period shown in Fig. B.10.

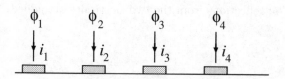

Fig. B.10. Infinite grating structures

Define the following new variables $Q(s)$ and $\Phi(s)$ by

$$Q(s) = \sum_{n=-\infty}^{+\infty} q_n \exp(+2\pi jns), \tag{B.79}$$

$$\Phi(s) = \sum_{n=-\infty}^{+\infty} \phi_n \exp(+2\pi jns), \tag{B.80}$$

where q_n and ϕ_n are the total charge and electrical potential on the n-th electrode, respectively.

Since q_n and ϕ_n can be regarded as the Fourier expansion coefficients of $Q(s)$ and $\Phi(s)$, respectively, one can rewrite (B.79) and (B.80) in the form

$$q_n = \int_0^1 Q(s)\exp(-2\pi jns)ds, \tag{B.81}$$

$$\phi_n = \int_0^1 \Phi(s)\exp(-2\pi jns)ds. \tag{B.82}$$

Equations (B.81) and (B.82) show that q_n and ϕ_n are expressed as a sum of contributions from various grating modes having wavenumbers $2\pi s/p$, where $0 \le s \le 1$. From the result described in Sects. B.2 and B.3, it is clear that $Q(s)$ and $\Phi(s)$ are not independent. That is, from (B.11), they are related to each other by the strip admittance [1, 2] $Y(s,\omega)$:

$$Q(s)/\Phi(s) = -2jY(s,\omega)/\omega \equiv 2\sin(s\pi)\epsilon_g(s,\omega), \tag{B.83}$$

Figure B.11 shows the calculated $\epsilon_g(s,\omega)$ on $128°$YX-LiNbO$_3$ with an Al grating of 2% p thickness at $f = 0.45V_B/p$, where $V_B = 4025$ m/sec is the slow-shear SSBW velocity. Two types of discontinuities are seen. At $s \cong 0.46$, there is a pole corresponding to the radiation of the Rayleigh-type SAW on the short-circuited grating. In addition, $\Im[\epsilon_g(s,\omega)]$ is large at $s < 0.23$. This is due to the radiation of longitudinal BAWs.

Substitution of (B.83) into (B.81) gives

$$q_n = 2\int_0^1 \sin(s\pi)\epsilon_g(s,\omega)\Phi(s)\exp(-2\pi jns)ds. \tag{B.84}$$

One may then obtain the following relation by substituting (B.80) into (B.84):

$$q_k = \sum_{\ell=-\infty}^{+\infty} G_{k-\ell}\phi_\ell, \tag{B.85}$$

where G_k is the newly defined discrete Green function [5] given by

$$G_k = 2\int_0^1 \sin(s\pi)\epsilon_g(s,\omega)\exp(-2\pi jks)ds. \tag{B.86}$$

As shown in Fig. B.12, G_k represents the induced charge q_k when unit potential is applied on the 0-th electrode, while the potential on the other

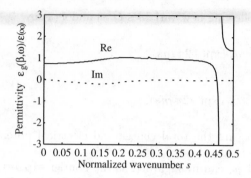

Fig. B.11. Effective permittivity for grating structure on 128°YX-LiNbO$_3$ at $h = 0.02p$ and $fp/V_B = 0.45$

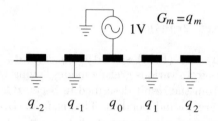

Fig. B.12. Discrete Green function

electrodes is zero. Because of reciprocity, the relation $G_k = G_{-k}$ holds, and this results in $\epsilon_g(1-s,\omega) = \epsilon_g(s,\omega)$.

The complex power P supplied through the electrodes is also estimated by $\epsilon_g(s)$ as follows

$$P = \frac{j\omega W}{2} \sum_{k=-\infty}^{+\infty} \phi_k^* q_k = j\omega W \int_0^1 |\Phi(s)|^2 \epsilon_g(s,\omega)\sin(s\pi)ds, \qquad (\text{B.87})$$

where W is the aperture.

B.4.2 Evaluation of Discrete Green Function

A method for evaluating the discrete Green function G_n is described. Here, several simple analytical results are derived by using the complex integral. It is shown that by using the technique developed for the evaluation of the effective permittivity using the complex integral, other characteristics, for example, the SSBW excitation strength, can be retrieved.

From the Cauchy–Riemann theorem, the integration path in (B.86) and (B.87) is modified as shown in Fig. B.13:

$$\int_0^1 ds = \int_0^{0-j\infty} ds + \int_{1-j\infty}^1 ds + \oint_{\Gamma_g} ds + \sum_{i=1}^3 \oint_{\Gamma_{Bi}} ds. \tag{B.88}$$

In this equation, Γ_g is the path rotating clockwise around the pole s_g, which is the solution of the equation $\epsilon_g^{-1}(s_g, \omega)=0$. Γ_{Bi} is the path along the branch cuts starting from $s_{Bi} = fp/V_{Bi}$, where V_{Bi} is the SSBW velocity. Here, the subscript i $(= 1, 2, 3)$ designates the type of SSBW, that is, slow-shear, fast-shear and longitudinal SSBW, respectively. The integration along the path of $s = [-j\infty, 1 - j\infty]$ vanishes because the integrand is zero.

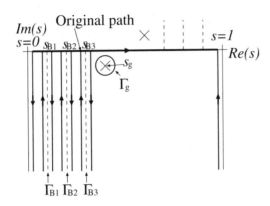

Fig. B.13. Integration path for discrete Green function

The first two terms on the right-hand side in (B.88) give the contribution from electrostatic coupling. The third term represents the contribution from grating mode radiation, and the fourth term from SSBW radiation.

Electrostatic Coupling. The contribution G_{en} from the electrostatic coupling in G_n is given by

$$G_{en} = 4 \int_0^{-j\infty} \sin(s\pi)\epsilon_g(s, \omega) \exp(-2\pi jns)ds$$

$$\cong 4\epsilon_g(0, \omega) \int_0^{-j\infty} \sin(s\pi) \exp(-2\pi jns)ds$$

$$\cong -\epsilon_g(0, \omega)/\{\pi(n^2 - 1/4)\}. \tag{B.89}$$

Grating Mode Excitation. Applying the residue theorem to (B.86), one obtains the contribution G_{gn} from the grating mode radiation as

$$G_{gn} = -j\frac{4}{\pi}\epsilon(\infty)K_g^2 \exp(-2\pi js_g|n|). \tag{B.90}$$

In this equation, K_g^2 is the effective electromechanical coupling factor of the grating mode given by

$$K_g^2 = \pi^2 \sin(s_g\pi) \left\{ \epsilon(\infty) \frac{\partial \epsilon_g(s,\omega)^{-1}}{\partial s} \Big|_{s=s_g} \right\}^{-1}. \tag{B.91}$$

Similarly, the application of the residue theorem to (B.87) gives the acoustic power $P_g^{\pm}$ of the grating mode radiated toward $\pm x_1$ as

$$P_g^{\pm} = \frac{\omega K_{rmg}^2 \epsilon(\infty) W}{\pi} \left| \sum_{k=-\infty}^{+\infty} \phi_n \exp(\mp 2\pi j n s_g) \right|^2. \tag{B.92}$$

Denote the charge carried by the $\pm x_1$-propagated grating mode as $q_{gn}^{\pm}$. When unit potential is applied to the 0-th electrode and the potential of the other electrodes is zero, $q_{gn}^{\pm}$ and $P_g^{\pm}$ are given by G_{gn} and $8^{-1}\pi\omega\epsilon(\infty)WK_g^2$, respectively. From this, one obtains the following relation

$$P_g^{\pm} = \eta_g^{-2} W |j\omega q_{gn}^{\pm}|^2, \tag{B.93}$$

where

$$\eta_g = 4\sqrt{\omega\epsilon(\infty)K_g^2/\pi}. \tag{B.94}$$

Figure B.14 shows the change in SAW velocity $V_p = fp/\Re(s_p)$ and the attenuation $\alpha_p = -2\pi\Im(s_p)/\Re(s_p) \times 8.686$ as a function of fp/V_B. There is a stopband due to Bragg reflection of the Rayleigh-type SAW at $0.482 < fp/V_B < 0.488$, and back-scattering to the BAW occurs when $fp/V_B > 0.5$.

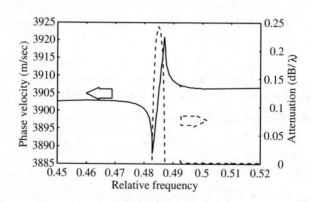

Fig. B.14. Velocity and attenuation of Rayleigh-type SAW as a function of fp/V_B

Figure B.15 shows the change in K_g^2 as a function of f. It is seen that K_g^2 changes very smoothly except at the stopband. The frequency dependence is due to that of the element factor described in Sect. 3.2.3. Near the stopband, K_g^2 changes rapidly due to the influence of multiple reflection, and it becomes imaginary within the stopband because most of radiated energy returns to the excitation point. Note that K_p^2 becomes complex at $fp/V_B > 0.5$ because of the coupling between the SAW and BAW. That is, the excitation source can receive BAWs generated by SAW reflection and vice versa.

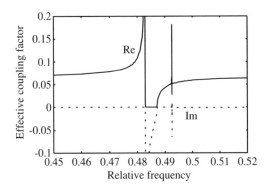

Fig. B.15. Electromechanical coupling factor K_g^2 for Rayleigh-type SAW as a function of fp/V_B

B.4.3 Delta-Function Model

The idea of the discrete Green function is applied to extend the delta-function model described in Sect. 3.3.1 [5]. Note that the effects of internal reflection are included in this analysis.

Assume that the propagation surface is completely covered with an SC grating. If the width and spacing of both the grating and IDT electrodes are the same, the analysis described in the previous sections is applicable.

If there exist no floating electrodes, the input admittance Y_{kk} of the k-th IDT consisting of M_k electrodes is simply given by

$$Y_{kk} = j\omega W \sum_{m=0}^{M_k-1} \sum_{n=0}^{M_k-1} S_{km}S_{kn}G_{m-n}. \tag{B.95}$$

The transfer admittance $Y_{k\ell}$ between the k-th and ℓ-th IDTs is given by

$$Y_{k\ell} = j\omega W \sum_{m=0}^{M_k-1} \sum_{n=0}^{M_\ell-1} S_{km}S_{\ell n}G_{D_{k\ell}+m-n}, \tag{B.96}$$

where $D_{k\ell}$ is the normalized distance between the two IDTs, and

$$S_k = \begin{cases} 1 & \text{for hot electrode} \\ 0 & \text{otherwise.} \end{cases} \tag{B.97}$$

Thus, if S_k is specified, the electrical characteristics of arbitrary IDT configurations can be determined numerically.

Although (B.95) and (B.96) are completely identical with those assumed in the simple delta-function model analysis, the present analysis takes account of the effects of electrical and acoustic interactions amongst the electrodes.

If there exist floating electrodes, the IDT behavior is characterized by the following procedure: (i) determination of the voltages of floating electrodes by applying unit potential to hot electrodes of an input IDT and by short-circuiting the output IDT, (ii) calculation of the total charges induced on hot electrodes by substituting the voltages of floating electrodes into (B.85), and (iii) determination of input and transfer admittances by summing the induced charges on the input and output IDT electrodes individually and by multiplying by $j\omega$.

By applying this procedure, the input admittance of an FEUDT [19] (see Fig. 3.5) on 128°YX-LiNbO$_3$ was analyzed.

Figure B.16 shows the calculated result, where the horizontal axis is the frequency normalized by the resonance frequency f_r and the vertical axis is the admittance normalized by $2\pi f_r W \epsilon(\infty)$. The result is in good agreement with that obtained by BEM [20]. It may be of practical importance to note that compared with BEM, the computational time and memory size required for the present analysis were reduced by 1/125 and 1/5, respectively.

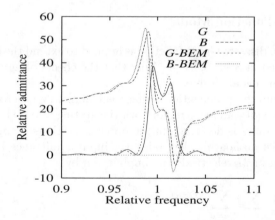

Fig. B.16. Input admittance characteristics of FEUDT on 128°YX-LiNbO$_3$

B.4.4 Infinite IDTs

Here a very simple but quite effective technique is described to evaluate SAW excitation properties in the metallic grating [21].

Let us consider the original grating structure shown in Fig. B.10. When $\phi_n = (-1)^n V_0/2$, the structure is equivalent to the infinite single-electrode-type IDT with periodicity $p_I = 2p$, and V_0 corresponds to the applied voltage.

From (B.80), $\Phi(\beta) = 2\pi V_0/p_I \times \delta(\beta - 2\pi/p_I)$. Then substitution of this relation into (B.81) and (B.83) gives

$$i_n = 2^{-1}(-1)^n WY(2\pi/p_I, \omega)V_0.$$

This means that the input admittance $\hat{Y}(\omega)$ per period for single-electrode-type IDTs with infinite length is given by [22]

$$\hat{Y}(\omega) = 2^{-1}WY(2\pi/p_I, \omega). \tag{B.98}$$

Note that calculation of $Y(2\pi/p_I, \omega)$ is considerably easier and faster than finding poles and/or zeros in $Y(\beta, \omega)$.

Figure B.17 shows the calculated $\hat{Y}(\omega)/\omega W\epsilon(\infty)$ on $(0, 47.3°, 90°)$ $\text{Li}_2\text{B}_4\text{O}_7$ with Al film of 2% p_I thickness as a function of relative frequency fp_I/V_B where $V_B = 3347$ m/sec is the velocity for the fast-shear BAW. Two resonances and antiresonances for the Rayleigh-type SAW are clearly seen. In this frequency region, $\Re[\hat{Y}(\omega)] = 0$ because of the cut-off for the BAW radiation.

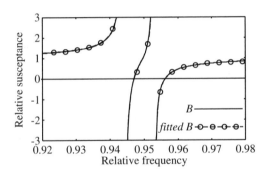

Fig. B.17. Input admittance of infinite IDT on $(0, 47.3°, 90°)$ $\text{Li}_2\text{B}_4\text{O}_7$

By using the technique described in Sect. 7.3.3, the COM parameters are retrieved. The figure also shows $\hat{Y}(\omega)$ calculated by using the COM. The agreement is excellent.

It should be noted that this substrate is known to support longitudinal leaky SAW [23], and also exhibits natural unidirectionality [24] for the Rayleigh-type SAWs [25]. The response due to the longitudinal leaky SAW appears in a much higher frequency range.

Next, as an example for the multi-finger case, let us consider the IDT with three fingers per period shown in Fig. B.18. This corresponds to the unit cell for the EWC/SPUDTstructure described in Sect. 3.1.2.

In this case, $\Phi^{(1)}(\beta) = 2\pi V_0/p_I \times \delta(\beta - 2\pi/p_I)$ and $\Phi^{(2)}(\beta) = \Phi^{(3)}(\beta) = 0$. Then the input admittance $\hat{Y}(\omega)$ per period for IDTs with infinite length is given by $\hat{Y}(\omega) = WY_{11}(2\pi/p_I, \omega)$.

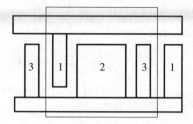

Fig. B.18. Unit cell for EWC/SPUDT structure with reflection and transduction

Fig. B.19 shows the calculated $\hat{Y}(\omega)/\omega W\epsilon(\infty)$ for the structure shown in Fig. B.18 on AT-cut quartz with Al electrodes of 4% p_{I} thickness. Due to the directivity caused by asymmetry of the IDT structure, two resonances and antiresonances are clearly seen. In the figure, $\hat{Y}(\omega)$ calculated by using the COM is also shown. It agrees well with the result of direct computation.

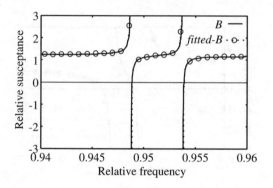

Fig. B.19. Input admittance of EWC/SPUDT on ST-cut quartz

The software SYNC and MSYNC based on this algorithm is now freely distributed through the Internet from the author's lab [16].

References

1. K. Bløtekjær, K.A. Ingebrigtsen and H. Skeie: A Method for Analysing Waves in Structures Consisting of Metallic Strips on Dispersive media, IEEE Trans. Electron. Devices, **ED-20** (1973) pp. 1133–1138.
2. K. Bløtekjær, K.A. Ingebrigtsen and H. Skeie: Acoustic Surface Waves in Piezo-electric Materials with Periodic Metallic Strips on the Substrate, IEEE Trans. Electron. Devices, **ED-20** (1973) pp. 1139–1146.
3. K.C. Wagner and O. Männer: Analysis of Obliquely Propagating SAWs in Periodic Arrays of Metal Strips, Proc. IEEE Ultrason. Symp. (1992) pp. 427–431.

4. Y. Zhang, J. Desbois and L. Boyer: Characteristic Parameters of Surface Acoustic Waves in a Periodic Metal Grating on a Piezoelectric Substrate, IEEE Trans. Ultrason., Ferroelec. and Freq. Cont., **UFFC-40**, 3 (1993) pp. 183–192.

5. K. Hashimoto and M. Yamaguchi: Analysis of Excitation and Propagation of Acoustic Waves Under Periodic Metallic-Grating Structure for SAW Device Modeling, Proc. IEEE Ultrason. Symp. (1993) pp. 143–148.

6. H.P. Reichinger and A.R. Baghai-Wadji: Dynamic 2D Analysis of SAW Devices Including Massloading, Proc. IEEE Ultrason. Symp. (1992) pp. 7–10.

7. P. Ventura, J. Desbois and L. Boyer: A Mixed FEM/analytical Model of the Electrode Mechanical Perturbation for SAW and PSAW Propagation, Proc. IEEE Ultrason. Symp. (1993) pp. 205–208.

8. P. Ventura, J.M. Hodé, and M. Solal: A New Efficient Combined FEM and Periodic Green's Function Formalism for the Analysis of Periodic SAW Structure, Proc. IEEE Ultrason. Symp. (1995) pp. 263–268.

9. G. Endoh, K. Hashimoto and M. Yamaguchi: SAW Propagation Characterisation by Finite Element Method and Spectral Domain Analysis, Jpn. J. Appl. Phys., **34**, 5B (1995) pp. 2638–2641.

10. K. Hashimoto, G. Endoh, M. Ohmaru and M. Yamaguchi: Analysis of Surface Acoustic Waves Obliquely Propagating under Metallic Gratings with Finite Thickness, Jpn. J. Appl. Phys., **35**, Part 1, 5B (1996) pp. 3006–3009.

11. K. Hashimoto, G.Q. Zheng and M. Yamaguchi: Fast Analysis of SAW Propagation under Multi-Electrode-Type Gratings with Finite Thickness, Proc. IEEE Ultrason. Symp. (1997) pp. 279–284.

12. T. Aoki and K.A. Ingebrigtsen: Equivalent Circuit Parameters of Interdigital Transducers Derived from Dispersion Relation for Surface Acoustic Waves in Periodic Metal Gratings, IEEE Trans. Sonics and Ultrason., **SU-24** (1977) pp. 167–178.

13. T. Aoki and K.A. Ingebrigtsen: Acoustic Surface Waves in Split Strip Metal Gratings on a Piezoelectric Surface, IEEE Trans. Sonics and Ultrason., **SU-24** (1977) pp. 179–193.

14. K. Hashimoto and M. Yamaguchi: Discrete Green Function Theory for Multi-Electrode Interdigital Transducers, Jpn. J. Appl. Phys., **34**, 5B (1995) pp. 2632–2637.

15. K. Hashimoto and M. Yamaguchi: Free Software Products for Simulation and Design of Surface Acoustic Wave and Surface Transverse Wave Devices, Proc. IEEE International Frequency Control Symp. (1996) pp. 300–307.

16. http://www.sawlab.te.chiba-u.ac.jp/~ken/freesoft.html

17. Y. Suzuki, H. Shimizu, M. Takeuchi, K. Nakazawa and A. Yamada: Some studies on SAW resonators and multi-Mode Filters, Proc. IEEE Ultrason. Symp. (1976) pp. 297–302.

18. R.F. Milsom, N.H.C. Reilly and M. Redwood: Analysis of Generation and Detection of Surface and Bulk Acoustic Waves by Interdigital Transducers, IEEE Trans. Sonics and Ultrason., **SU-24** (1977) pp. 147–166.

19. M. Takeuchi and K. Yamanouchi: Field Analysis of SAW Single-Phase Unidirectional Transducers Using Internal Floating Electrodes, Proc. IEEE Ultrason. Symp. (1988) pp. 57–61.

20. K. Hashimoto and M. Yamaguchi: Derivation of Coupling-of-Modes Parameters for SAW Device Analysis by Means of Boundary Element Method, Proc. IEEE Ultrason. Symp. (1991) pp. 21–26.

21. K. Hashimoto, J. Koskela and M.M. Salomaa: Fast Determination of Coupling-Of-Modes Parameters Based on Strip Admittance Approach, Proc. IEEE Ultrason. Symp. (1999) to be published.

22. P. Ventura, J.M. Hodé, M. Solal and L. Chommeloux: Accurate Analysis of Pseudo-SAW Devices, Proc. 9th European Freq. & Time Forum (1995) pp. 200–204.
23. T. Sato and H. Abe: Longitudinal Leaky Surface Waves for High Frequency SAW Device Application, Proc. IEEE Ultrason. Symp. (1995) pp. 305–315.
24. P.V. Wright: Natural Single-Phase Unidirectional Transducer, Proc. IEEE Ultrason. Symp. (1985) pp. 58–63.
25. M. Takeuchi, H. Odagawa and K. Yamanouchi: Crystal Orientations for Natural Single Phase Unidirectional Transducers (NSPUDT) on $Li_2B_4O_7$, Electron. Lett., **30**, 24 (1994) pp. 2081–2082.

Index

π-type filter 149–152, 155
p matrix 67, 74–76, 211
Q 124–126, 130, 167, 259
$(0°, 140°, 24°)$ $La_3Ga_5SiO_{14}$ 184, 185
$(0, 47.3°, 90°)$ $Li_2B_4O_7$ 113, 182, 183, 261, 263, 317
-3 dB bandwidth 113, 117, 119, 143, 146, 148, 150, 159
$128°YX$-$LiNbO_3$ 41, 42, 164, 169, 171, 177, 178, 218–220, 225, 226, 231, 232, 254, 263, 264, 306–309, 311, 312, 316
3-port circuit 76
$36°YX$-$LiTaO_3$ 113, 114, 237–240, 242–248, 251–254, 257, 260, 264
$41°YX$-$LiNbO_3$ 238–240, 242, 246
$42°YX$-$LiTaO_3$ 146, 159, 245
$45°YZ$-$Li_2B_4O_7$ 81, 182, 213, 220–222
$4mmm$ 274
$64°YX$-$LiNbO_3$ 238, 239, 241, 257, 258
$6mm$ 274, 276, 277, 284, 287
$90°$off-X AT-cut quartz 259

abbreviated notation 272–274, 284
absolute -5 dB bandwidth 159, 160
absorber 237
acoustic impedance 26, 33, 36, 65, 66, 215
acoustic length 208, 294, 297
acoustic power 314
acoustic wave 107, 293, 304, 305
acousto-optic 3
admittance chart 71, 72
admittance matrix 92
AlN 164, 184
aluminium nitride 184
angular frequency 297
anisotropic 2, 5, 9, 20, 79, 163
anisotropic material 2, 18
anisotropy 3, 4, 8, 21, 82, 133, 182, 241
antiresonance 226, 317, 318
antiresonance frequency 124, 150, 151, 155, 158, 210

antisymmetric mode 132, 134, 137, 145, 193, 195
Aoki's theory 294, 297, 308
aperture 8, 52, 58, 88, 89, 92, 103–105, 131, 134–137, 167, 168, 213, 224, 312
apodization 87, 103, 116, 131, 195
apodization loss 92, 94, 109, 113
apodize 88, 89, 91, 92, 107
apodized weighting 88
array factor 56
as cut 107
AT-cut quartz 171, 172, 174, 176, 265, 318
attenuation 2, 41, 42, 314
attenuation coefficient 36

back-scattered BAW 243, 250, 253
back-scattering 29, 30, 41, 220, 243, 248, 307, 314
balanced input and output 152
balun 152
bandpass filter 32, 117
bandwidth 73, 143, 146, 148, 151, 155–158, 186, 239
BAW 1, 6, 17–19, 22, 26, 29, 30, 41, 61, 69, 79–84, 94, 108–110, 138, 140, 167, 170–172, 174, 195, 220, 227, 237, 239, 241, 248–250, 267, 268, 314, 315
BAW beam 83
BAW efficiency 82
BAW excitation 79, 138
BAW power 17, 81
BAW radiation 79, 81, 83, 138–140, 248, 317
BAW resonator 10, 56, 65, 174
BAW scattering 30
BAW spurious 195
BAW transducer 79
BAW velocity 3, 18, 26, 30, 64, 165
BAW wavelength 79
BAW wavenumber 79
beam steering 5, 164, 166, 180

BEM   293, 316
Bessel function   58
BGS wave   19, 21, 237, 266, 285, 287, 289
BGS wave excitation   288–290
BGS wave velocity   289
bidirectional   52, 227
bidirectional IDT   47, 76
bidirectional loss   77, 94, 112, 113
bidirectionality   48, 77, 112
Bløtekjær's theory   293, 294, 296, 297, 304, 306
Blackmann window   99
Blackmann-Harris window   99
Bleustein-Gulyaev-Shimizu wave   19, 237
bonding pad   111
bonding wire   230, 255
boundary condition   83, 110, 132, 193, 197, 250, 275, 277, 280, 282, 285
boundary element method   293
Bragg condition   28–30, 35, 197
Bragg frequency   29, 30, 34–36, 38, 140, 215
Bragg reflection   27–29, 33, 34, 41, 47, 48, 196, 314
branch cut   288, 313
bulk acoustic wave   1
bus-bar   48, 50, 51, 130, 131, 133, 137, 201, 204, 215, 225, 309

capacitance coefficient   224
capacitance ratio   124, 129, 143, 144, 150, 158, 179
cascade-connection   34, 229, 230
Cauchy-Riemann theorem   288, 312
cavity   127, 130
cavity length   127, 249
center frequency   159
characteristic impedance   75
charge density   277
Chebyshev polynomial   57
circuit Q 74, 143, 149, 150
circular wave   103
closed waveguide   12, 14
COM   67, 191, 196, 214, 222, 227, 317, 318
COM analysis   213, 225
COM equation   192, 197, 200, 202–205, 208, 211, 214, 221, 223, 226, 228, 251, 252, 267
COM model   216
COM parameter   205, 213, 214, 216, 217, 219, 220, 222, 224, 226, 227,

231–234, 248, 253, 254, 257, 258, 261–265, 317
COM simulation   259
COM theory   200, 218, 219, 221, 248, 254, 267, 268, 293
complex integral   312
conductance   207, 212, 213, 233
constant-voltage source   128
contamination   127
convolution   99, 105, 169, 171, 283
counter-electrode   184–187
coupling length   194
coupling-of-modes   67, 191
coupling-of-modes theory   191
critical angle   6, 16, 18, 20
crossed field model   65, 66
crystal axis   163
crystallographic asymmetry   51
crystallographic symmetry   274
cut angle   163, 164, 184
cut-off   15, 41, 69, 80–83, 140, 171, 187, 267, 268, 307, 317
cut-off frequency   13, 15, 80, 82–84, 131, 138, 243, 248, 250
cut-off waveguide   14
Czochralski method   176, 179, 183

DBAW   108–110, 173, 241
decay constant   18, 266
deep bulk acoustic wave   108
deformation   271
delay line   3
delay time   241
delta-function   89
delta-function model   61, 65, 66, 74, 87, 91, 93, 107, 108, 110, 111, 207, 208, 217, 315, 316
detuning factor   197
diamond   185, 187, 188
dielectric constant   174, 176, 179, 184, 276
differential operator   193
diffraction   8, 87, 90, 103, 104, 108, 130, 167, 243
directionality   227
directivity   76, 78, 79, 115, 206, 207, 211–213, 217, 318
discrete Green function   311–313
dispersion   40–42, 248, 249, 260, 306, 308
dispersion characteristics   198, 205, 206
dispersion equation   266, 267

dispersion relation   13–16, 41–43, 45,
    131, 187, 216–220, 224, 225, 243, 244,
    248–252, 260, 266, 267, 285, 287, 293,
    296, 301, 306, 310
displacement   2, 271, 275
displacement vector   271, 272
distortion   115
dogleg weighting   87, 103
double rotated cut   164
double-electrode grating   297, 302,
    303, 309, 310
double-electrode-type IDT   47, 48,
    51–53, 56, 57, 111, 223, 224
double-mode resonator filter   145–147,
    152, 153
drift   127
dummy electrode   61, 111
durability   149
DUT   49, 69

ECRF   117
effective acoustic admittance matrix
    282, 283, 287, 305
effective permittivity   41, 52, 81, 82,
    108, 109, 165, 167–169, 171, 172,
    174–178, 180–184, 237–239, 282, 283,
    285, 287, 295, 296, 304, 306, 312
eigenmode   17, 19, 22, 27, 136, 191, 259
elastic constant   274
elasticity   10, 123, 271
electric field   275, 276
electric flux density   275, 277
electric potential   276, 277
electric regeneration   39
electrical boundary condition   18, 19
electrical feedthrough   94, 95, 111, 256
electrical matching   70–72
electrical mismatching   74, 114
electrical reflection   39, 89, 94, 218, 220
electrical regeneration   77, 94, 112,
    115, 218
electromagnetic wave   61, 69
electromechanical coupling factor   44,
    53, 54, 73, 110, 124, 137, 164, 165, 170,
    172–174, 176, 177, 179–182, 184–188,
    215, 238, 259, 266, 287, 289, 313, 315
electrostatic approximation   247
electrostatic coupling   61, 288, 313
electrostatic field   279, 280
element factor   55–58, 61, 62, 218, 314
end effect   59–61
energy storing effect   14, 15, 30–32, 36,
    66, 138, 214, 215, 248, 249, 267

energy trapping   130
equivalent circuit   14, 15, 26, 30, 32,
    35, 65, 66, 123, 124, 128, 135, 138,
    141, 145, 150, 158, 204, 215, 216, 232,
    255–257
equivalent circuit model   56, 65, 66, 91,
    93, 215
equivalent mirror   34
Euler angle   163, 164
evanescent   11, 14, 18, 33, 38, 133, 199,
    280
evanescent mode   14, 15, 136
EWC/SPUDT   49, 50, 119, 317, 318
excitation center   48, 50, 51, 56, 78, 224
excitation efficiency   201, 203, 216, 252
externally coupled resonator filter
    117, 118

Fabry-Perot model   127, 143, 213
Fabry-Perot resonator   31, 32, 117
fast-shear BAW   237, 243, 317
fast-shear SSBW   243, 313
fast-shear wave   3, 240
FEM   56, 293, 294, 304, 308
FEMSDA   218–220, 224, 226, 248,
    250–252, 294, 306, 316
ferroelectric   164, 176, 179
FEUDT   50, 316
fiber   191–194
figure of merit   127, 129, 156
finger resistivity   204
finger-pair   47–49, 60–63, 66, 82, 87, 89,
    98, 101, 113, 210, 213, 246
finite element method   56, 293, 304
finite impulse response filter   89
FIR   89
floating-electrode type UDT   50
Floquet theorem   27, 28, 196, 297, 300,
    305
Fourier component   200
Fourier expansion   28, 97, 99, 100, 295,
    298, 311
Fourier integral   139, 200
Fourier integral form   104
Fourier transform   55, 89, 97–100, 139,
    168, 169, 281, 282, 299
fractional -3 dB bandwidth   63, 74,
    142, 165
Fraunhofer zone   8
free surface   19, 43, 54, 55, 61, 165, 168,
    169, 172, 173, 176, 177, 179, 180, 214,
    229, 238–243, 281, 284, 287, 289
free surface approximation   104, 107

frequency dispersion   40, 44, 164, 166,
    168, 186, 187, 250, 308, 309
frequency dispersive   115, 186
frequency drift   126
frequency response   60, 82, 87, 89,
    90, 94, 95, 97, 98, 101, 104, 109, 113,
    116–120, 141, 142, 146–148, 150–153,
    155–157, 159, 160, 184, 209, 245, 246,
    255, 258, 263
frequency stability   127
Fresnel zone   8
front-end filter   184
fundamental mode   187
fundamental resonance   138, 167

gap length   208
Gibb's phenomenon   98
grating   20, 22, 25, 26, 28, 30, 32,
    34–38, 40, 132, 133, 138–140, 148, 176,
    198, 199, 211, 218, 225, 248–250, 259,
    266, 293, 294, 296, 301, 304, 306, 309,
    310, 312
grating electrode   293
grating mode   27, 36, 44, 198, 205, 247,
    295, 296, 301, 305, 311, 313, 314
grating reflector   32, 34, 51, 117, 123,
    127–130, 138, 141, 144, 258
grating vector   28
Green function   168–170, 284
group delay   90, 115, 119, 126, 186
group delay ripple   117, 118
group velocity   5, 12–14, 166, 280
group-type UDT   49
guard electrode   111

half-width   232, 234
Hamming window   99, 100, 107
Hankel function   110, 172
higher mode   13, 146, 187
higher-order resonance   130
Hilbert transform   81
Hooke's law   274

IDT   22, 25, 47, 48, 52, 53, 55–66,
    69–71, 73, 74, 76–79, 81, 82, 87–89,
    91–95, 97, 99–101, 103–114, 117, 123,
    127–132, 134, 138, 140–143, 145, 148,
    165–168, 170, 184, 185, 195, 200, 201,
    203–214, 217, 221–224, 226, 227, 234,
    239, 241, 246, 247, 249, 250, 256, 258,
    259, 284, 315–318
IDT periodicity   140
IF   184, 185
IIDT   113

IIDT filter   113, 114, 263
image connection   114
image parameter   154, 155
immittance chart   72, 73
impedance chart   72
impedance element filter   149
impedance matching   70, 73, 74, 77, 78,
    92, 112, 113, 167
impedance matching condition   114
impedance mismatch   26
impulse response   89, 94, 95, 113, 115,
    126, 128, 239–241
in phase   12, 28, 51, 103, 113, 118, 220,
    256
in-line field model   65, 66
inertia   123
infinite grating   310
inharmonic spurious resonance   130
input admittance   64, 67, 77, 78, 128,
    143, 209, 226, 246, 315–318
input impedance   149, 184
insertion loss   69, 74, 77, 78, 87, 93, 94,
    96, 97, 110–114, 116–119, 143, 146,
    148, 149, 151, 157, 159, 160, 165, 167,
    245, 249, 255, 259
integration path   287, 288, 312, 313
interdigital transducer   25, 47
interdigitated-interdigital transducer
    113
interference   200
internal impedance   128
internal reflection   65, 127, 206–209,
    315
internal resonance   115
inverse Fourier transform   89, 283
isotropic   1, 13, 17, 83, 133, 274, 275
isotropic material   8, 19

Kronecker's delta   136

L   17, 18, 20, 21, 174, 182
L-type BAW   7, 17, 19, 83, 171, 243
$La_3Ga_5SiO_{14}$ 183
ladder-type filter   153–156, 158–160,
    245, 246, 255–257
Lamb wave   187
langasite   183
lattice disorder   164
lattice-type filter   152, 153
law of energy conservation   17, 31
law of superposition   55, 61, 105
layered structure   168
leaked BAW   20, 21, 109, 241, 242

leaky mode   16, 136
leaky SAW   20, 21, 109, 113, 167,
   172–175, 177, 179, 180, 182, 227,
   238–240, 242–249, 252–258, 263, 280
least square error   97
Legendre function   40, 52, 295
$Li_2B_4O_7$ 4, 21, 182
$LiNbO_3$ 164, 176–178, 180
linear phase   90
linear programming   101, 102
$LiTaO_3$ 164, 179, 180
loaded Q 143, 146, 151
long-term stability   127
longitudinal BAW   311
longitudinal leaky SAW   317
longitudinal mode   130, 148
longitudinal mode resonance   128
longitudinal SSBW   41, 313
longitudinal wave   1–3, 6, 7, 110, 275,
   277, 279, 307
longitudinal-type leaky SAW   22, 113,
   182, 183, 261
longitudinally coupled double-mode
   resonator filter   145, 148
lossless   33, 192
Love wave   19
lowest mode   187
LST-cut quartz   174, 175

main response   94
mainlobe   62, 113
Mason's equivalent circuit   65
mass density   275
mass loading effect   293, 294, 306–308
matching condition   146, 149
matching element   230
material constant   174, 176, 179, 182,
   183, 232, 261
Maxwell equation   276, 278
mechanical boundary condition   18, 19
mechanical reflection   39, 47, 48, 50,
   51, 66, 77, 89, 95, 115, 217, 218, 220
metal strip   195
metallic grating   26, 39, 40, 218, 224,
   293, 294, 296, 297, 301, 304, 307, 308
metallization ratio   56, 254
metallized surface   19, 43, 54, 55, 165,
   168–170, 173, 174, 176, 177, 179, 180,
   229, 238–243, 245, 247, 266, 282, 287,
   289
microwave wafer-probe   234
mid-term stability   127
mini-max   101

mini-max method   100
minimum phase   90
mismatching loss   94, 149
mobile communication   177, 180, 184,
   237
mode   13, 14, 20, 131–137, 147, 191,
   194, 196, 200
mode amplitude   191, 196, 203, 226,
   267
mode profile   138
modulation   28
motional admittance   145
motional capacitance   123, 129, 137,
   215
motional inductance   123
motional resistance   123
MSC   167, 194, 195
MSYNC   226, 318
MULTI   224–226, 294, 308
multi-mode resonance   128
multi-mode resonator filter   145, 146,
   148
multi-phase UDT   49
multi-strip coupler   167, 194, 195
multi-strip grating   225
multi-track filter   118–120
multiple reflection   32, 117, 314
multiple resonance   128
mutual coupling coefficient   191, 196

natural unidirectionality   51, 203
network analyzer   68, 75, 94, 152
Newton equation   274, 278
nonleaky   180
nonleaky mode   16
nonleaky SAW   20, 21, 167, 174–177,
   179, 180, 182, 184, 218–221, 225, 226,
   280
nonlinear   115, 126
nonpiezoelectric   187
nonresonant SPUDT   115
normal mode   191
normalized frequency   41, 250
Nyquist filter   101, 102

object function   101, 102
OBLIQ   294, 308
oblique reflector   117
observation point   200
ohmic loss   49
one-dimensional analysis   136, 137
one-port SAW resonator   123, 124,
   126, 127, 141–144, 147, 150, 151, 158,
   208, 210, 211, 231, 232, 234, 245, 258

open waveguide   15, 16
open-circuited   301, 309
open-circuited grating   39–45, 50, 195,
  205, 206, 216–218, 220, 225, 252, 260,
  296, 306, 307
orthogonal relation   136
oscillation   126
oscillation condition   126
oscillation frequency   126, 127, 130
oscillator   141, 142
out of phase   48, 51, 113
out-of-band   114, 157
out-of-band rejection   32, 103, 111,
  113, 114, 146–149, 151, 153, 155–157,
  160, 245, 256

parabolic approximation   105, 133
paraelectric   174, 182, 183
parallel resonance   149
parasitic capacitance   232
parasitic effect   110
parasitic element   230
parasitic impedance   230, 234, 256
partial wave   279
particle displacement   304
passband   34, 37, 38, 97–103, 107,
  111, 113–117, 131, 141, 146, 151, 153,
  155–157, 182, 248, 256, 257
passband shape   98, 115, 117, 119
passband width   113, 115
penetration depth   14, 18, 137, 249,
  251, 254, 260
periodic grating   293
periodic structure   196
periodicity   210, 214
peripheral circuit   68, 74, 167, 230
permittivity   211, 280
perturbation   39, 191, 259, 307
perturbation method   67, 214, 216
perturbation theory   214
perturbed mode   197, 202, 223, 229,
  252
phase delay   69
phase difference   195
phase lag   12, 89
phase matching   80
phase matching condition   28, 29, 79
phase shift   12, 14, 78, 138, 214, 215,
  249
phase velocity   1–3, 5, 12–14, 29, 41–
  43, 133–135, 166, 180, 182, 186–188,
  195, 275, 278
phase weighting   113

photolithography   47, 49
photomask   245
piezoelectric   2, 19, 164, 180, 186
piezoelectric constant   276
piezoelectric film   185
piezoelectric material   18, 20, 48, 79,
  163, 275, 277, 290
piezoelectric medium   277
piezoelectric substrate   19, 26, 39, 48,
  109
piezoelectric thin film   184, 186
piezoelectricity   8–10, 19, 21, 146,
  164, 171, 174, 176, 177, 179, 182, 183,
  185–187, 237, 275, 277
plane wave   5, 8, 12, 103, 110, 275, 277,
  278, 294, 297
plasticity   271
Poisson's ratio   21, 182
polarization   2–4, 7, 164, 176
pole   311, 313, 317
polycrystalline   7
positive and negative reflection type
  grating   40
power conservation   202
power density   17
power dissipation   68, 126
power flow   5, 80, 104, 166, 180
power flow angle   5, 166, 241
Poynting theorem   304
pressure wave   1
propagation angle   3
propagation direction   2, 4, 6, 20, 163,
  164, 166, 180, 184
propagation loss   2, 20, 109, 128, 167,
  172–174, 177, 179, 180, 182, 202, 232,
  234, 237, 238, 245, 253, 259
propagation path   119
propagation surface   176, 247
pseudo SAW   20

QARP   141
quality factor   124
quartz   22, 117, 174, 175, 183, 211, 261
quasi-electrostatic approximation
  276, 278
quasi-longitudinal wave   2
quasi-shear wave   2, 3

radiation admittance   64, 65, 82, 91, 92
radiation angle   80, 108
radiation condition   280
radiation conductance   82, 91, 92, 109,
  129, 165, 170, 171
radiation pattern   17

radiation susceptance   82
Rayleigh-type SAW   18–22, 55, 180,
    186, 187, 237, 238, 243, 248, 249, 254,
    263–265, 284, 285, 306, 308, 309, 311,
    314, 315, 317
reciprocity   7, 62, 68, 77, 92, 296, 312
rectangular window   98, 99
reflection   36, 111, 195, 200, 210, 214,
    216, 309
reflection angle   7
reflection center   26, 27, 50, 51, 78, 224
reflection coefficient   6, 7, 26, 31,
    34–38, 67–69, 77–79, 95, 128, 144,
    197–200, 202, 208, 215, 309
reflectivity   78, 211
reflector   25, 52, 113, 117–119, 123,
    138, 140, 141, 195, 208–210, 213, 214,
    227, 237, 249, 250, 258, 259
reflector grating   138
reflector stopband   213
regeneration   77, 94
rejection band   101, 103, 119, 131, 153,
    155–157, 256, 257
rejection band level   167, 182
Remez exchange method   100–102
residue theorem   170, 289, 313, 314
resonance   56, 115, 123, 124, 139–141,
    151, 197, 210, 211, 213, 226, 233, 234,
    249, 256, 317, 318
resonance Q 74, 124, 129, 138,
    140–144, 150, 245, 246
resonance characteristics   123
resonance circuit   73, 142
resonance condition   48, 128, 144
resonance frequency   36, 48, 50, 61,
    62, 64, 73, 74, 81, 97, 108, 109, 113,
    123–125, 128, 135, 138, 140, 141, 145,
    150, 151, 155, 158, 164, 165, 167, 171,
    210, 215, 216, 224, 232, 233, 248, 250,
    259
resonant cavity   127
resonant mode   145, 146
resonant SPUDT   115, 116
resonator   65, 127–130, 137, 138, 140,
    148, 149, 151, 155, 156, 167, 179, 182,
    209–211, 237
resonator filter   195
return loss   69, 70, 81
RF filter   237, 248
ripple   38, 94, 98, 99, 101, 131, 199, 211
rotated Y-cut LiNbO$_3$ 176, 177
rotated Y-cut LiTaO$_3$ 179, 245
rotated Y-cut quartz   174

rotation angle   21, 176, 179, 180, 182,
    245, 246
RSPUDT   115

saggital plane   82, 279
sand-blasting   107–110, 237, 242
sapphire   185
SAW   17, 20, 21, 25–30, 39–42, 44, 47,
    48, 50, 54, 61, 62, 64, 76, 77, 81, 82, 87,
    91, 92, 94, 97, 103–105, 107–112, 117,
    118, 128, 130, 133, 138, 139, 144, 155,
    163–172, 180, 182, 185, 186, 194, 195,
    198–200, 202, 203, 205, 206, 216, 217,
    224, 225, 238, 246, 250, 251, 259, 266,
    267, 278, 281, 282, 284, 293, 294, 296,
    297, 306–308, 310, 315
SAW amplitude   26, 88
SAW detection   207
SAW device   60, 66, 68, 77, 79, 89, 92,
    94, 101, 102, 108, 109, 163, 164, 166,
    167, 176, 177, 184, 185, 194, 200, 227,
    290, 293
SAW energy   18, 130
SAW excitation   48, 51, 61, 87, 105,
    115, 170, 171, 180, 185, 206, 214, 316
SAW field   18, 48, 92, 103, 105, 136,
    139, 140, 186, 187, 199
SAW filter   184, 185
SAW oscillator   126
SAW propagation   18, 26, 32, 39, 40,
    45, 54, 63, 89, 293, 296, 297, 309
SAW propagation loss   25, 94
SAW property   184, 186
SAW radiation   69, 82
SAW reflection   77, 129, 144
SAW regeneration   94
SAW resonator   10, 34, 127–130, 140,
    144, 149, 153, 155, 156
SAW response   94, 110
SAW transversal filter   107
SAW velocity   18, 20, 30, 34, 38, 41,
    43–46, 55, 62, 64, 65, 107, 127, 133,
    138, 164, 165, 173, 184, 187, 213, 214,
    232, 299, 306–308, 314
SAW wavelength   40, 168, 186
SBAW   22, 108
scalar potential theory   131, 132
scattered BAW   30, 31, 249
scatterer   2
scattering coefficient   68, 70, 74, 75, 78,
    93
scattering loss   2
scattering matrix   142
SDA   293, 294

second-order effect   66
self-coupling coefficient   214
semi-infinite   17, 19
series resonance   135
series weighting   87, 88
Sezawa mode   187
SH   18–21, 182, 242
SH wave   4, 6, 7, 13
SH-type BAW   19–21, 266
SH-type leaky SAW   238, 257
SH-type SAW   19–21, 174, 180, 237,
    238, 249, 250, 259, 285
SH-type SSBW   22, 174
shallow bulk acoustic wave   22
shape factor   116
shear BAW   167
shear wave   1–3, 6, 19, 110, 259, 275,
    277, 279, 287
shear-horizontal wave   4
shear-vertical wave   4, 7
short-circuited   301, 309
short-circuited grating   39–46, 49, 50,
    61, 62, 195, 205, 206, 208, 214–218,
    220, 223, 224, 229, 243, 245, 247, 248,
    250, 252, 260, 267, 296, 306, 307, 311
short-term stability   126, 127
sidelobe   62, 89, 98, 99, 116
sidelobe level   99
sidelobe suppression   99, 117
single rotated cut   164
single-electrode-type IDT   47, 51–53,
    56–58, 60, 223, 224, 316, 317
single-phase   202
single-phase UDT   65
single-phase unidirectional transducer
    49
slow-shear   20, 177, 226, 237, 245
slow-shear BAW   20, 21, 171
slow-shear SSBW   41, 177, 181, 184,
    218, 220, 225, 306, 311, 313
slow-shear wave   3, 187, 240
slowness   3, 6, 79, 82, 165, 168, 283
slowness curve   308
slowness surface   3–8, 80, 82, 83, 107,
    133, 279
Smith chart   70–72
Smith's equivalent circuit model   65
Snell's law   6
solid-electrode-type IDT   47, 59
spatial frequency   4, 28
specific impedance   68
spectral domain analysis   293
split-electrode-type IDT   47

SPUDT   49–51, 114–119, 225
spurious   131, 148
spurious BAW response   107–109, 138,
    180
spurious DBAW response   110
spurious level   109
spurious resonance   149
spurious response   22, 79, 108, 109,
    130, 167, 172, 181
spurious SSBW response   110, 167
SSBW   22, 41, 108–111, 167, 172, 173,
    176, 177, 226, 238–240, 243, 246–250,
    259, 260, 313
SSBW excitation   167, 176, 182, 288,
    290, 312
SSBW propagation   259
SSBW radiation   171, 177, 313
SSBW velocity   286, 313
ST-cut quartz   107, 116, 148, 174, 183,
    184, 318
standing wave   38, 45, 133, 140
static capacitance   52, 56, 58, 59, 64,
    65, 111, 123, 142, 170, 202, 215, 223,
    224, 255, 256
stationary condition   305
steepest descent path   290
stiffness constant   2, 274
stopband   34–38, 41, 43–46, 113, 117,
    197–200, 205, 211, 213, 216–218, 220,
    225, 232, 243, 245, 248–251, 260, 266,
    307, 309, 310, 314
stopband width   260, 307, 309
stored energy   11
strain   48, 274–276
stress   9, 53, 273–276, 304
strip admittance   40, 296, 311
strong-resonant SPUDT   115, 116
STW   20, 176, 250, 259–262
STW resonator   259
substrate orientation   3, 48, 174
substrate surface   166, 168
surface acoustic wave   4, 17, 19, 20
surface boundary condition   17, 18, 20,
    83, 110, 167, 173, 245, 249
surface electrical boundary condition
    19, 164, 165, 174, 176, 180, 240, 290
surface electrical potential   58, 104,
    110, 195, 203
surface skimming bulk wave   22
surface transverse wave   20, 22, 176,
    259
SV   17, 18, 20, 21, 174, 182, 237
SV wave   4, 7

SV-type BAW   7, 17–21, 83, 237
symmetric mode   132, 134, 137, 145,
    193, 195
SYNC   226, 318

TCD   166
TCF   165, 166, 174, 176, 180–182, 184,
    185
TCV   174, 176, 177, 179
temperature coefficient of delay   166
temperature coefficient of frequency
    165, 181
temperature coefficient of velocity
    174, 176, 177, 179
temperature stability   165, 166, 183,
    185
TeO$_2$ 3, 5
tetragonal   9
thermal expansion   127
thermal expansion coefficient   166,
    174, 176, 179, 182
thermal noise   126
thickness resonance   13, 15, 109
thickness vibration   65, 232, 266
time constant   126
time delay   12, 34, 94, 166
TiO$_2$ 4
total reflection   6, 7, 11, 12, 14–16, 130
transduction coefficient   201, 222, 223
transduction loss   79
transfer admittance   62, 63, 65, 87,
    110, 128, 141, 144, 229, 315, 316
transfer admittance matrix   301
transfer function   88, 92, 97, 98, 100,
    109, 195
transition band   103, 111, 119, 152,
    153, 157
transition bandwidth   99, 158
transmission coefficient   6, 7, 31, 67,
    68, 95, 199
transmission line   69
transmission response   12, 244
transversal filter   91–94, 97, 102, 103,
    109–115, 118, 141, 167, 182, 195
transverse mode   130, 131, 134, 135,
    137, 147, 148
transverse mode resonance   130, 135
transverse resonance condition   12, 16
transverse vibration   65
transverse wave   1
transversely coupled resonator filter
    147, 148
transversely-coupled resonator filter
    148

trigonal   9
trimming   160
triple transit echo   94
triple-electrode grating   301, 302
TTE   94–96, 112–114
TTE level   112
TTE ripple   96, 97
tunneling   11
TV-IF filter   179, 184
two-port matrix   32, 154
two-port SAW resonator   141–145,
    148, 149, 213, 229

UDT   48–50, 117, 224
unapodize   89, 91, 92, 97
unbalanced input and output   152
unidirectional IDT   48, 77, 78, 224
unidirectionality   48, 49, 52, 217, 225,
    226
uniformity   164
unitary   68, 75, 192, 267
unitary condition   68, 76, 78, 115, 192,
    197, 202
unloaded Q 143
unperturbed SAW   267

variational analysis   216
varicap   127
VCAL   107
VCO   127, 131
velocity dispersion   187, 227, 307
voltage-controlled oscillator   127

Wagner's theory   296
wave equation   275
waveform   89
wavefront   1, 4, 6, 8, 163, 166, 275
waveguide   12–15, 130, 259
waveguide mode   12, 14
wavelength   2, 4, 28, 29, 48, 56, 134,
    164, 186, 187, 242, 294
wavenumber   3, 4, 6, 11, 14, 16, 18,
    26–29, 33, 40, 45, 54, 61, 66, 69, 128,
    140, 144, 191, 193, 196, 197, 200, 202,
    215, 218, 229, 251, 252, 266, 281, 282,
    295–297, 305, 311
wavevector   4–6, 12, 29, 30, 80, 81, 105,
    131, 139, 279
weak-resonant SPUDT   115, 116
weighting   87, 89, 98, 113, 115, 117
weighting function   89, 90, 97, 98
weighting loss   128
width weighting   87, 88

window function   98–100
withdrawal weighting   87, 88, 91, 92,
    103, 113, 117

X-112°Y LiTaO₃   107, 110, 119, 164,
    180, 234, 264, 265
X-cut LiTaO₃   180, 181

Z-path filter   117–119
zero TCF   174, 177, 182, 259
zero-reflection   114
zero-reflection condition   78
zinc oxide   184
ZnO   164, 184, 186–188

Printing: Mercedes-Druck, Berlin
Binding: Buchbinderei Lüderitz & Bauer, Berlin